ACCELERATOR INSTRUMENTATION

CONFERENCE PROCEEDINGS NO. **229**

PARTICLES AND FIELDS SERIES 44

ACCELERATOR INSTRUMENTATION

SECOND ANNUAL WORKSHOP

BATAVIA, IL 1990

EDITOR:
ELLIOTT S. McCRORY
FERMI NATIONAL
ACCELERATOR LABORATORY

American Institute of Physics   New York

Authorization to photocopy items for internal or personal use, beyond the free copying permitted under the 1978 U.S. Copyright Law (see statement below), is granted by the American Institute of Physics for users registered with the Copyright Clearance Center (CCC) Transactional Reporting Service, provided that the base fee of $2.00 per copy is paid directly to CCC, 27 Congress St., Salem, MA 01970. For those organizations that have been granted a photocopy license by CCC, a separate system of payment has been arranged. The fee code for users of the Transactional Reporting Service is: 0094-243X/87 $2.00.

© 1991 American Institute of Physics.

Individual readers of this volume and nonprofit libraries, acting for them, are permitted to make fair use of the material in it, such as copying an article for use in teaching or research. Permission is granted to quote from this volume in scientific work with the customary acknowledgment of the source. To reprint a figure, table, or other excerpt requires the consent of one of the original authors and notification to AIP. Republication or systematic or multiple reproduction of any material in this volume is permitted only under license from AIP. Address inquiries to Series Editor, AIP Conference Proceedings, AIP, 335 East 45th Street, New York, NY 10017-3483.

L.C. Catalog Card No. 91-55347
ISBN 0-88318-832-1
DOE CONF-9010267

Printed in the United States of America.

Contents

Preface vii

Schedule viii

INVITED TALKS

Instrumentation and Diagnostics Used in LEP Commissioning, with Accent
on the LEP Beam Orbit Measurement System 1
    Jean Borer
The AGS Booster Radiation Loss Monitor System 35
    Ed Beadle
Electronic Systems for Beam Position Monitors at CEBAF 48
    W. Barry, Jay Heefner, and J. Perry
Panel Discussion on Beam Position Monitors Around the Labs 75
    Scott Miller, Moderator
Wire Scanner Systems for Beam Size and Emittance Measurements at SLC 88
    Marc Ross
Ion Profile Monitors Using Microchannel Plates 107
    John Krider
Schottky Signal Monitoring at Fermilab 108
    David Peterson
Panel Discussion on Where Does the Instrumentation Engineer's Function
End and That of the Computer Engineer Begin 131
    Scott Miller, Moderator
Advanced, Time-Resolved Imaging Techniques for Electron-Beam
Characterizations 151
    Alexander H. Lumpkin
Fiber Optic Links for Instrumentation 180
    Ralph Pasquinelli
Instrumentation Issues at SSC 195
    Don Martin

POSTER PRESENTATIONS

High Resolution, Position Sensitive Detector for Energetic Particle Beams 213
    E. P. Marsh, M. D. Strathman, D. A. Reed, and R. W. Odom
Anaylsis of the Beam Position Measurement with Button-type
Pickups in APS 218
    Y. Chung
A High-Frequency Schottky Detector for Use in the Tevatron 225
    D. A. Goldberg and G. R. Lambertson
A Wall Current Beam Position Monitor Built On a Ceramic Chamber 235
    Yan Yin

A Fast Beam Loss Monitor System for the
KEK Proton Synchrotron Complex.. 243
    J. A. Holt, J. Kishiro, D Arakawa, and S. Hiramatsu
Tuned-Antenna Driver for Microstrip Probe Sensitivity Testing...................... 244
    R. B. Shurter and J. D. Gilpatrick
A Controlled Master Frequency Oscillator for the SSC Low Energy
Booster.. 246
    L. K. Mestha
A New Scheme for Measuring the Length of Very Short Bunches
at CEBAF 254
    C. G. Yao
Instrumentation at the Bates Linear Accelerator Center.................................... 260
    J. B. Flanz, E. E. Ihloff, K. D. Jacobs, and T. Russ
Cornell Synchrotron Tune Correction.. 267
    C. Dunnam, J. Byrd, R. Meller
A Beam Position Monitoring System for Brookhaven's Linac
to Booster Transfer Line 273
    T. J. Shea, C. M. Degen, D. M. Gassner, and V. LoDestro
A Real Time Bunch Length Monitor in the TRISTAN AR............................... 280
    Takao Ieiri
Discharge Phenomena in the Button Electrodes of the Beam Position
Monitors of TRISTAN MR 287
    M. Tejima, T. Ieiri, H. Ishii, T. Shintake, K. Mori, and Y. Mizumachi
The Beam Loss Monitors at CEBAF 294
    J. Perry
A Longitudinal Emittance Measurement Program
for the Fermilab Booster... 300
    V. Bharadwaj and M. Popovic
Log-Ratio Circuit for Beam Position Monitoring.. 308
    F. D. Wells, R. E. Shafer, J. D. Gilpatrick, and R. B. Shurter
Imaging Micron-Sized Beams with Optical Transition Radiation..................... 315
    D. W. Rule and R. B. Fiorito
A Tune Measurement System for the Tevatron Using
a Phase-Locked Loop.. 322
    J. Fitzgerald and R. Gonzalez
Progress Towards a Turn-by-Turn Beam Profile Monitor
for the Fermilab Booster... 328
    J. B. Rosenzweig, V. Bharadwaj, J. Lackey, and P. Zhou

Other Presentations... 337

List of Participants 339

Preface

The second annual Workshop on Accelerator Instrumentation was held at the Fermi National Accelerator Laboratory on October 1–4, 1990. The emphasis this year was on the engineering aspects of operational instrumentation systems and on those systems which are actively under development. There were 143 participants from five countries, marking a notable increase in foreign participation over the first Workshop.

With the advent of the Superconducting Super Collider project in Dallas, LEP at CERN, CEBAF in Virginia, and ongoing development at the existing labs, the interest in beam instrumentation is high. The modern Instrumentation Engineer is challenged to meet the expectations of accelerator designers in this age of ever-more-powerful and simple-to-use high-tech machines.

These Proceedings contain twenty-eight contributions. This year, the program results in three types of contributions. The bulk of the contributions come from the ten invited talks. Transcripts from the two discussion sessions, held at the end of the first days of the Workshop, are also included. The third type of contribution comes from the poster sessions, a first for this Workshop. Of the twenty-three posters presented at the Workshop, we have nineteen contributions in these Proceedings.

We are grateful for the support, financial and otherwise, of Fermilab, especially the Director, John Peoples (a former accelerator instrumenter himself!) and Gerry Dugan, the Head of the Accelerator Division at Fermilab.

A special thanks goes out to the organizing committee which is listed below. We want to thank Don Martin for organizing the vendor participation. As the editor, I would personally like to thank last year's editor Vincent Castillo, along with Bob Webber, Ralph Pasquinelli, and Frank Cole for helpful discussions during the editing process. Also, thank you to the speakers and to the panel discussion participants.

And what conference could possibly go off smoothly without local organizers? Our exemplary staff was headed up by Pat Smith and Cynthia Sazama. Thanks also to Terry Morris, Marion Richardson, and Margie Harvey for help during the workshop. Our thanks go out to them.

Accelerator Instrumentation Workshop Organizing Committee

Marc Ross	Stanford Linear Accelerator Center
Keith Jobe	
Richard Witkover	Brookhaven National Laboratory
Gerry Bennett	
Jim Hinkson	Lawrence Berkeley Laboratory
Greg Stover	
Olin van Dyck	Los Alamos National Laboratory
Bob Shafer	
Antanas Rauchas	Argonne National Laboratory
Walter Barry	CEBAF
Robert Rossmanith	
Ralph Pasquinelli	Fermilab
Pat Smith	
Bob Webber	Fermilab/SSCL
Don Martin	SSCL

INSTRUMENTATION WORKSHOP SCHEDULE
OCT 1-4, 1990
FERMI NATIONAL ACCELERATOR LABORATORY

MONDAY OCTOBER 1, 1990

8:00	REGISTRATION -- Ramsey Auditorium Lobby
9:00	Opening remarks -- (Gerry Dugan, Accel. Div. Head)
9:15	First session -- Instrumentation and Diagnostics Equipment Used in LEP Commissioning With Accent on Beam Orbit Measurement System (Jean Borer, CERN)
10:45	Break
11:15	Second session -- Beam Loss Monitor Development for the BNL Booster (Ed Beadle, BNL)
12:30	Lunch and informal discussion
1:30	Third session -- BPM Development at CEBAF Using Pseudo-Random Modulation and Detection (Jay Heefner, CEBAF)
2:30	Panel Discussion -- Beam Position Monitoring Around the Labs
3:30	Break
4:00	Poster Session and Vendor Displays
7:30	Light catered dinner at Fermilab, 2nd Floor Crossover

TUESDAY, OCTOBER 2, 1990

9:00	Fourth session -- Wire Scanner Systems for Beam Size and Emittance Measurements at SLC (Marc Ross, SLAC)
10:30	Break
11:15	Fifth session -- Ion Profile Monitor Using Microchannel Plate (John Krider, FNAL)
12:30	Lunch and informal discussion
1:30	Sixth session -- Schottky Signal Monitors at Fermilab for Tune and Emittance Measurement (David Peterson, FNAL)
2:45	Break
3:30	Panel discussion -- Where Does the Instrumentation Engineer's function End and that of the Computer Group Begin?
5:00	Leave for Banquet at Columbia Yacht Club. Board buses in front of hi-rise.

WEDNESDAY, OCTOBER 3, 1990

9:00	Seventh session -- Advanced Time-Resolved Imaging Techniques for Electron Beam Characterizations (Alex H. Lumpkin, LANL)
10:30	Break
11:00	Eighth session -- Applications of Optical Fiber Systems in Beam Instrumentation and Feedback (Ralph Pasquinelli, FNAL)
12:30	Lunch and informal discussion
1:30	Ninth session -- Beam Profile Measurements Using Synchrotron Light (Ron Nawrocky, BNL)
2:40	Break
3:00	Tenth session -- Instrumentation Issues at SSC (Don Martin, SSC)
4:10	Closing Remarks
4:30	Organizers Meeting

THURSDAY, OCTOBER 4, 1990

9:00	Fermilab tours -- Sign up sheets are available for tours of P-Bar Source Area and Booster Area. Please sign up if you are interested in either tour.

*Some of the participants
in the*
1990 Workshop
on
**Accelerator Instrumentation
Fermi National Accelerator Laboratory
October 1-4, 1990
Batavia, Illinois
USA**

Invited Talks

SL/BI/jb   CERN/SL/90-107 (BI)

INSTRUMENTATION AND DIAGNOSTICS USED IN LEP COMMISSIONING, WITH ACCENT ON THE LEP BEAM ORBIT MEASUREMENT SYSTEM

J. Borer
European Organization for Nuclear Research (CERN) CH-1211 Geneva 23, Switzerland

ABSTRACT

The LEP machine is equipped with a very complete instrumentation [1,2] for beam position, intensity and profile, tune and optic parameters, interaction rate, from which a summary is first presented. Second, the largest instrument for beam orbit measurement is described in more depth.

The Beam Orbit Measurement (BOM) System with its 504 Beam Position Monitors and 40 Processing Electronics Racks distributed along the 27 km of the LEP tunnel and linked by the Control System's Token Ring has been installed and pre-tested with simulated beam signals. The signal processing equipment must guarantee high reliability and precision despite high γ-irradiation and long cable transmission. In addition most of the electronics will not be accessible during operation. The analog signal processing is based on normalizers using phase modulation and on Flash ADC's. Local memories allow the recording of data at each bunch passage for more than 1000 revolutions. It is followed by digital signal processing in local VME crates equipped with MC68000 microprocessor. The BOM system will be able to acquire data of up to 8 bunches from which injection trajectories, average orbits, integer and fractional part of Q, β function and post event analysis will be processed.

A first test with beam was successfully performed on a single station equipped for 12 monitors during the injection test of July 1988. It did validate the hardware and processing design. But complete results could only be reached in July 1989 when the proper triggering of all electronic stations could be adjusted on the circulating beam, via the Beam Synchronous Timing System. Most of the data analysis is done simultaneously in 40 microprocessors which communicate with a data Collector. Operational results are presented. The BOM System has been a key instrument for the success of LEP performance.

1. SUMMARY OF LEP-BI MONITORS

1.1 INJECTION SPLIT FOILS

Four Split Foil monitors [3] are installed: one horizontal(BPFH)-vertical(BPFV) pair on each injection channel close to Interaction Point 1. They are made of two Ti strips of 7 μm thickness and with a gap of 40 μm. These position monitor present little interference with the injected beam and allow for many beam revolutions. Their sensitivity is about 30 pC per foil for a bunch of $8 \cdot 10^9$ particles. They have been used for injection tuning, mainly at LEP start-up while the BOM System was run-in and beam synchronised.

1.2 INJECTION SCREENS (with TV transmission)

The Luminescent Screens [3] are strongly perturbing position monitors, but they belong to the basic equipment of accelerators mainly for their start-up. They present the advantage to "see" the beam with its position, size and intensity. They are the most practical instrument to thread the beam for the first time along 28 km. They are made of Cr doped Al_2O_3 plates with a 2 mm thick Cu shield for synchrotron radiation protection. The screen picture is transmitted by a 45° stainless steel mirror and a CCD TV camera to video monitors in the control room. The 24 LS installed in LEP are:

- three successive LS after injection points
- two at each side of beam stoppers near QS10s
- two at each side of the safety stopper near Interaction Point 1

1.3. CURRENT TRANSFORMERS

Four assemblies (BCTR), on the same vacuum chamber with ceramic gap, comprising average current (DC), individual bunches (BB) and single turn (ST) current transformers [4,5], are installed in LEP: two at injection points and two at 163m from IP1.

The DC transformer is a parametric magnetic amplifier which presents less noise than former Zero Flux CTs from which more than 500 are used for precision monitoring of magnets and klystrons current. The toroidal cores are made of wound amorphous Vitrovac 6025 tape with combined longitudinal and cross-field (perpendicular to the tape axis) annealing yielding an optimized BH curve with best dynamic properties and smaller magnetic domains for minimum Barkhausen noise. The Parametric CT covers a large dynamic range and the noise level is < 0.2μA rms (1 s integration time). The DC transformer specification is a range of +-10 mA with a resolution of +-1 μA.

The BB and ST transformers are of the Integrating Current Transformer type. They are made with two toroidal cores also from Vitrovac 6025 tape. They are housed in a toroidal copper box with a gap forming a single turn winding. The integrating capacitor is formed of 36 capacitor chips of 100 pF forming a total capacity of 3800 pF. The charge induced I_b by each bunch is temporarily stored in this capacitor. The equivalent circuit is presented on Fig.1 and its tranfer function correspond to a band-pass filter stretching I_b by a factor of about 1000. This analog signal is applied

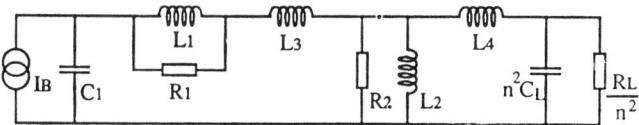

- I_B : beam current (current source)
- C_1 : storage capacitor of secondary loop
- C_L : storage capacitor of tertiary loop
- L_1 : inductance associated with core 1
- L_2 : inductance associated with core 2
- L_3 : leakage inductance in secondary loop
- L_4 : leakage inductance in tertiary loop
- R_1 : core losses of core 1
- R_2 : core losses of core 2
- R_L : load resistor (coaxial cable, 50 ohms)
- n : number of turns, winding of core 2 (10 ... 20)

Fig.1 : Simplified equivalent circuit of ICT

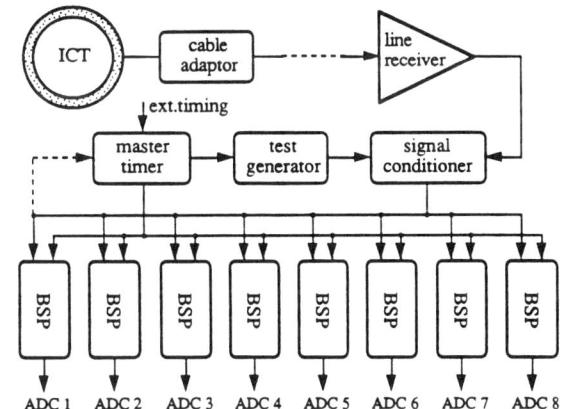

Fig.2 : Block diagram of the BB Monitor

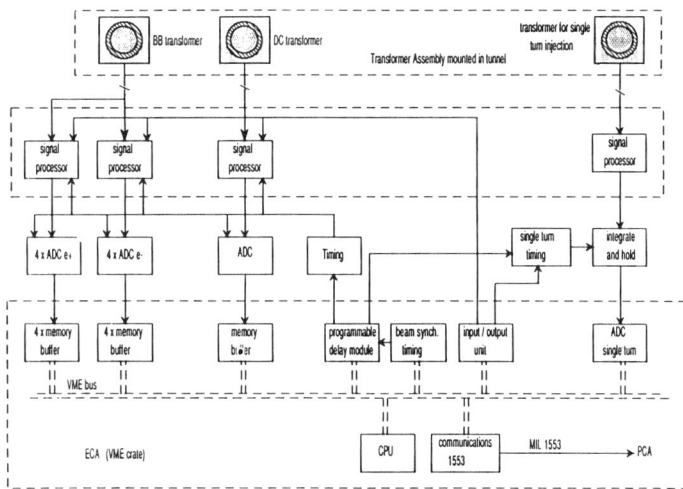

Fig.3 : Detailed diagram of the acquisition system

4 LEP Beam Orbit Measurement System

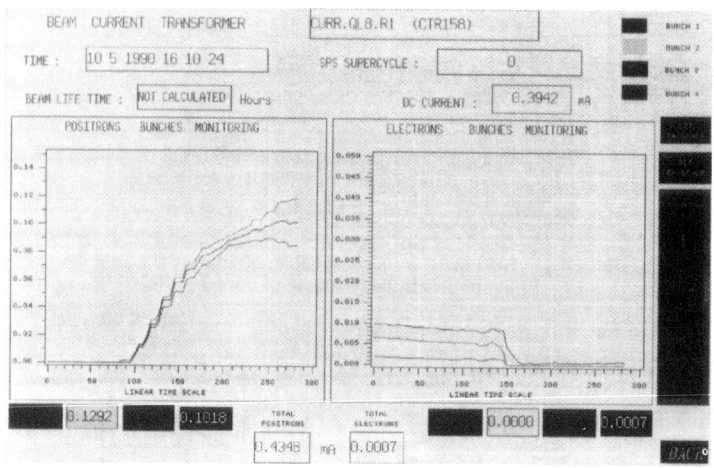

Fig.4 : Display of bunch currents with a beam loss

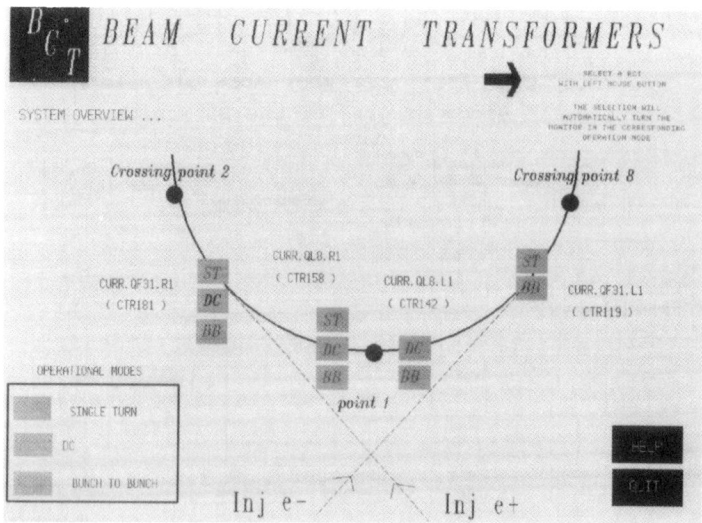

Fig.5 : LEP Current monitors layout

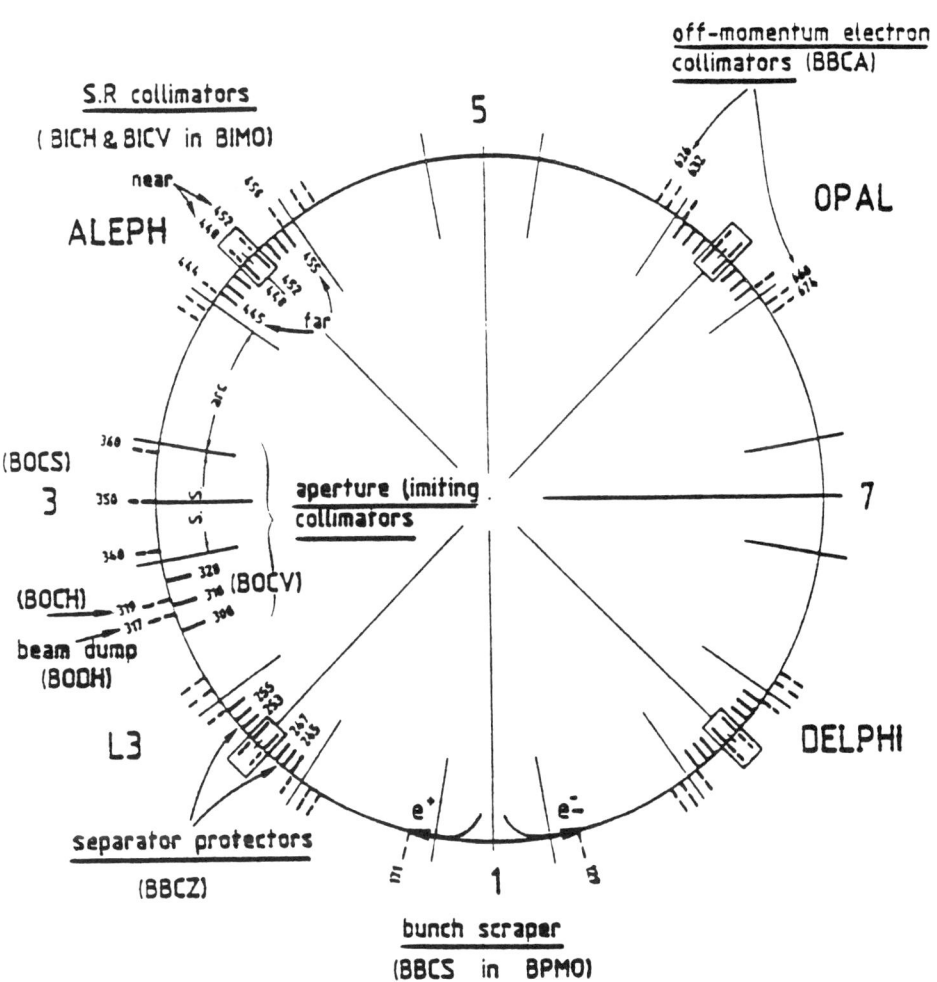

Fig. 6a: LEP Main Ring Collimators

6   LEP Beam Orbit Measurement System

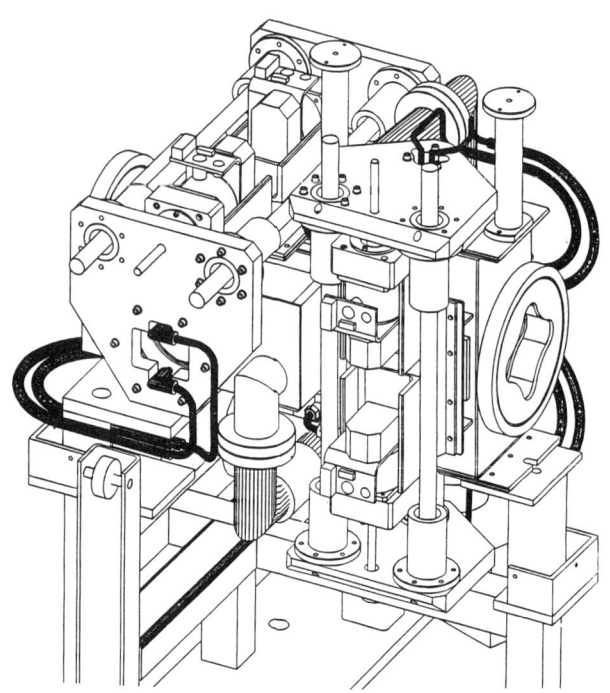

Fig.6 : Total aperture collimator (BIMO)

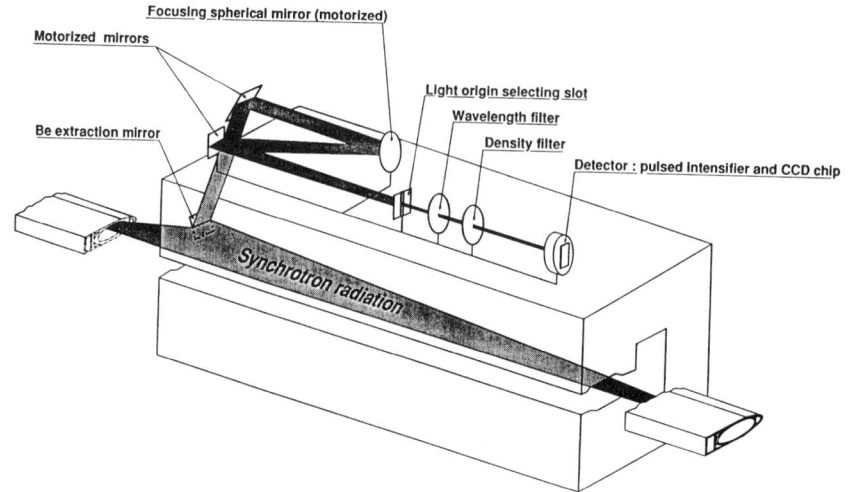

Fig.7a : Telescope BEUV setup

to 8 parallel analog processing channels (Fig.2) for bunch selection and Integrate-Hold function with DC level compensation and very low noise. The output signal is proportional to the charge, expressed as current, of the selected bunch and averaged for a given number of revolutions. Fig.3 present the acquisition system and Fig.4 an example of bunch currents display on a console. The BB and ST transformer specification is a range of +- 2.5 mA with a resolution of 10 nA. The BB and ST transformers use the Beam Synchronous Timing control for bunches and turns definition.

The transformers disposition in LEP is presented on Fig.5. The DC and BB channels are essential for the machine operation to the point that the picture of digital Ameters directly connected to the analog preprocessors is transmitted by TV in the control room ! All the filling process is followed with the BB transformers. The ST channel was intensively used for the injection test of july 1988 and at start up of LEP for the measurement of the number of particles injected. It is not as usefull once the beam can be accumulated with RF trapping.

1.4. COLLIMATORS

The collimators [6] have 3 different functions and their diposition around LEP is given on Fig.6a : Aperture Definition collimators, Total Aperture collimators (Fig.6b) for experiments protection and Total Aperture collimators for separator protection. The arcs H-aperture defining collimators could not sustain (overheating) direct impact of full beam if they were made of high z material. Hence they are made low z block (18 cm Al) with high z thin plating (10 μm Au). The other collimators are made of 13 cm W. They are well matched to the vacuum chamber with Cu transition blocks for RF losses or beam loading reasons. Their positioning is driven by stepping motors with a 5 μm resolution and 100 μm setting accuracy.

1.5. STOPPERS with SCREENS

Two beam stoppers (BCSC or BCSE) made of copper blocks [3] are installed near QS10s on both side of experimental Interaction Points. In addition another one has been placed near QL7 close of IP1 for safty reasons. Operationnaly they are either used for safety and personnal protection or for single turn injection necessary to verify and optimise injection and Beam Synchronous Timing functionning.

1.6. BEAM SCRAPERS

Two scrapers (BBCS) have been installed [3] and they are incorporated in the split-foil injection monitors (BPMO). They are foreseen for bunch charges equalisation but are practically not used.

1.7. PROFILE MONITORS:

Many instruments have been installed for the analysis of transverse and longitudinal profiles and emittances. The most used one is the synchrotron radiation UV telescope since it does not interfere with the beams and allows for life monitoring of the beam stability or cross-section on a TV screen in the control room. It gives a "feeling" about beam health to the operators while stacking and ramping.

1.7.1 U.V. TELESCOPES (BEUV)

Abundant synchrotron radiation, in the UV range, is available in LEP and is compatible with usual UV components and some air path, whilst minimizing the broadening of the beam profile through diffraction [7]. The extraction mirror is a delicate piece of equipment since the ultimate optical quality (+-25 nm flatness) must be maintained under heavy power deposition. It is made of beryllium and is situated at 21m from the source and 75mm from the beam as can be seen from Fig.7a . Best performance

8 LEP Beam Orbit Measurement System

is reached with a system at QD giving ε_z with high resolution and a system at Q12 for ε_x. $\Delta p/p$ can also be estimated from σ_x measured at QD and Q12. Hence 4 telescopes are installed on either sides of IP8. The beam image is captured by a pulsed intensifier and a CCD camera with 288x384 pixels. The precision on the profiles varies from 2 to 4%. Digital processing allow for 3 modes of operation: i) The normal TV mode for observation in the control room, ii) Acquisition of a single shot with computation of the sigmas and data compaction for display on a console, iii) Acquisition of a series (10) of two dimensional profiles stored on the CCD before processing and spaced in time as close as one revolution period apart. Operational results are shown on Fig.7b.

1.7.2 X-RAY vertical BEXE and longitudinal BEXE

Another profile instrument (Fig.8a) [8], for the X-ray range, is placed very close to the beam. A special vacuum enclosure allows direct exposure of solid state detectors to the main dipole synchrotron radiation, through a window of 400 mm beryllium. The CdTe photoconductors have been chosen for their very short carrier lifetime (10 ps). A vertical line array of photoconductors (Fig.8b) measures the vertical profile with a pitch of 100 mm, at each bunch passage. The detector and its movable support for alignment and in-out motions are inside a tank with a vacuum of 10^{-6} Torr for high voltage and corrosion reasons. The processing scheme is shown on Fig. 9. The multiplexed analog signal presenting the vertical profile is then acquired for each bunch and revolution by BOM NB-ADC modules and parallel buffer memory for further digital processing and display. One detector has been installed in LEP and has survived months of irradiation. This is a crucial and encouraging result because the power deposited by X Rays in the 4 mm of CdTe can be estimated to 40 mW/mm^2 for a beam of 1 mA, corresponding to a dose of 10^{11} Grays. No damages could be noticed, which is quite exceptional for a semiconductor device.

On the same support a longitudinal profile or bunch length detector is mounted. Its layout is presented on Fig.10. The folded delay lines with different length are connecting a row of CdTe photoconductors. Once all photoconductors are illuminated by the intense X-ray pulse the P_o ones polarized with a large DC bias send through the delay lines a pulse proportional to the light pulse to reach the other photoconductors P_i with staggered delays. At P_i the out-going pulses charge become points of an autocorrelation profile of the light pulse. The readout electronics is the same as for the the vertical profile. The detector prototype has been a long venture in collaboration with the D.LETI Institute C.E.A. Grenoble,F. The detector has not yet been fully tested with beam but a test with a subpicosecond laser has given an autocorrelation pulse of 19 ps (FWHH). The acquisition hardware and software is also not yet completed.

Such instruments have very interesting features: non intercepting, real time single shot measurements, measuring at rate of 44 kHz, the vertical resolution is not diffraction limited and there is no critical timing needed. Two detectors are installed for e+ and e-.

1.7.3 WIRE SCANNER

Two sets of wire scanners [9], each for measuring the horizontal and vertical profiles, are installed in LEP in a straight section close to IP1 where the horizontal and vertical dispersion is zero. A carbon fibre with a diameter of 36 μm moves through the beam with a speed of about 0.5 m/s. When the bunch is traversing the fibre, the electons or positrons produce bremsstrahlung photons with an energy spectrum of up to the beam energy. The bremsstrahlung photons are detected with a scintillator placed some ten meter downwards at the start of the bending section. The detector is shielded against low energy synchrotron radiation. This allows a measurement of the beam profiles which is practically background free.

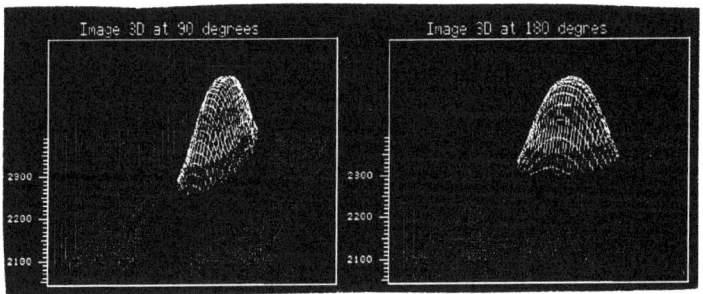

Fig. 7b: Example of BEUV display

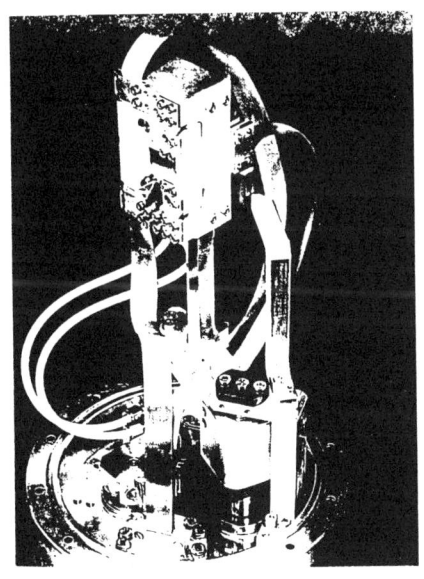

Fig. 8a : Motorized support in vacuum or BEXE

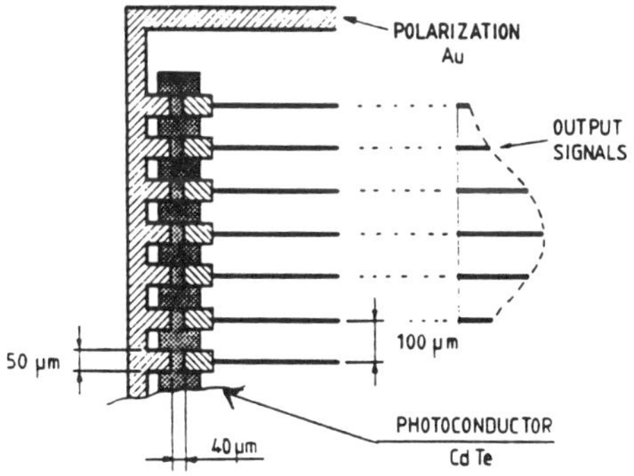

Fig. 8b : Vertical detector layout

10 LEP Beam Orbit Measurement System

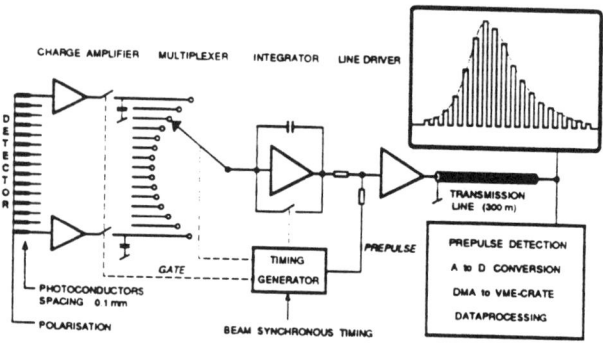

Fig. 9 : Detector synoptic diagram

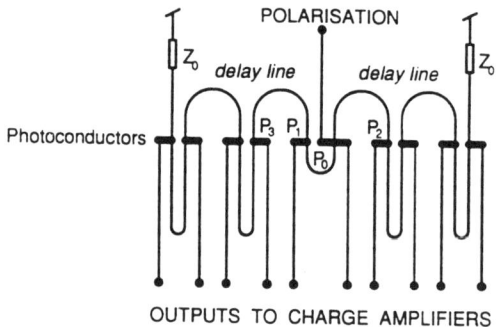

Fig. 10 : Autocor. circuit for longitudinal profile

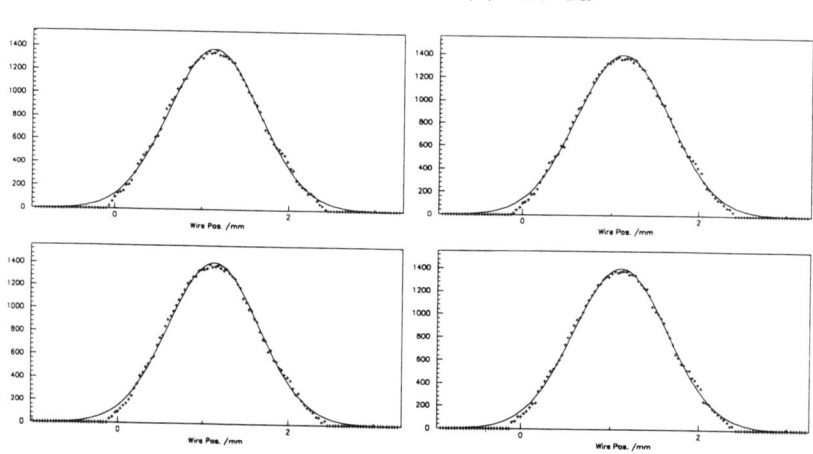

Fig. 11 : Wire scanner display

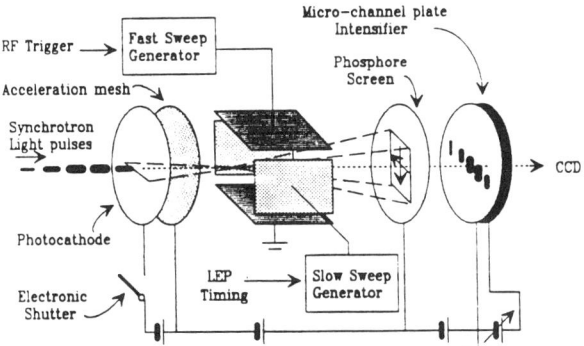

Fig. 12 : Streak camera principle

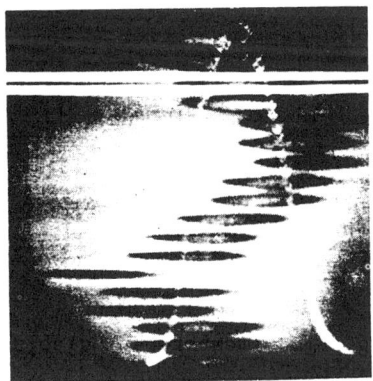

Fig. 13 : Display of synchrotron oscillation

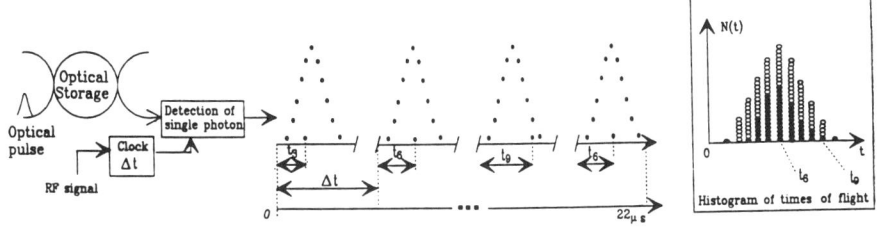

Fig. 14a : Principle diagram of a single pulse storage and stochastic sampling

12 LEP Beam Orbit Measurement System

By a precise measurement of the position of the fibre and of the pulse height induced by the bremsstrahlung photons whenever a bunch hits the fibre, the profile is measured with a resolution of less than 10 micron. During the LEP run some thousand scans were done, during machine development periods as well as for optimization of the luminosity by minimizing the beam size. Very small changes of the beam height could be detected by the instrument and the calibration of other (nondestructive) monitors measuring the profiles could be done.

Some problems were encountered when the scanner was used to measure the vertical beam size at the 20 GeV injection energy. In this case, the beam was substantially blown up during the scan because of multiple coulomb scattering. This lead to an asymmetric profile since the blow up increases during the time the fibre traverses the beam. A reduction of it was achieved by increasing the speed of the scanner. In addition, the emittance calculating software corrects for the asymmetry, which further reduces the error.

During the first month of LEP operation, fibres broke occasionally. This is probably due to heating of the fibre by the coupling of the fibre with the electromagnetic fields of the beam. A second mechanism to heat the fibre is due to the energy loss from the beam particles traversing the fibre. A beryllium fibre which was tried out was even more fragile. In order to increase the lifetime of the fibres, different measures were taken : the speed of the fibre passing the beam was increased, the control of the motion was improved. After that no fibres were destroyed anymore. Nevertheless the problem of the fibre heating by the beam remains a topic to be studied in more details.

The detector signal processing and display programs are commissioned and operational. Fig.11 presents an Apollo display in the control room.

1.7.4 STREAK CAMERA

An industrially developed Streak Camera [10], with two orthogonal deflections allowing for the stacking of up to 50 streaks on the screen of a CCD readout, is installed in the IP1 optical laboratory where it receives the visible light deviated by Be mirrors. The slow sweep can be varied for all bunches recording or interspacing for up to 20 synchrotron periods. Selected bunches are gated with a pulsed bias of the photocathode. The camera set-up is presented on Fig.12 . The time resolution has been measured with a picosecond laser pulse and is better than 6 ps FWHH. Fig.13 presents a beam longitudinal profile and synchrotron oscillation as recorded from the light produced by the low field bending magnets. The light source is being intensified with mini-wigglers. Synchronizing the Streak Camera and BEUV detector should allow for synchro-betatron oscillation observation. This instrument and its computer control and CCD camera image processing is also in the running-in phase.

1.7.6 PHOTON STORAGE RING

Another new technique for longitudinal profile from synchrotron radiation is in development but has not yet reached the operational level: the photon storage ring [11] with random sampling. Fig.14a explains the principle and Fig.14b describes the setup.

1.8. TUNE CONTROLLER or Q-METER

One Q-meter instrument [12] is installed in LEP near IP1. Its beam hardware is one standard button BPM and two single turn ferrite kickers installed around a ceramic vacuum chamber. It measures the coherent non-integer part of the betatron frequencies with a resolution $<10^{-3}$ for electron and positron beams. Fig. 15 presents the principle of the instrument. The beam position acquisition uses the modules of the WB processing of the BOM system.

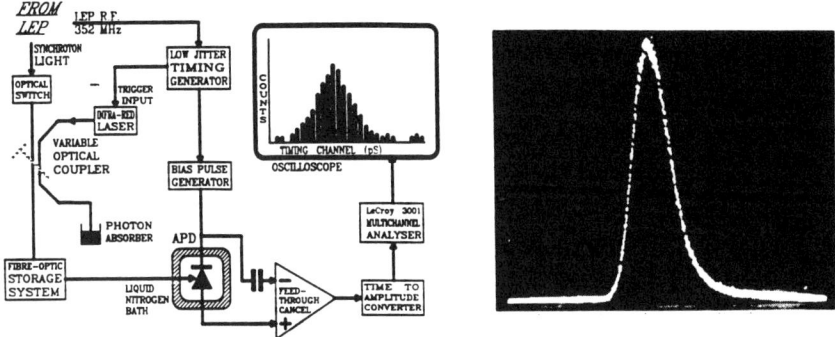

Fig. 14b : Measuring setup and random sampling of a bunch signal

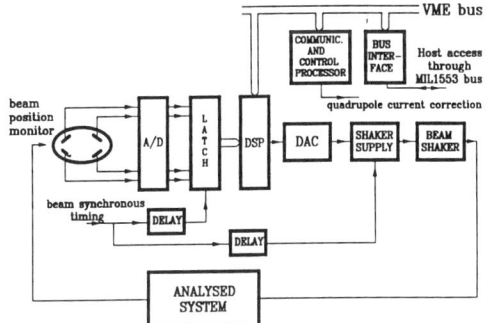

Fig. 15 : Q-Meter block diagram

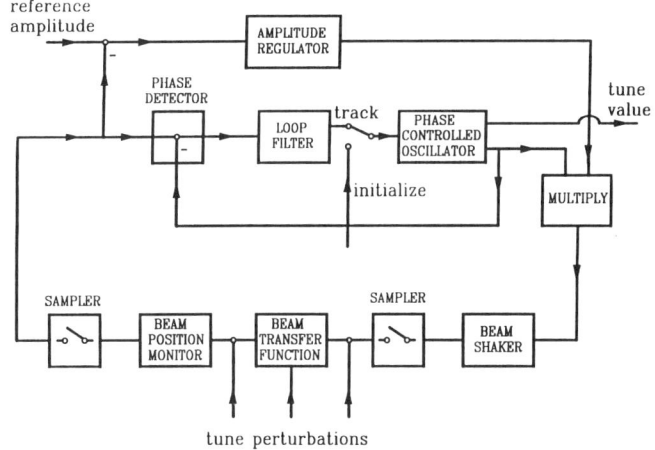

Fig. 16 : Signal flow diagram for PLL mode

14 LEP Beam Orbit Measurement System

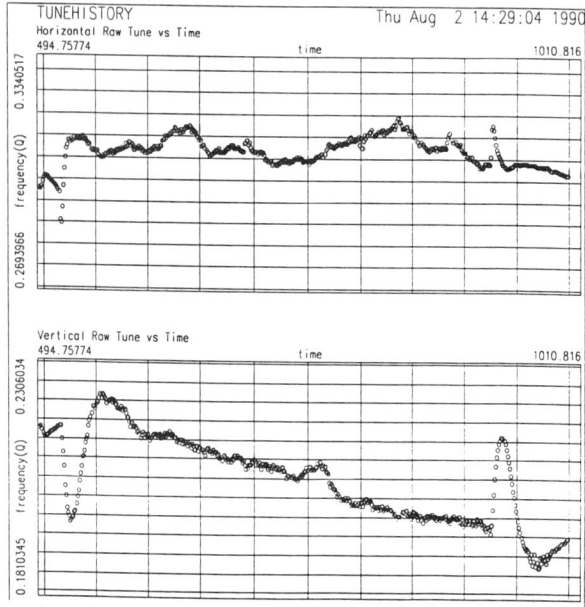

Fig. 17 : Tune history record in PLL mode

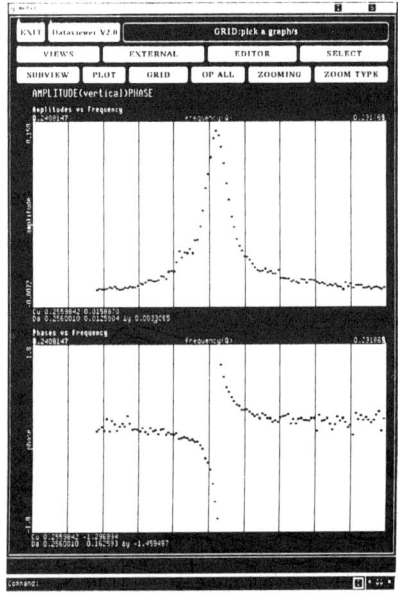

Fig. 18 : Tune measurement display

The digital signal processing is done with two VME processors MOTOROLA 68020. Due to the flexibility offered by digital signal processing the Q-meter offers three modes of operation:
- FFT Mode: The beam is excited with random noise (constant spectrum) and the resulting motion (0.05 to 0.5 mm) is acquired for 256 to 4096 revolutions which are processed by FFT in a time of 170 to 860 ms. The resolution $\Delta q = 1 / n_s$, which is related to the number of samples, has been further improved by an interpolation algorithm. For continuous q monitoring the q_H and q_V data are stored in a circular buffer.
- Swept Frequency Mode: as in usual frequency analysers the beam excitation frequency is swept according a staircase function. The beam response is processed by Harmonic Analysis with an average of 1000 samples per point. The obtained accuracy is 2.10^{-4} which is smaller than the stability of LEP over 10 min.
- PLL Mode: the Phase Locked Loop mode block diagram is given in Fig.16. With a correctly chosen phase shift between the PCO output and the beam shaker, the PLL will track and even capture the q resonnance. An amplitude regulation is necessary to avoid over excitation and beam loss. This mode is a delicate one since it depends on machine state but it is faster than possible tune changes due to variation of current in the magnets. Fig.17 represents a tune record where tune changes were introduced and where the small peaks are due to parasitic influence of SPS cycles.

The use of LEP Q-meter has proved to be a precious tool for both operation and precise accelerator studies. One example is the observation of the σ and π mode of oscillation during collision which gives information releted to luminosity, effective emittance and aspect ratio at I.P.. A closed loop tune regulation during ramping process was difficult to optimise but is now operational. Fig.18 is a tune diplay on APPOLO console.

1.9. INTERACTION RATE BHABHA MONITORS

Each H collimator-block of the IP collimators has been equiped with a compact silicon tungsten calorimeter [13] in order to be at very small angle for Bhabha monitoring. Four monitors per experimental IP, 15 m away, are forming the Lep luminosity monitoring system. Silicon detectors as active material and tungsten as absorber and shower generator form a compact design presented on fig.19a .The calorimeter is separated from the vacuum by a 1 mm stainless steel window and the back side is protected from synchrotron radiation by 30 rad. length of tungsten and a 2 mm lead shield and protective sandwich all around. During commissioning, the silicon detectors suffered from an alarming increase of their diode reverse current. Since these currents recovered during machine stop suggest phenomena of positive charge accumulation by SR and e+/e- background in the silicon oxide. This was mostly overcome by the shielding and by biasing the detectors only during measurements. The calorimeter assembly with its front end electronics is shown on Fig.19b and luminosity measurement at IP4 with beam steering on Fig.20 . The monitors have shown their capability of measuring the relative luminosities but the potentialities are still not fully exploited (software).

1.10. TRANSVERSE POLARIMETER

A laser polarimeter [14] based on spin-dependent Compton scattering of circularly polarized photons from polarized electrons has been installed in the LEP straight section LSS1. The principle is shown on Fig.21 . The vertical angular distribution of the recoil high energy γ rays, as measured at a detector 247m downstream the laser interaction region (LIR), shows an up-down asymmetry depending on the right - left helicity of the incident laser photons and proportional to the e+e- transverse polarization level. The installation of the laser polarimeter has been completed in 1989 (Fig.22) with its Nd-YAG laser ($\lambda_\Phi = 532$ nm) placed in the optical lab 125 m away from the LIR. Evidence of polarized beams has been recorded during the last run of 1990, on Aug. 30, as can be seen from Fig.23 .

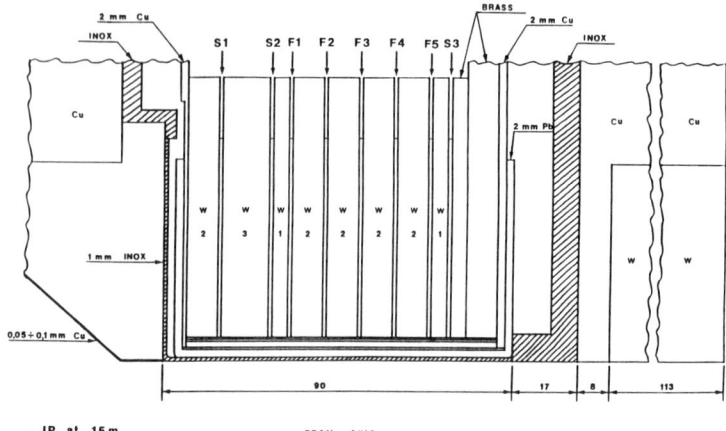

Fig. 19a : Construction of the calorimeter in the block

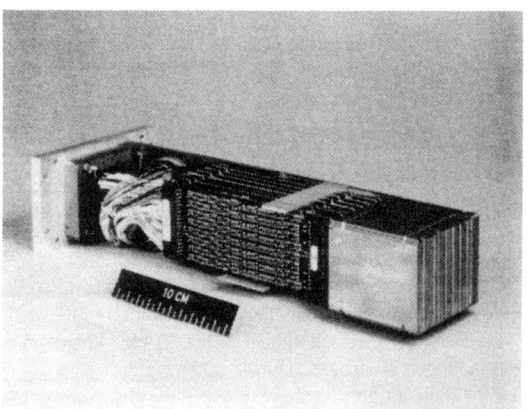

Fig. 19b : Calorimeter with front-end electronics

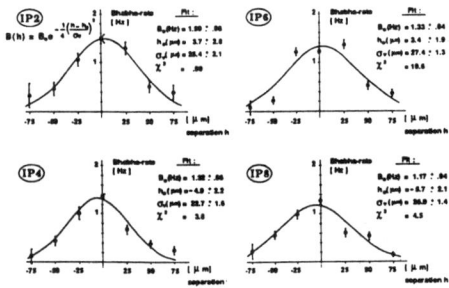

Fig. 20 : Relative luminosity rates versus vertical steering in the 4 IPs

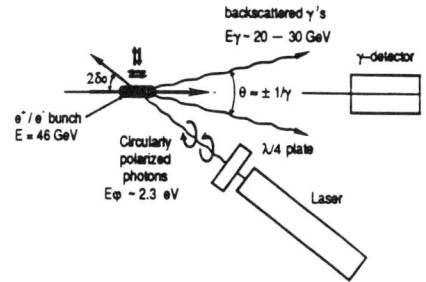

Fig. 21 : Principle of the laser polarimeter

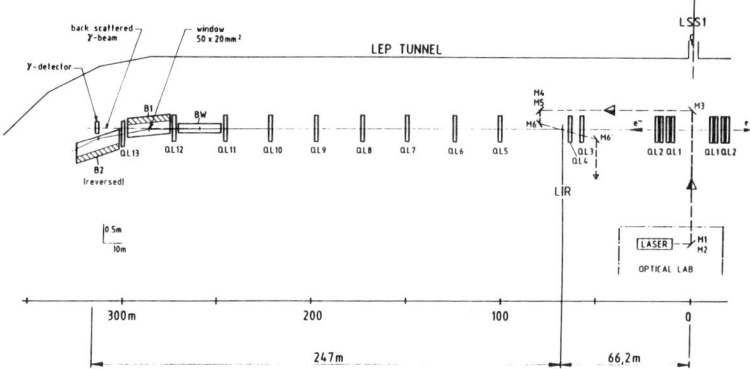

Fig. 22 : General layout of the LEP polarimeter

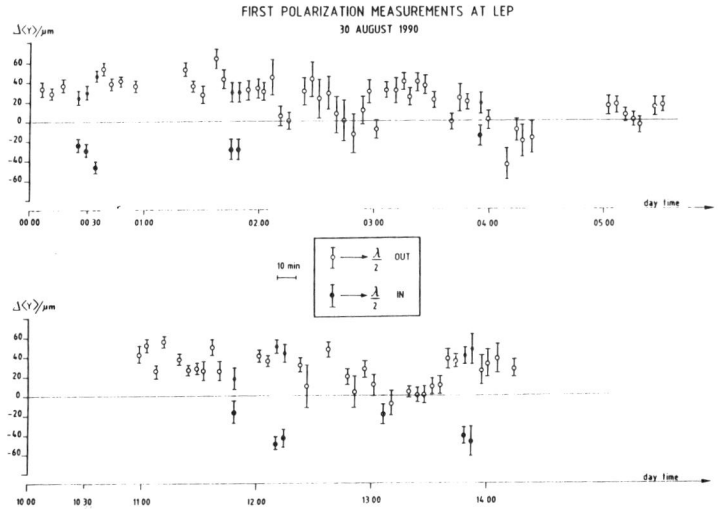

Fig. 23 : First natural polarisation measurement as function of time

18   LEP Beam Orbit Measurement System

2. BEAM ORBIT MEASUREMENT SYSTEM

2.1. INTRODUCTION

The BOM system has been designed to measure with a fast response the beam positions from 504 Beam Position Monitors (BPM) located near QD's along the 28 km of the LEP machine [15]. Since the radiation level in the LEP tunnel will be high, the processing electronics has been grouped in 24 shielded underground areas, resulting in long cable connections between monitors and electronic stations (up to 800 m) and restricted access during machine operation. The latter implies the need for very good remote diagnostic means and high reliability as well as easy maintainability.

The BOM system (Fig.24) has to be able to measure first turn trajectories (single passage) for local problem detection or closed orbits (average over many bunches and turns). This implies a synchronizing mean for unambiguous capture of the same bunch signal all around the 24 stations. A special Beam Synchronous Timing (BST) system has been implemented [4] which is also used as a synchronous serial link for the control of the BOM system and other instruments. This requirement being satisfied, the data can be stored in the local processor's memories. Since the average time between bunches is 11 ms, all signals are recorded and the memory has been tailored to contain up to 1000 revolutions.

Thanks to their development for TV applications, cheap 8-bits-FADCs could be chosen for the fast acquisition circuitry. These raw data will then be further processed by a local microprocessor to derive closed orbits or lattice parameters, or to make a post event analysis (transient recorder mode).

The resulting volume of data to be transmitted is very much reduced and does not create a heavy load to the control network. The overall BOM software organization is presented in Fig.35 . It shows that many users can make request to the BOM Server which will organize the BOM activity, set priorities and collect the results from the various points. The response time for a closed orbit request should be within seconds.

The radiation level in the LEP tunnel implies special design of critical parts of the BOM system. Careful attention has been devoted to the BPM button electrodes, signal cables and front-end electronics.

2.2. BEAM POSITION MONITORS

The design criteria for the LEP monitors were the following [15,34]:

- short length, about 10 cm,
- to be welded to the vacuum chamber,
- skewed sensor positions in order to avoid direct inpact of synchrotron radiation,
- minimum RF loading and higher modes coupling to the beam,
- high precision for interchangeability,
- no further tests after welding,
- flanges for sensor exchange,
- resistance to corrosion and to baking up to 300°C,
- no difficult administration of test parameters, i.e. full interchangeability.

The best solution is a capacitive monitor with button-like electrode, as also used in most other electron machines [16,17,18,20]. The button electrode (Fig.25) is held by its 50 Ω central conductor and positioned by a ceramic Al_2O_3 washer. Both the electrode capacity and its distance to

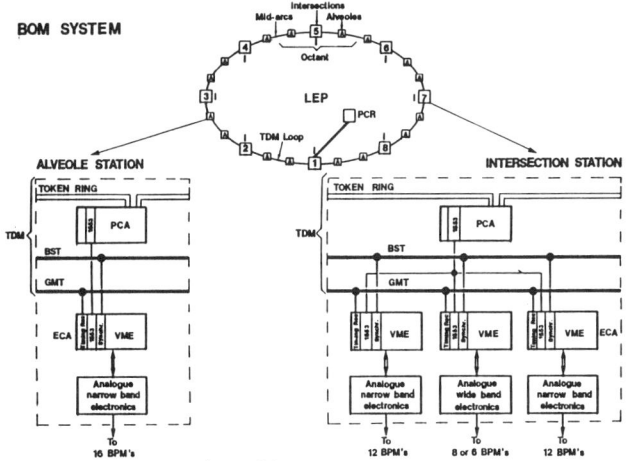

Fig. 24 : BOM system

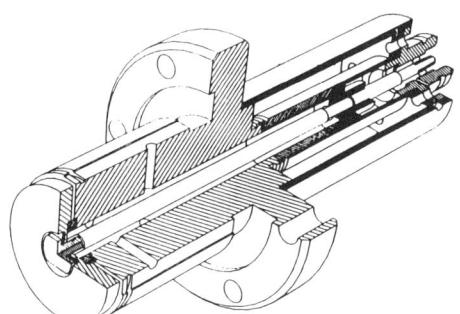

Fig. 25 : Button electrode

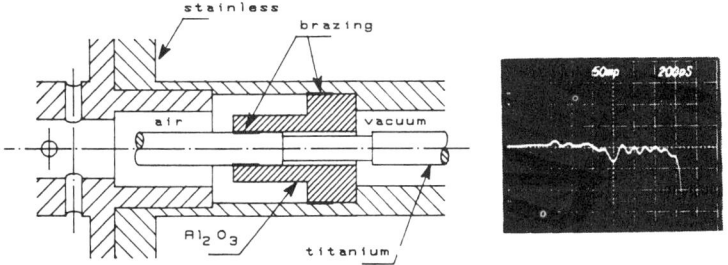

Fig. 26 : Feedthrough structure with its TDR response

the beam are critical elements for the sensitivity and for the zero offset of each monitor. This requires tight mechanical tolerances for both the buttons and the monitor block, in order to insure an overall transverse position tolerance of 0.15 mm.

The structure of the feedthrough (Fig.26) with its mixed vacuum(air)-ceramic dielectric and shifted inner and outer brazing has been inspired from the KEK design [19,20]. This design, as compared to the simple disc design, gives to the central rod a larger diameter and a better mechanical strength. The LEP design has been further improved for extending its frequency bandwidth in shifting the diameter transitions to achieve a local capacity compensation. The impedance matching is maintained well up to 12 GHz. The feedthrough TDR shows a performance comparable to the one of an SMA connection.

The central rod is made of titanium and the outer conductor as well as the flange with its plug are made of stainless steel 304L. Gold-nickel braze has been chosen for its corrosion resistance to nitric acid vapors [26] produced by the ionized moist air. Most of the LEP vacuum chamber is made of an extruded aluminium pipe which would not have provided an accurate nor stable enough support for the four buttons of a BPM. Aluminium blocks (Fig.27) have been designed to be welded onto normal vacuum chambers. But the button electrode flange (70 mm diameter) being made of stainless steel (304L) an important differential expansion will be present during baking at 150°C. Thorough testing has shown that the metallic seal (LEP type or Helicoflex) can take the expansion difference by its deformation and the seal leak tightness is very reliable. Some BPM blocks in experimental straight sections have different cross-sections (Figs. 28,29) to match the chamber geometry and are made of stainless steel to allow for bakings up to 300°C.

The BPM sensitivity has been carefully measured on a high precision test bench (Fig.30). The beam is simulated by a thin wire antenna held under tension by a frame which position is controlled by stepping motors with a resolution of 5 microns. The overall position reproducibility is within 30 microns. The HP 8505A network analyser is both used to feed the antenna (70 MHz) through an amplifier and to process the four reference button signals which are multiplexed by relays. The overall bench is fully computer controlled.

On reception from series production the button electrodes are first mechanically tested with a jig and then tested on the same bench for their coupling factors. Their capacity (about 8.7 pF) is finally measured with a capacimeter of 0.01 pF resolution. In order to insure the BPM electrical center precision both parameters are stored for each button and quads are formed of buttons with very close values, in steps of 0.01 dB and 0.1 pF.

Once the transverse electrical characteristics of each type of block have been established and recorded, the series blocks are only tested mechanically to be within the given tolerances. The blocks are produced with computer controlled milling machines. On installation the blocks are rigidly bound to machine quadrupoles with an accuracy of 0,1 mm relative to their magnetic axes constitute thereafter fixed points of the vacuum chamber.

Taking into account high mechanical precision for interchangeability, precision supports and alignment, the overall absolute precision is 0.3 mm rms.

The same BPMs are also installed for other applications: Experiments timing (8 BTPC), Instrumentation timing (2 BTPE), RF Phase control (20 BAPC), Q-meter (3 BQPE), Analog observation (2 BQPE) and Transverse Feedback (10 BAPE).

Fig. 27 : Aluminium block
(456 BPPE 131 x 70 mm)

Fig. 28 : Stainless steel block
(24 BPPC ⌀ 100 mm,
8 BPPS ⌀ 120 mm)

Fig. 29 : Stainless steel large
aperture block
(16 BPPM ⌀ 160 mm)

Fig. 30 : Electrical test bench

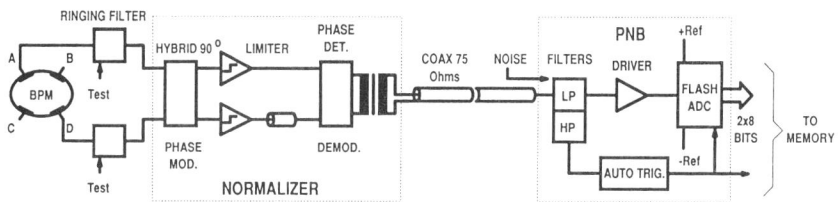

Fig. 31 : Narrow band processing scheme

22 LEP Beam Orbit Measurement System

2.3. RADIATION CONSIDERATIONS AND CABLE SELECTION

The material used for the blocks, the button electrodes and the brazes have very high radiation and corrosion resistance. The button coaxial line has been lengthened in order to set the end N-connector as far as possible from the chamber. The BPM is coated with lead in order to reduce the radiation leak through the button hole toward the connected cable. At LEP maximum energy it is expected that those coax cable jumpers may have to be replaced by mineral coax cables.

The choice of coaxial cables, (of total length about 300 km) for signal transmission was of great concern since they have to run inside the LEP tunnel.

The cables were first selected according to their best possible immunity to EMC effects and minimum transmission attenuation. The influence of pulsed ionizing radiation was then tested at DESY in the PETRA machine. Different cables were laid against the unshielded vacuum chamber inside the magnet gap. The radiation peak rate was about 7×10^5 Rad/s. The result was that ionization effects could be observed in cables with air dielectric but none in the ones with foam or solid polyethylene. Cable samples were finally irradiated with a cobalt source up to a dose of 10^8 Rad. Different foam densities, 0.4 and 0.25, were tested as well as different foaming methods like using freon, nitrogen, or a chemical reaction. The result of the test was that the main factor for radiation resistance is the foam density.

Finally the selected cables are 75 W and 50 W CATV cables with solid aluminium extruded outer conductors and 0.4 density polyethylene foam produced with pure nitrogen (for inner corrosion reason). Their price is low and their electrical performance is quite close to the one of high quality air coax cables.

2.4. ANALOG SIGNAL PROCESSING

The bunch induced pulse on the button electrode has a range of 1 to 2500 V (8.7-pF capacity, 25 mm excursion) and a short duration of about 100 ps at mid-height. Since the signal power collected by the button electrodes is not large [15,24] and has to be transmitted over long distances of up to 800 m, the analog preprocessing is not obvious. A solution inspired by previous designs [15,21] has been adopted in LEP for 448 BPM's connected to the so-called narrow band (NB) processing circuit for monitors wherever the time of passage between e^+ and e^- is <600 ns. Since the average time between bunches is 11 μsec. all signals can be recorded and stored in a local memory for up to 1000 revolutions.

2.4.1 Narrow Band processing scheme

This circuit is presented on Fig.31. It is characterized by a FADC of 8 bits resolution and 15 MHz maximum rate (Thomson EF 8308). Eight bits resolution is coarse for a BOM measurement and would not cope with the beam intensity variation of 200. An intensity analogue normalizing circuit has been added, which is based on a differential phase modulator with a $\pi/2$ hybrid followed by a linear phase demodulator made of an exclusive or gate. The resulting transfer function is :

$$\delta = \text{arctg}\, \Delta / \Sigma$$

But such a circuit has to be driven by a burst of sinusoidal signal. It is produced by ringing filters directly connected to the button electrodes (Fig.32). Their central frequency is 70 MHz and the burst last for about 150 to 400 ns. These filters must have very close and stable response to avoid phase errors very critical for the absolute precision of the BOM. The filters are double tuned circuits printed on Al_2O_3 substrate to withstand radiation. A capacitive test input allows for a complete test and

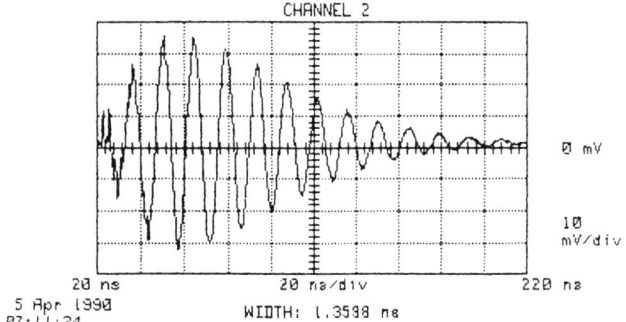

Fig. 32 : Beam induced signal at the output of the ringing filter

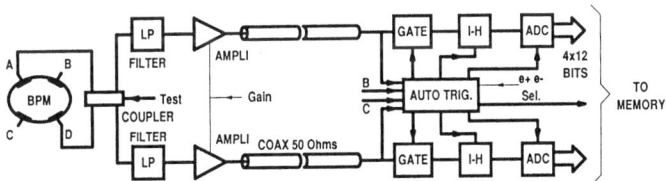

Fig. 33 : Wide band processing scheme

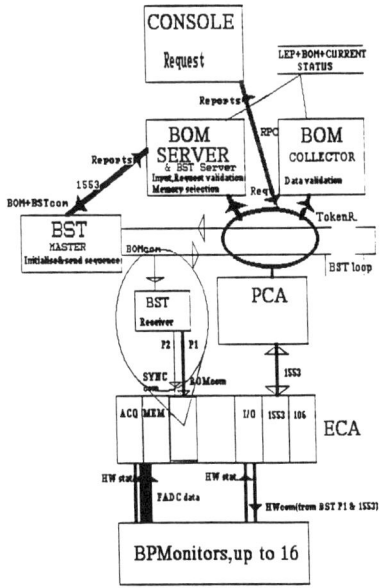

Fig. 34 : BOM system organisation

24 LEP Beam Orbit Measurement System

calibration of the processing chain since **its overall gain is directly proportionnal to the BOM scaling factor.**

The four signals are processed in diagonal pairs and transmitted to the electronic stations **by only two cables.** The transmitted pulses present a duty cycle modulation of 140 MHz carrier. The mean value is the position information and **the carrier is used as auto-trigger** for the FADC acquisition. Two acquisitions are programmed, one before the pulse to monitor the base-line and a second one 150 ns later to record the magnitude of the low-pass filtered pulse. The NB system needs then about 450 ns for the filter signal to decay before being ready for the next measurement. Therefore, it cannot be used for monitors near to the intersections, where e^+ and e^- bunches are closer together.

The eight bits of the FADC limit the readout resolution to:

$$\delta X = +- 85 \, \mu m \quad \text{and} \quad \delta Y = +-130 \, \mu m \quad \text{for } +-1/2 \text{ LSB.}$$

But it can be improved by the use of either following two verniers:

i) Statistical interpolation : if a random noise is added to the signal, the measurements are distributed around the steps defined by the LSB and the result of the average of these measurements is a smoothed transfer function. The optimum result is when the sigma of the input noise distribution equals 1/2 LSB. This noise signal is to be produced by noise generators which are not yet installed.

ii) Zoom: if the FADC references can be reduced for a smaller input voltage excursion, then the resolution can be improved by the same factor (not yet implemented).

Both vernier methods are limited by the FADC non-linearities. These are due to some coupling between input and output registers (not Grey code) inside the chip. A good part of these non-linearities are systematic and are compensated during the digital processing, thus improving the resolution by a factor four. When the amplitude of a coherent oscillation is measured over many revolutions the resulting BOM resolution should reach:

$$\delta X_{rms} = 7 \, \mu m$$
$$\delta Y_{rms} = 11 \, \mu m$$

Since any calibration procedure of an electronic processing chain presents also some error, the estimated r.m.s. absolute position error is 0.1 mm, which has to be added to the mechanical ones.

2.4.2 Wide Band processing scheme

As mentioned before, the BPMs close to intersections (56 units) require a faster acquisition electronics [22] which is called the Wide Band system (WB). It is represented in Fig. 33. Its characteristic is linear and therefore simpler to process. But its circuitry is more complex and requires gain changes according to intensity and a 12 bits ADC for larger signal magnitude dynamics. It is also beam self-triggered, can distinguish e^+ from e^- by their polarity, but can only treat one kind of particles at a time.

The button electrodes are directly connected to 50 Ω cables and low pass filters. This pre-integrated pulse is still asymmetric and the surface of the first part is sampled by an Integrate & Hold circuit. Its output value is then converted by the ADC in 3 μs. The WB system treats the four button signals separately. Two double directional couplers are used to inject the test and calibration signal before the low pass filters.

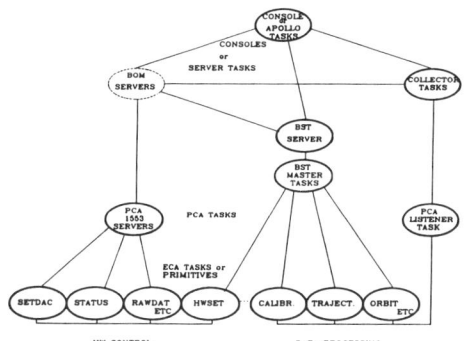

Fig. 35 : BOM software organisation

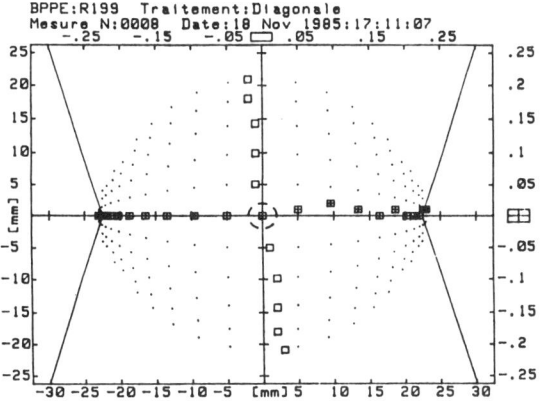

Fig. 36 : First order electrical transform

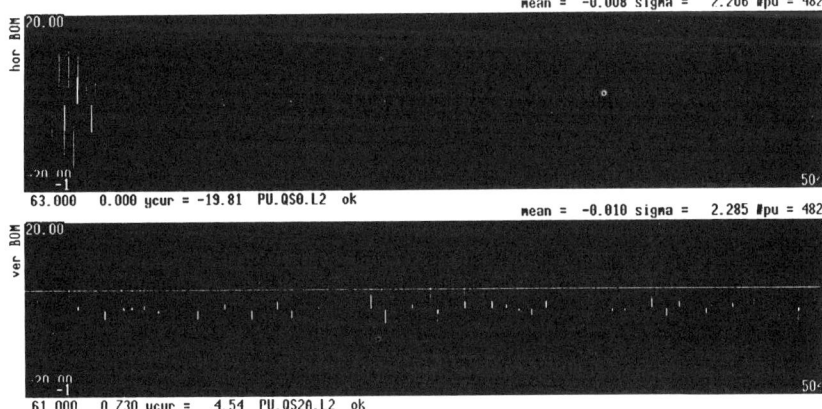

Fig. 37 : Orbit display with a closed horizontal local kick in arc 1-2

26 LEP Beam Orbit Measurement System

2.5. BOM SOFTWARE DESCRIPTION

The software for the control of the BOM system and data processing (Fig.35) did also represent a large effort (30%) relative to the hardware realization. Its concept is strongly influenced by the system distribution (Fig.24) and by the BOM control scheme [28], which depends fully on the LEP Control System [29]. Its development took place while the Control system features were progressively frozen. This was quite a burden for the instrument specialists. The local software (ECA-VME bus) is written in pascal within RMS68K system.

The request from the user (Fig.34) on an Apollo console is processed by the BOM Server (not yet fully implemented) which asks the BST Master to initiate a measurement sequence. It will define the BOM operating parameters like choice of particle type, gain setting, zoom condition, number of acquisitions etc. These data are forwarded to the Master of the Beam Synchronous Timing [27]. The task commands and the parameters are then transmitted to all stations through a command field of the BST message. All stations are synchronized by an acquisition command in the same command field and a turn number to be used as label for the self-triggered acquisition of up to 8 bunches position within a turn. The BST is synchronized to the RF clock. The turn period of 88,9 ms is further subdivided in fine time periods for bunch numbering purpose and rejection of parasitic signals outside the normal bunch sequence.

The BOM hardware is set according to the request parameters and raw beam position data are acquired and stored in the local ECA memory under beam synchronous commands. At each acquisition the base line data (NB Syst.) and position data are transmitted to a double multi-input acquisition memory situated in VME ECA crates (Equipment Controller Assembly). This acquisition memory has a depth of about 1000 revolutions. At the end of an acquisition sequence the Master sends a task command which starts the digital processing.

The BOM system has also the possibility to work in post trigger mode for transient capture. In this case the BOM system acquires permanently until it receives a stop command related to a defined event. The purpose of the double acquisition memory with its inputs in parallel is to allow for monitoring without perturbing the course of the running measurements.

When the processing task is started, the local CPU retrieves off line the data from the acquisition memory. The position calculation out of the raw data is done in two steps: first, the compensation and corrections of the analog acquisition errors and second the correction of the monitor non-linearities. It processes the data with the use of calibration and linearisation tables in order to correct for errors and non linearity of electronics, i.e. for the NB system: the Normalisers inverse transfer function (distorted arctan) as established by laboratory measurements and that of the FADCs. The first order electrical transform of a BPM of elliptical cross-section calculated from the diagonal differences [23,24] is shown in Fig.36 .

$$x = K_H (\Delta_1 - \Delta_2)$$
$$y = K_V (\Delta_1 + \Delta_2)$$

These measured positions x and y show strong non-linearities influenced by the particular geometry of the vacuum chamber equipotential. Polynamials P(x,y), with only six coefficients have been fitted to the four types of BPM geometries and will serve to reconstruct real positions X and Y to the utmost accuracy:

$$P(x,y) = ax+bx^3+cx^5+dxy^2+exy^4+fx^3y^2$$

$$X = P_j(x,y)$$
$$Y = P_j(y,x)$$

The positions will be further processed (Fig.35) for first turn trajectories (up to 16 turns), closed orbits, and FFT analysis. This last process means that magnitude and phase of the transverse oscillations of a permanently excited beam will be calculated for each monitor.

The resulting X and Y position file for 4 to 16 BPM is sent to the local PCA where a "Listener" routine sends it to the BOM "Collector" PCA via the Control System. The BOM Collector receives 40 files, checks for consistency, creates the BOM measurement output file and warns the user. The orbit output file is transmitted to an APOLLO display program (Fig.37) for visual analysis and futher use [32].

The BOM software is also designed for hardware diagnostic, calibration and closed orbit simulation in the absence of beam.

Due to time shortage, only a stripped-down development version of the software could be supplied for the start-up, called version 0, which has been successfully used until 1990 for LEP beam control and the evaluation of BOM performance. It does not yet offers the beam and bunch selection and parasitic rejection according to the data time label. This version allows for two first revolutions trajectories measurement (e+/e- and first bunch) and closed orbit measurement averaged over 10 or 50 revolutions and all bunches (no separation of e+/e- beams or single beam measurement).

2.6. BOM FABRICATION AND INSTALLATION

The injection test of July 1988 validated the BOM hardware design and series production was launched.

The subsequent last year of LEP construction and installation was very loaded. All BPMs were ready and installed as part of the quadrupole vacuum chambers [25] by the Vacuum Group. About 15 leaks on button electrode feedthroughs (total 2200) appeared after transport and tunnel installation, but no more since then. The origin of the leaks (tot. about 40) appearing at the central titanium conductor (Fig.37) has been investigated with the manufacturer and found to be due to combination of gold-nickel braze (imposed by corrosion resistance) with titanium forming a very hard alloy and too large a chamfer filled with braze and hence reinforcing the thin central condutor tube. A new model is studied. About 10 electrodes presented short-ciruits. They were due to very thin Al fibers produced by scraping the machining grooves at electrode introduction in the BPM block. The shorts could be removed by electrical discharge and current.

The BOM system did require an almost industrial production, testing and calibrating of its 3000 electronic modules and 150 crates. The series production and wiring was executed by industry, but the testing and calibration was done at CERN on especially developed and computer controlled test benches, which are also needed for BOM maintenance. Just after chamber installation and once the cables were pulled, a first team connected all the front-end electronics next to the BPMs. Two to three other teams assembled and tested the WB and NB stations hardware. The two months foreseen for overall BOM system check-out, with BST and Control system interface, were reduced to nothing due to late delivery of the tunnel under the Jura. It resulted in a tremendous debugging effort in the first weeks of LEP start-up.

28 LEP Beam Orbit Measurement System

2.7. BOM PERFORMANCE AND PERSPECTIVES

The first overall BOM system test [31] was only possible with a beam passing all around LEP and hence at LEP start-up (July 1989). The first two weeks were mainly dedicated to communication software debugging and to BST synchronisation relative to beam signals in all stations. **This made it difficult to use BOM trajectory measurement to guide the beam all around.** The BOM performance and reliability improved quickly, closed orbits could be corrected and the first events from colliding beams could be recorded after a week. Thanks to hardware stability and its few failures or bugs during the first weeks the effort could be concentrated on the software. After two months "An r.m.s. orbit distorsion of about 1 mm was achived routinely" [30].

During the January to March 1990 shut-down the BOM HW has been further tested and it was found that high voltage command cards created disturbances and that the test generators were desynchronised resulting in noisy calibration. These problems were cured and the BOM calibration precision was improved by a factor of 4 as verified by simulated orbit test. The BOM system was recalibrated for the 1990 running period which resulted in a better vertical orbit distortion of sigma 0.7-mm and reduced emittance [32] using 100 correctors.

Different methods have been used for BOM precision and resolution estimation. Since the average of position measurements either vertically from all BPMs or horizontally from those (112 BPMs) of straight sections presenting no dispersion should be zero, any discrepancy is an instrumental offset. The NB part of the system presents an horizontal offset of -0.8 for e- to -1.6 mm for e+ relative to its calibration and to other position measuring instruments. This offset is due to the sensitivity of the NB analog processing chain to the difference of length between the simulated beam signal (6 ns FWHH) used for calibration and the beam signal (100 ps, two polarities). In addition the double tuned circuit, chosen for faster decay reason, presents frequency variation during the burst (Fig. 3). Since the beam or simulation pulse width has an influence on the autotriggering of the measurement point, values may fluctuate by a few bits as it can be inferred from Fig.39 . For the time being the NB Horizontal offset is corrected by a program (BOMCOROFFSET) which subtract the offset. This NB circuit problem influences also the scale of the measurements. The scale has been checked by the dispersion method (Fig.40) : changing the RF by ±100 Hz and comparing the horizontal position in the arcs with its calculated value or by the bump method: beam position is recorded with induced open or closed bumps both in straight section or arcs and a scattergraph [33] is plotted of the measured values relative to the MAD calculated ones. Fig.41 shows the comparison of NB and WB monitors for closed bump across an intersection point. All methods have shown that the NB scale is too small and with variation from BPM to BPM: factor 0.8±10%. It is not the case for the WB system.

This NB problem is being cured during the 1990-91 shut-down: the test generators are modified, the ringing filters are replaced by a single tuned circuit which does not present frequency variation and an improved normalizer generation is under development.

The WB part presents instabilities due to the degradation of the gain relay contact. Laboratory investigation shows that these instabilities are due to the aging of a contact lubricant used in this type of coaxial relay. The replacement is progressing and should be ready for next start-up.

Except for these two points the BOM HW performance present good reliability and also has the advantage of its redundance. Calibration rate is of the order of two months. Special attention has been given to the cables for their EMC noise rejection and insensitivity to large bursts of synchrotron light. There has been no problems with cross-talk or picked-up noise.

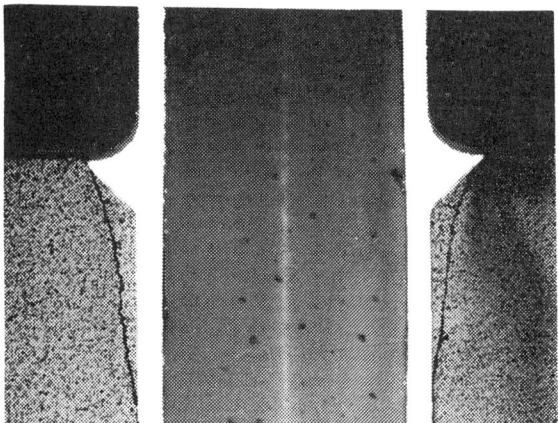

Fig. 37 : Feedthrough ceramic crack due to braze ring

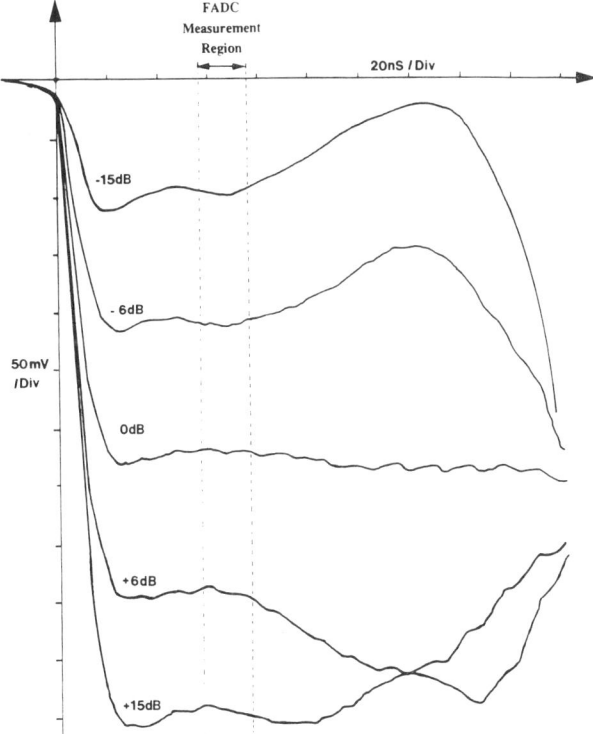

Fig. 38 : NB Processor input signal as seen by the FADC at acquisition for different signal ratios from a pair of electrodes

30 LEP Beam Orbit Measurement System

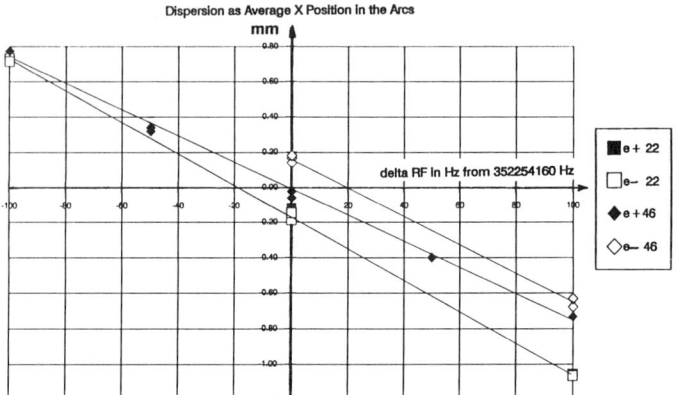

Fig. 40 : Dispersion measurement

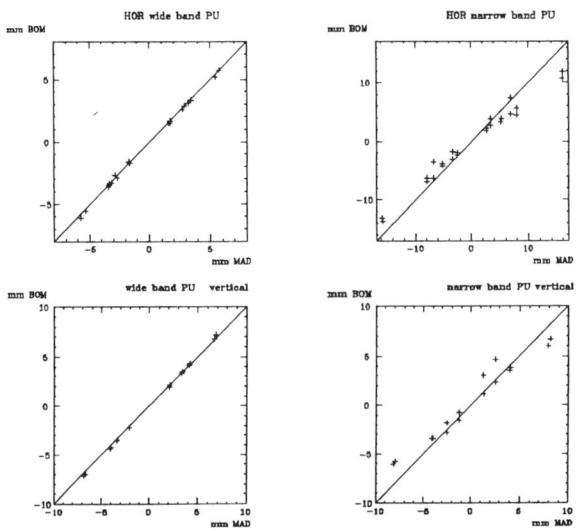

Fig. 41 : Scale measurement WB/NB, measured beam positions relative to calculated ones

Finally the orbit resolution is better than 65 µm and its absolute precision is 0.4 mm rms, as estimated from orbits analysis, even so the noise generators for interpolation are not yet operational.

The overall BOM system software reliability is insufficient and directly related to that of the LEP control system and to the present minimum diagnostic means. The version 0 software execution speed is too slow to allow for orbit averaging over a large number of revolutions. A version 1 has been installed and tested in the local processors (ECA). It not only offers additional BOM measurement facilities, like separate orbits for e+/e- in presence of the two beams, but also four times faster processing speed, i.e. 27 s for processing the average orbit over 30 turns and 8 bunches. Further improvement is in progress in order to reach the original design aims like acquiring an orbit in a few seconds for which the hardware and BST have been designed. A version 2 software based on OS9 system and C language and also on the use of direct Ethernet/Token Ring bridge connection to the Apollo console, is in preparation first for the WB monitors. Its purpose is to improve the response speed and performance.

2.8. CONCLUSION about BOM System

The main goal of the BOM system has been achieved, i.e. to allow for an orbit correction to less than 1 mm r.m.s. The absolute accuracy of the overwhole system was recently demonstrated when the vertical r.m.s. value was reduced to 0.5 mm (Polarisation run) with both results of increased luminosity of 20% and evidence of polarisation. Shortcoming in the normaliser NB system and gain switching of the WB system are being eliminated for the next running period by redesign of the involved modules.

Once version 2 of the software will be fully implemented more sophistications will be introduced to take advantage of the large acquisition capabilities of the BOM system like 1000 turns memory and second memory allowing to run suveillance and background tasks.

3. ACKNOWLEDGEMENT

The LEP Instrumentation has been quite an accomplishment requiring not only clever design but also tedious work and a lot of motivation for the 28 km of LEP and this from the scientific to the execution staff. This report should be a tribute to their merit.

I would like to thank my colleagues for their contribution to the preparation of this document. Their names are all quoted in the references.

REFERENCES

[1] C. Bovet, Beam Diagnostics for LEP, invited paper 13th Int.Conf. on High Energy Accelerators, Novosibirsk, USSR (1986), CERN/LEP-BI/86-16.

[2] M. Placidi, Status of LEP Instrumentation, Copies of transparencies presented at the 24th LEP Machine Advisory Committee 9th Feb. 1988.

[3] R. Jung, Private communication.

[4] K.B. Unser, Measuring Bunch Intensity, Beam Loss and Bunch Lifetime in LEP, Paper presented at the European Particle Accelerator Conference 90, Nice, F., June 12-16, 1990, CERN/SL/90-27 (BI)

[5] G. Burtin, R. Colchester, C. Fischer, J.-Y. Hemery, R. Jung, M. Vanden Eyden, J.M Vouillot, Mechanical design,Signal Processing and Operator Interface of the LEP Current Transformers, Paper presented at the European Particle Accelerator Conference 90, Nice, F., June 12-16, 1990, CERN/SL/90-30 (BI)

[6] F. Bertinelli and R. Jung, Design and Construction of LEP Collimators, IEEE Particle Accelerator Conference, Washington DC, March 1987. Proceedings 87CH2387-9 vol.3.

[7] C. Bovet, R. Jung, Emittance Measurement by Means of UV Synchrotron Light from LEP Main Dipoles, LEP Note, July 17th 1986. unpublished.

[8] E. Rossa, C. Bovet, B. Jenny, J. Spangaard (CERN), E. Jeanclaude (THOMSON,PARIS), M. Cuzin, M:C: Gentet, C. Ravel (CEA D:LETI;GRENOBLE), X-Ray Monitors to Measure Bunch Length and Vertical Profile at LEP, Paper presented at the European Particle Accelerator Conference 90, Nice, F., June 12-16, 1990, CERN/SL/90-23 (BI)

[9] R. Schmidt, C. Bovet, A. Burns, F. Ferioli, Q. King, J. Koopman, J. Mann, H. Michel, L. Vos, Experience with the LEP Wire Scanners, LEP Commissioning Note 24, April 12, 1990.

[10] E. Rossa, N. Adams (CERN), F. Tomasini, J.-M. Roth (ARP, STRASBOURG;F), Double Sweep Streak Camera for LEP, Paper presented at the European Particle Accelerator Conference 90, Nice, F., June 12-16, 1990, CERN/SL/90-24 (BI)

[11] C. Bovet, E. Rossa, N. Adams, A. Simonin, Single Shot Bunch Length Measurement at LEP by Stochastic Sampling of Synchrotron Light Photons, Paper presented at the European Particle Accelerator Conference 90, Nice, F., June 12-16, 1990, CERN/SL/90-25 (BI)

[12] K.D. Lohmann, M. Placidi, H. Schminkler, Design and Functionality of the LEP Q-Meter, Paper presented at the European Particle Accelerator Conference 90, Nice, F., June 12-16, 1990, CERN/SL/90-32 (BI)

[13] G.P. Ferri, M. Glaser, G. von Holtey, F. Lemeilleur, Commissioning and Operating Experience with the Interaction Rate and Background Monitors of the LEP, Paper presented at the European Particle Accelerator Conference 90, Nice, F., June 12-16, 1990, CERN/SL/90-26 (BI)

[14] M. Placidi, G.P. Ferri, M. Glaser, R. Jung, L. Knudsen, F. Lemeilleur, D. Rabault, R. Schmidt, (CERN), B. Dehning, J. Badier (MAX-PLANCK-INST.,MUNICH;G), Design and First Performance of The LEP Laser Polarimeter, Paper presented at the European Particle Accelerator Conference 90, Nice, F., June 12-16, 1990, CERN/SL/90-28 (BI)

[15] J. Borer et al., Beam Orbit Measurement for correction in LEP, LEP Note 437 (Febr. 1983).

[16] DESY, PETRA, updated version of PETRA Proposal (Febr. 1976).

[17] F. Peters, PETRA Monitors, private communication (Febr. 1978).

[18] J.L. Pellegrin, General Information about PEP BPM System, privat communication (Aug. 1982).

[19] Hajime Ishimaru, Al-Alloy-Ceramic Ultra-high Vacuum and Cryogenic Feedthrough, useful from DC to 6.5 GHz, Vacuum, Vol. 32, 753 (1982).

[20] Takao IEIRI et al., Performance of the Beam Position Monitors of the TRISTAN Accumulation Ring, Proceedings of the 5th Symposium on Accelerator Science and Technology, KEK (Sept. 1984).

[21] V. Rossi, Quadrature Normaliser for Position Measurement of continuous and single bunched Beam, SPS/ABM/VR/nbl/85-230 Report.

[22] J. Borer et al., Système de mesure de Trajectoires en TT6-TT1 et d'Orbites en ISR-R2 pour les Antiprotons,
Partie I, CERN/ISR-RF/82-08 (1982),
Partie II, ISR-RF/JB/DC/CP/cb/TN 6 (1982), unpublished.

[23] J. Borer, C. Bovet, Computed response of four Pick-up Buttons in an elliptical Vacuum Chamber, LEP Note 461 (July 1983), unpublished.

[24] J.-Borer, C.Bovet, D. Cocq, H. Kropf, A. Manarin, C.Paillard, M.Rabany, G.Vismara, The LEP Beam Orbit Measurement System, Proceedings of the IEEE Particle Conference, March 16-19, 1987.

[25] J-C. Brunet, Industrial Fabrication of Ultrahigh Vacuum Quadrupole Chambers for LEP, 11th Int. Vacuum Congress, K_ln, Sept. 25-26,1989. CERN/LEP-VA/89-60

[26] J-P. Bacher, N. Hilleret, Corrosion of Accelerator Vacuum System: Risks and Prevention, 11th Int. Vacuum Congress, Köln, Sept. 25-26, 1989, CERN/LEP-VA/89-52

[27] G. Baribaud, D. Brahy, A. Cojan, F. Momal, M. Rabany, R. Saban, J-C. Wolles, The Beam Synchronous Timing System for the LEP Instrumentation, Int. Conference on Accelerator and Large Experimental Physics Control Systems, Vancouver, Oct.30-Nov.3,1989, CERN/LEP-BI/89-66.

[28] G. Baribaud, D. Brahy, A. Cojan, F. Momal, M. Rabany, R. Saban, A Thys, J-C. Wolles, A Distributed DATA Acquisition and Monitoring System for the Beam Instrumentation of LEP, Int. Conference onAccelerator and Large Experimental Physics Control Systems, Vancouver, Oct.30-Nov.3,1989, CERN/LEP-BI/89-66.

[29] P-G. Innocenti, The LEP Control System: Architecture, Features and Performances, August 1989, CERN/SPS-ACC/89-35.

[30] D. Brand et al., A Description of the Orbit Correction Procedure used for the LEP Start-up, LEP commissioning Note 11, Sept. 29th, 1989.

[31] G. Bribaud, J. Borer, C. Bovet, D. Brahy, D. Cocq, H. Kropf, A. Manarin, F. Momal, C. Paillard, M. Rabany, R. Saban, G. Vismara, The LEP Beam Orbit Measurement System : Status and Running-in Results, European Particle Accelerator Conference, Nice France, June 12-16, 1990, CERN/SL/90-31 (BI).

[32]   D. Brandt, P. Defert, W. Herr, J-P. Koutchouk, J. Miles, T. Risselada, R. Schmidt, Orbit Processing for LEP, European Particle Accelerator Conference, Nice, France, June 12-16, 1990, CERN/SL/90.

[33]   H. Moshammer, BOM Calibration, private communication, July 31, 1990.

[34]   A video tape (PAL) presenting some aspects (mainly the button electrode) of the fabrication, testing and installation is also available.

THE AGS BOOSTER RADIATION LOSS MONITOR SYSTEM

E.R. Beadle and G.W. Bennett
AGS Department
Brookhaven National Laboratory
Upton, N.Y. 11973

Introduction

Loss monitor systems detect and measure the radiation produced by particles lost from the primary beam. These systems are used to measure beam loss, residual activation, beam size, and extraction efficiency. In addition, they facilitate accelerator tuning and aid in equipment protection[1,2,3]. One broad function is to help minimize activation caused by beam loss, and thus reduce personnel radiation exposure. At Brookhaven National Laboratory loss monitor systems are used at the 200 MeV proton linac, and at the Alternating Gradient Synchrotron (AGS) to identify the temporal and spatial character of beam losses[4,5].

The Booster Radiation Loss Monitor System (BRLM) described here features adaptive dynamic range, ionization chamber detectors spanning the transfer lines as well as the ring, precision analog integrators, a VME based "smart" analog to digital converter, and a microprocessor based controller-interface to the host computer. In addition, a complex scheme allows selected detectors to provide a variety of protective features, and the capability to isolate any of these detectors from the protection circuitry. The BRLM system is an essential diagnostic tool for commissioning when losses can be large, yet has sufficient sensitivity to detect small losses during normal running.

The Booster (Figure 1) has a circumference of 201.7 m and is designed to accelerate protons, polarized protons, and ions from Carbon to Gold. For unpolarized protons the Booster is designed to operate at 7.5 Hz with an intensity of 1.5×10^{13} protons/pulse, and an extraction energy of 1.5 GeV. Polarized protons will be accumulated in the Booster from 20 or more linac pulses, requiring a 3 second cycle to deliver 10^{12} protons. Protons are injected into the Booster via the Linac-to-Booster (LTB) transfer line, and all species are extracted using the Booster-to-AGS (BTA) transfer line. Heavy ions are injected into the Booster from the Tandem Van de Graaff using a separate transport line. The Heavy Ion intensities are planned to range from 50 to 3×10^9 per pulse with a 700 ms cycle duration.

AGS Booster Radiation Loss Monitor System

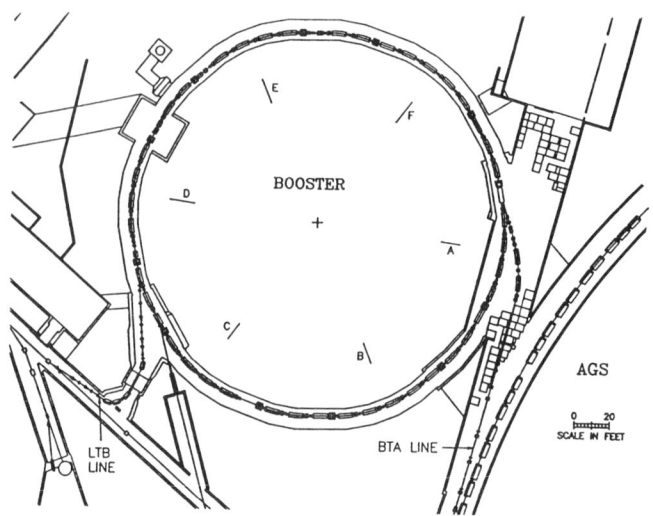

Figure 1. Layout Drawing of the AGS Booster

The AGS Booster Loss Monitor System is similar to the AGS Loss Monitor System in use since 1979[6]. That system has 120 detectors providing complete coverage of the AGS ring. Each detector's output current is converted to a voltage which is digitized by a voltage to frequency converter. The V/F outputs are accumulated in 16 bit counters which are interfaced to an 8085 microprocessor. The data is processed to provide numeric and graphic displays for one chosen time interval. The Booster system incorporates the most useful features of the AGS system and provides new capabilities. One feature of the BRLM is its <u>adaptive dynamic range</u>. This means that the upper limit of the dynamic range is automatically shifted to a higher level, and the lower limit is fixed. Therefore, losses that would otherwise saturate the signal processing electronics can be tracked. Other improvements include increased sensitivity to beam loss, better graphic displays, and support for at least three simultaneous users with up to seven time windows for data accumulation per user per Booster cycle.

Architecture

As shown in the hardware block diagram (Figure 2), the detectors are biased by floating supplies. The bias supplies output to analog integrators, which accumulate the charge from the detectors. The integrators were originally designed by BNL and are now commercially available (ATL Inc.). Following the integrators is a commercial VME bus based data acquisition system (Datel Inc.). This system consists of multiplexing sample and hold cards (Datel DVME-645) and an intelligent analog to digital converter board (Datel DVME-601). Using preprogrammed firmware the A/D card locally controls the entire acquisition process via the analog expansion bus.

The scan trigger causes the A/D card to output the hold command and address each channel for digitization. After the S/Hs have settled, the integrators are reset to zero and then resume integrating. The circulating beam monitor is used to provide the BRLM user with a measure of the circulating beam intensity at the times integrated loss data is acquired.

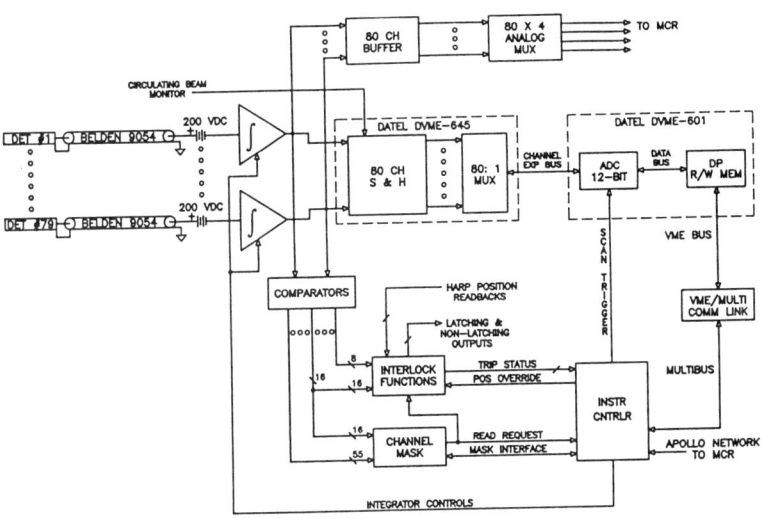

Figure 2. BRLM System Block Diagram

Each channel is digitized to 12 bits and written to the dual-ported memory on-board the A/D card with a throughput of 140 KHz. The dual-port memory is shared with the Multibus based Instrument Controller which provides control of the BRLM hardware and communication with the host computer. Communication between the Multibus and VMEbus systems is achieved using a commercial bus translator (Bit-3 Corp.). The bus translator makes the data transfer transparent to the processors on both sides.

The comparator block provides threshold detection for the integrator outputs. The comparator outputs are used by the channel mask and interlocking functions. The channel mask prevents selected channels from requesting ADC triggers. Some channels, primarily those located in the transfer lines, are used for beam interlocks by various systems in the accelerator to either turn off linac injection promptly, or prevent Booster acceleration in subsequent cycles, preventing damage to accelerator components from lost or missteered beam.

The 80 x 4 analog mux enables the main control room to monitor a subset of the analog signal outputs.

Adaptive Dynamic Range

The data acquisition process is started in response to a scan trigger. The trigger is caused by either a programmatic request from the host computer, or by the BRLM hardware from the comparators. The combination of software and hardware generated scan requests is used to implement an adaptive dynamic range limit. If the scan and reset process were to only be initiated by the software, one or more channels might exceed the ADC full scale range between requests, and some data would be lost. The comparators are used to flag the condition that at least one signal is approaching the ADC limit. This flag is processed by the channel mask to establish that it originates from a detector designated to generate read requests. If so, the flag is forwarded to the instrument controller, which arbitrates between software and comparator sources before producing a scan trigger. In either case, the scan trigger causes all the channels to be digitized and stored, and then the BRLM timing system resets all the integrators so they can continue to accumulate data. When the machine cycle is complete, the software assembles the individual scans into the time intervals requested by the users. Although the ADC limits the instantaneous dynamic range on each scan, the total dynamic range is bounded only by the word size of the processor summing the individual scans. Additionally, the minimum signal that can be measured is equal to 1 LSB of the A/D convertor, and this signal level defines the lower limit of the dynamic range.

Typically, the standard ring channel comparator trips are set to 80% of the ADC full scale, 8V. Auxiliary ring channels and those originating in the transfer lines are primarily for equipment protection, with the trip levels adjusted for a small margin above normal operational losses.

The mask also allows shifting of the dynamic range for increased precision measuring low level signals in the presence of high losses. Additionally, the mask also allows disabling malfunctioning channels which could "lock-up" the ADC subsystem. Although a channel is disabled from requesting service its data is still digitized by the ADC. The mask concept is that of an AND-OR array. Each AND gate receives an input from one comparator and a disable line. The outputs of the AND gates are ORed together to generate a read request. In this way any or all of the channels can be disabled in software from the main control room. This provides the system with the convenience of switching detector channels on and off from the control room in a controlled and documented manner.

The mask is designed for up to 72 detector channels and is expandable to 96. Only 72 channels are required because the 8 injection line detectors are automatically read after injection by a background logging task; and multiple ADC scans cannot be performed on the timescale of the linac injection.

Detector Design and Mounting

The detector is a coaxial ionizaton chamber made from lengths of RG 318/U style Heliax cable (Andrews Corp.), pressurized to 10 psig with an Argon-CO_2 mixture. The bias voltage of 200 VDC has been selected so that the detector operates near the middle of the ion saturation region. In this region the detector behaves as an ideal current source. The relatively low bias voltage does not require guard rings on the signal connector, simplifying the design and lowering the cost of the detector. The ring detectors are each 5 m long; other locations have detectors 3.6, 5, or 6.4 m in length. At one end is a rexolite insulated UHF connector for signal output and bias input. Rexolite is used because of its radiation resistance. Both ends have fittings to allow gas flow when desired.

The detectors are mounted close to the median plane on the tunnel wall cable tray approximately 30 inches radially inside the reference orbit (Figure 3). Median plane mounting maximizes the detector sensitivity to beam loss. The cable tray provides constant geometry with respect to the central orbit while not blocking access to the magnets and beam pipe. In locations where the tunnel walls fall away from the beam pipe special mounting stands are used to maintain the detector geometry.

In the ring there are 48 detectors, one spanning each lattice half cell, with the ends of the chambers overlapping slightly at the middle of each dipole magnet reference location. In addition there are 16 more detectors, 8 in each transfer line. The transfer line detectors are positioned below the beam line on the magnet support. This position results in potentially less sensitivity than on the beam plane, but experience with the 200 MeV linac system shows the sensitivity is still adequate. There is also provision for mounting up to 16 auxiliary detectors at arbitrary locations in the Booster and transfer lines as required.

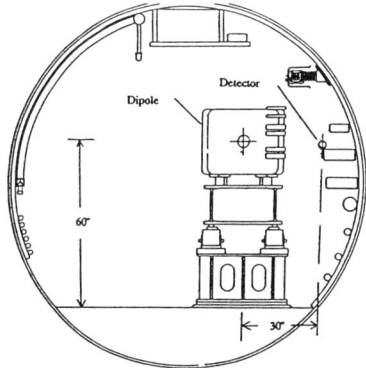

Figure 3. Tunnel Cut-Away View Showing Detector Mounting

Detector Biasing

The detectors are biased at 200 VDC and the output is processed by integrator circuits which are DC coupled to the detector output. The 200 VDC is generated using a DC to DC converter. A simplified schematic is shown in Figure 4. The circuit uses an unregulated 15 VDC supply chopped at 1 MHz with a power FET. The FET is schematically shown as a switch. Neither the value of the DC input voltage nor the switching frequency is critical because of the large operating region of the ion chamber and wide transformer bandwidth. The transformer is a 15:1 step-up (North Hills Co.). The secondary voltage is adjustable over a range of 180 - 225 V under no load conditions. The adjustment is achieved by varying the primary circuit resistance implemented with a potentiometer. The circuit is rated for use at up to 1.5 ma at 140 V. At currents greater than 0.5 ma the output voltage decreases due to transformer losses. The circuit is very stable, with the output regulated to within 2%, and an output temperature coefficient of less than 0.1% per degree Celsius.

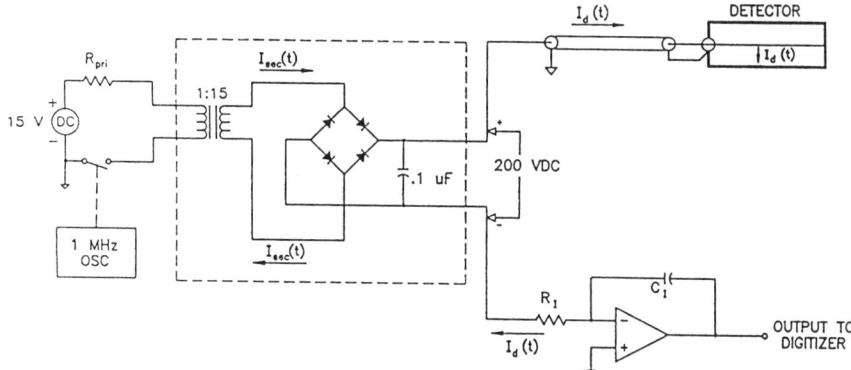

Figure 4. Simplified Schematic of the 200 VDC Bias Supply.

The supply output current is rectified and filtered by the diode bridge and capacitor. DC coupling to the detector is achieved by virtue of the secondary coil current. The current exiting the secondary and entering the detector must be "made up" and the integrator returns the load current. Therefore the detector current is directly measured by the integrator. Each transformer actually has four secondary coils that couple to the single primary coil. Therefore, up to 4 detectors can be accommodated by a single transformer assembly. The 6U eurocard packaging used allows two transformers per module to be mounted, and thus a single module supplies up to 8 detectors.

The circuit has been designed and fabricated to reduce leakage current to below 10 pA per channel, and provide high isolation between channels. To reduce the effects of leakage due to variations in ambient temperature and humidity the boards have been baked and conformally coated. Also, the bias and signal outputs are on two separate connectors to decrease leakage. The channel-to-channel isolation achieved is greater than 10^{14} ohms.

The cable connecting the bias supply and the detector is a Belden 9054 RG59/U type coaxial triple-shielded cable with 75 ohm impedance. The cable is constructed with dual 95% braids followed by 100% foil coverage on the outer layer. This cable was selected for its shielding effectiveness of 110 db. Adequate shielding is important when measuring low level signals in the hostile RF environment of accelerators. The leakage resistance of this cable is very high due to the polyethylene dielectric, which is also much better in radiation environments than Teflon dielectric cables.

Integrators

A simplified schematic of the integrator circuit is shown in Figure 5. The integrators have the capability to gate the integration time interval, hold or reset the output, and select gain. The integrate, hold, and reset modes are controlled by the hold and reset switches. The scale factor is selected using the gain switch, which selects the integrator feedback capacitance for either the high sensitivity (C1) or low sensitivity (C1+C2) value. The reset and hold switches are implemented with FETs because of speed requirements. The gain switch is realized with a relay because low resistance is required. When necessary, gain switching is performed between machine cycles, thus adequate settling time for the relay is available. All of the switches are controlled by TTL compatible circuits that interface to the instrument controller.

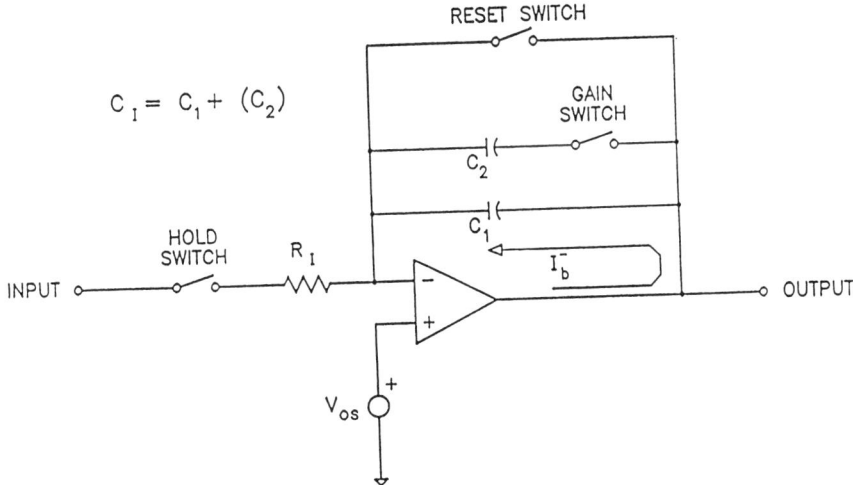

Figure 5. Simplified Integrator Schematic

In the integrate mode the gain switch is set to the position desired, the hold switch is closed connecting the input into the circuit, and the reset switch is opened. To enter the hold mode the input is disconnected by opening the hold switch. However, the hold function is implemented with separate hardware, the Datel DVME-645, allowing maximization of the integration time duty factor during a machine cycle. To reset the integrator, the best method is to open the hold switch and close the reset switch. However, in this application it is acceptable to keep the hold switch closed when resetting. The result will be a small residual voltage, much less than the A/D converter resolution, on the capacitor resulting from the detector signal at the end of the reset interval. The reset duration is set for 10 us by a monostable multivibrator. In the integrate mode, bias and offset errors affect the accuracy of the integrator output. The integrator output voltage error may be written as:

$$V_{error} = V_{out} + \frac{1}{C}\int I(t)dt = \frac{1}{RC}\int V_{os}dt + \frac{1}{C}\int I_{b-}dt + V_{os}$$

where V_{out} is integrator output voltage, $I(t)$ is the input current (assumed to flow into the integrator), V_{os} is the offset voltage, and I_{b-} is the opamp bias current. Because the detector acts as a current source the first error term is negligible. The next term is the integral of the inverting terminal bias current which causes the baseline to drift. To minimize the effect of this parameter a JFET input opamp (Burr Brown 3527BM) is used, and the bias current is on the order of 1 - 10 pA. Because the input stage is FET based, the amplifier input offset voltage is very high compared to bipolar units. The voltage offset directly affects the accuracy of the integrator, and the zero adjustment can be used for reducing this term. The zero adjustment can also be used to counteract the effects of the offset voltage in applications where low source impedances are used. Compensation for the charge injection caused by the JFET switches is also included. RC networks differentiate complements of the FET switch control lines, providing pulses of charge into the analog signal path that null the effects of charge injection (Figure 6).

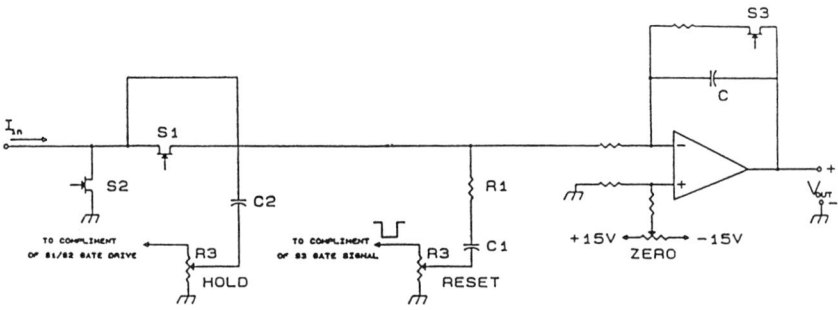

Figure 6. Charge Injection Compensation and Zero Adjustment

Selecting values for the integrator input resistance and feedback capacitance requires consideration of several factors. The input resistor limits the speed of response for the integrator circuit. To quantify the effect of the input time constant the system has been modeled using SPICE. The coaxial signal cable was modeled as resistance and capacitance shunting the integrator input. The detector was modeled as an ideal current source, and the opamp as an ideal voltage controlled voltage source to eliminate any model dependent effects. For simplicity, the integrator input current and output voltage were studied as a function of time in response to a square current pulse. The standard value of input impedance, 36 kOhm, with the typical cable capacitance forms an input time constant of 360 us, and requires more than 10 time constants to accumulate only 93% of the signal! Therefore for the integrators monitoring injection, which require response times on the order of 1 - 10 us, an input impedance of less than 1 kOhm is needed. However, voltage noise pick-up induces currents into the integrator, and 510 Ohms has been experimentally determined to be the lowest value consistent with a 12 bit digitizing system. For channels used in the extraction interlocks, the single turn ejection generates signals on a 1 us timescale. The hardware is not able to take corrective action so quickly due to the capacitance of the signal cables. Therefore, for lossy extractions the only choice is to inhibit further Booster cycles. The time between successive Booster cycles is at least 130 milliseconds, and large input resistances are acceptable, thus 36 kOhms is used. This high value has the benefit of good noise immunity and limits the loss of signal to less than 3% during the integrator reset. A 5 kOhm impedance has been selected for the detectors in the ring as a compromise between speed and noise immunity. The integrator capacitors also must be considered. Accurate, stable, and low leakage capacitors are required, therefore 1% mica capacitors are used. The minimum capacitor value is selected on the basis of the maximum drift rate and noise level tolerable. For this application 180 pf has been selected. At the high end, the amplifier output drive limits the charging rate of the capacitance and thus the exposure rate of the detector. A 1 nf value has been chosen in order to capture a 10% beam loss at full energy without saturating the integrator.

Data Acquisition System

A block diagram of the data acquisition system is shown in Figure 7. The system uses the Bit-3 communication link to transfer the data between the Multibus instrument controller and the VMEbus data acquisition hardware. The link transparently transfers the data between the controller and data acquisition system without requiring special VME drivers. The link maps the VME memory to the Multibus processor as local memory and allows both processors to operate at full speed. The Bit-3 link must be placed in slot 1 in the VME chassis because it provides the VMEbus arbitration logic. Although not shown on the diagram, there is an adaptor on the Multibus side as well. In response to a scan trigger the DVME-601 card automatically outputs the addresses to access the held values of the integrated detector outputs and the circulating beam monitor, digitizes these values, and places the data in the dual-port memory shared with the Bit-3 link. The 601 board has the capacity to address a total of 256 channels, with 16 being onboard channels. The onboard channels are multiplexed and then sampled, but some sampling skew will result if they are used. To eliminate this as source of error, only offboard channels have been used.

The buffer/driver card is provided to increase the bus drive ability of the 601 to match its addressing capacity. The buffer card uses two FAST series TTL chips (F541) as buffers. The low pass filter shown is for signal conditioning, and it attenuates radio frequency interference from the 1 MHz oscillator in the bias supply, and the 200 MHz RF in the linac.

The ADC output is coded as bipolar 2's compliment, however, other codes are selectable with jumpers. This application uses the 601's onboard instrumentation amplifier at unity gain. Other gains are jumper selectable in decade steps between 1 and 1000, but gains greater than unity require unacceptably long settling times for this application. The DVME-601 board includes a Motorola 68010 microprocessor, 32 kBytes of read/write memory with an expansion socket for additional memory, and preprogrammed firmware. Therefore this card provides considerable data preprocessing capability, however, the current design philosophy does not make full use of these features. The DVME-645 is a 16 channel simultaneous sample and hold with an output multiplexer, cascadable to 256 channels. The board is jumper programmable to set either single or double ended mode, and base address. The BRLM uses the single ended mode. The S/Hs are compatible with a 12 bit system, drooping less than 1 LSB in the 560 us required to complete the 80 channel scan.

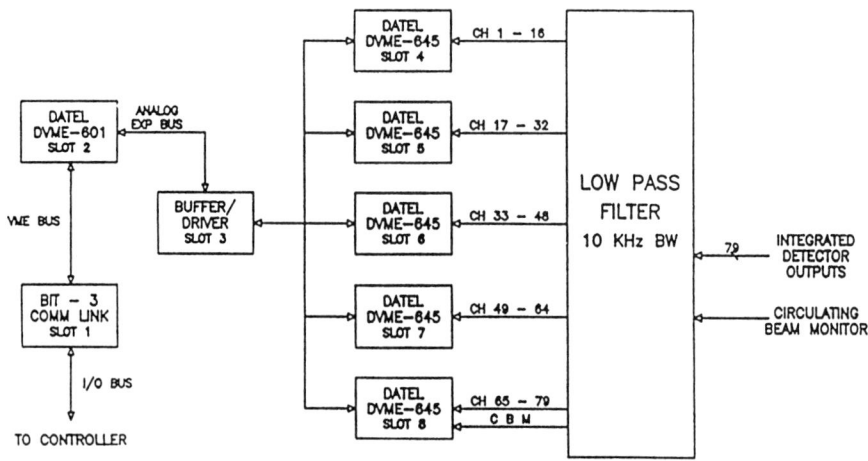

Figure 7. Block Diagram of Data Acquisition System

Figure 8 shows the data acquisition system timing. In diagram 8a the integrator output approaches a threshold (V_T). The comparator monitoring that channel will then raise its flag to the channel mask, which in turn requests the controller to generate a scan trigger. The controller outputs the trigger and the data acquisition hardware responds with a sample command to the S/Hs and commences the digitization process. The S/Hs then acquire all the integrator outputs, and after a 15 us delay that allows for acquisition and settling, all the integrators are reset to zero. The integrators then continue to accumulate the detector outputs. The digitizied data from each scan is stored in tables for post-processing to generate the displays for each user. As discussed above, depending on the integrator input time constant, the signal cable will store some of the signal occurring during reset. In Figure 8b the mode reflecting a programmed reset is depicted. In this case the integrators have accumulated a loss below threshold, and a read request is generated by the application code at the host level. A scan trigger is generated for the data acquisition system, and the timing proceeds as before.

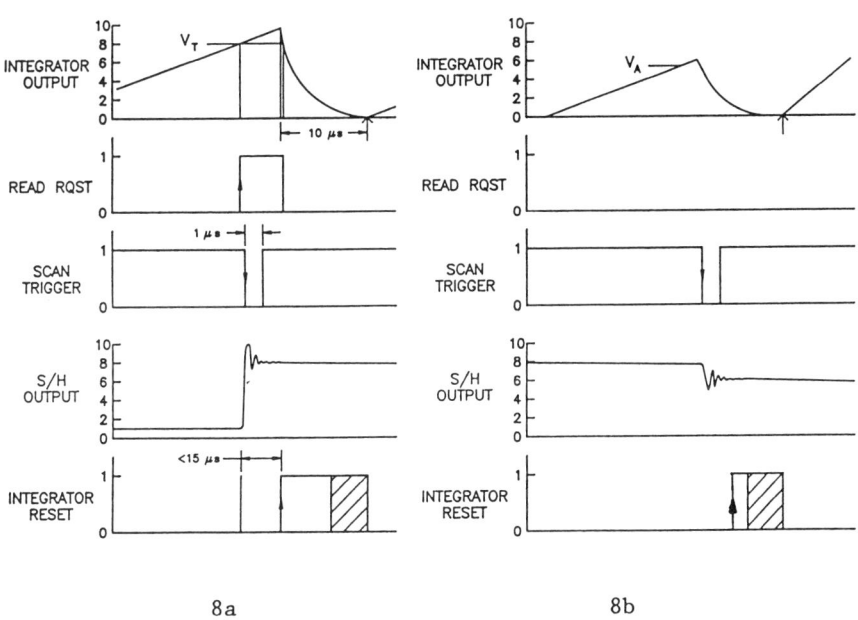

8a   8b

Figure 8. Data Acquisition Timing

<u>Beam Interlock Functions</u>

This hardware offers equipment protection by monitoring a set of selected detector channels for integrated losses exceeding some threshold. Primarily the transfer line channels are used, however, during injection and extraction the entire ring is also however, during injection and extraction the channel mask output is also monitored. The signals from this hardware are used as part of the Booster's non-latching Fast Beam Interrupt (FBI) and latching interrupt systems.

The non-latching system causes the linac injection pulse to be shut off at the ion source using the FBI circuits. The nominal proton injection duration is 200 us, and can be shut down in a few microseconds. The BRLM output to this system is formed by ORing the comparator outputs for the LTB and ring detector channels during injection. The BRLM output to this system is automatically reset at the start of a new Booster cycle. Latching interlocks trip an external hardware latch, inhibiting Booster injection until a manual key switch is cleared by a main control room operator. The BTA line detectors are used only in the latching interlock circuitry and can trigger the output at any time during the cycle. The ring and auxiliary detector channels can be used in this system also, but they are time gated to initiate latching interlocks only during injection and extraction times. Latching is used to allow somewhat complex logic schemes to be implemented, e.g. two trips out of any three successive cycles.

Conditions exist when high losses must be tolerated, such as invasive beam profile measurements using a multiwire device like a "harp". Harp insertion causes spray downstream that may trip one or more detectors, resulting in interlocks which turn the beam off, and thus prohibit the measurement. To prevent this situation, the capability to disable the offending detectors when a harp is inserted has been included. Position readbacks indicate each harp's position and hardware switches select the detectors to be disabled. In locations, such as the BTA line, where multiple harps can affect the same detector, the readbacks are encoded in an ORing circuit. The control room operators can override the readbacks and maintain the integrity of the BRLM portion of the interlocking systems regardless of a harp's position. This feature can be used as a safety check against unauthorized users actuating devices, and as a way to check the interlock circuits.

<u>CONTROLS INTERFACE</u>

Control for the system is provided by the Instrument Controller. The controller hardware is divided into input/output and communication sections. The I/O section is the direct interface to the BRLM. It consists of one half of the Bit-3 adaptor, TTL compatible custom interface modules, timing decoders, and a microprocessor module. The interface modules contain multiple Intel 8255s, and thus provide a high density of I/O lines. The timing decoder is used to interpret timing triggers available from the accelerator timing system. The processor module provides local intelligence which decodes and formats data for the communications section. The communication section consists of a dual-port memory, a processing module for instruction parsing, and a direct memory access (DMA) channel between the controller and the data network station. The station connects through a communication network to the Apollo host computers.

SOFTWARE

The software is the user's interface to the BRLM system. Accessible at this level is control over programmable read times, display of data, channel mask status, and the harp position overrides. Several displays for the detectors are available to the system user, showing information such as physical location, operational status, detector data, circulating beam monitor readings, and interlock system assignments. The displays for detector data include numerical representation of the outputs over specified intervals, and graphs of the numerical data with capability to overlay several cycles. For these displays, the software assembles the sampled detector data for each user and schedules the programmed resets to avoid conflicts. There is also a "standard" user called the background logging task. This task runs continually to generate a record of the machine operation. The software also provides a flight recorder feature, similar to an aircraft "black-box". Several acceleration cycles of loss data are buffered, with older data successively averaged over progressively longer time intervals. At any time the data can be played back to look for prognosticators of beam loss and to serve as a diagnostic aid in checking other systems. As an addition to this task the ability to generate and display the mean and standard deviation for losses both spatially and in time is being developed. These parameters can be used to inhibit the machine, in non-realtime, when loss conditions deviate significantly from established norms.

REFERENCES

1. M.A. Plum et al., "Fail-Safe Chamber Errant Beam Detector for Personnel Protection," Proc. IEEE Part. Acel. Conf., Chicago, Ill., 1989.
2. R.G. Jacobsen and T. Mattison, "Beam Loss Monitors in the SLAC Final Focus," Proc. IEEE Part. Acel. Conf., Chicago, Ill., 1989.
3. M. Fishman and D. Reagan, " The SLAC Ion Chamber System for Machine Protection," IEEE Trans. on Nucl. Sci., pp 1096 -1097, June 1967.
4. R.L. Witkover, "Microprocessor Based Beam Loss Monitor System for the AGS," IEEE Trans. on Nucl. Sci., pp 3313 - 3315, June 1979.
5. G.S. Levine and J.C. Balsamo Jr., "The AGS Beam Loss Monitoring System," BNL Report 19836.
6. R.L. Witkover, "Microprocessor Based Beam Loss Monitor System for the AGS," IEEE Trans. on Nucl. Sci., pp 3313 - 3315, June 1979.

ELECTRONIC SYSTEMS FOR BEAM POSITION MONITORS AT CEBAF*

W. Barry, J. Heefner (presenter), and J. Perry
Continuous Electron Beam Accelerator Facility
12000 Jefferson Avenue, Newport News, VA 23606

ABSTRACT

The Continuous Electron Beam Accelerator Facility (CEBAF), presently under construction, consists of a pair of .4 GeV recirculating linacs and will produce a 4 GeV CW beam with average currents of up to 200 μA. Because the accelerator is recirculating, multiple beams of different energies are present simultaneously in the linac beamlines. In these sections, it is required that the position of each different energy beam be measured separately. To achieve this, the electron beam itself must be modulated in time in such a way that each different energy beam in the linacs induces signals on the beam position monitors (BPMs) that can be separated using correlation techniques.

In this paper, two types of correlation receiving systems used at CEBAF for the linac BPMs are described in detail. These receivers are based on the pulsed and bi-phase pseudorandom modulated carrier methods for modulating the electron beam. In addition, the BPM receivers for the CEBAF arc regions, where the different energy beams are physically separated, are also described.

OVERVIEW

The Continuous Electron Beam Accelerator Facility (CEBAF) is a five pass recirculating electron accelerator with an energy gain of 800 MeV per pass.[1] The accelerator itself consists of a pair of 400 MeV CW linacs connected together by two sets of five recirculating arcs stacked on top of one another. Each linac consists of 20 superconducting accelerating modules with energy gains of 20 MeV each. The beam characteristics relevant to the Beam Position Monitor (BPM) system are as follows:

1) Fundamental RF frequency $f_o = 1.497$ GHz
2) Bunch length (1° of RF) $\tau_b = 1.85$ psec
3) Average beam current $1\ \mu A \leq I_{av} \leq 200\ \mu A$
4) Recirculation time $\tau = 4.2\ \mu sec$
5) CW with 1 bunch per RF cycle

As indicated in figure 1, the five different energy beams reside in separate vacuum chambers in the arcs. However, in the linacs, all five energies (passes) reside in the same beampipe. In order to tune up the machine and perform

---

* This work was supported by the U.S. Department of Energy under contract DE-AC05-84ER40150.

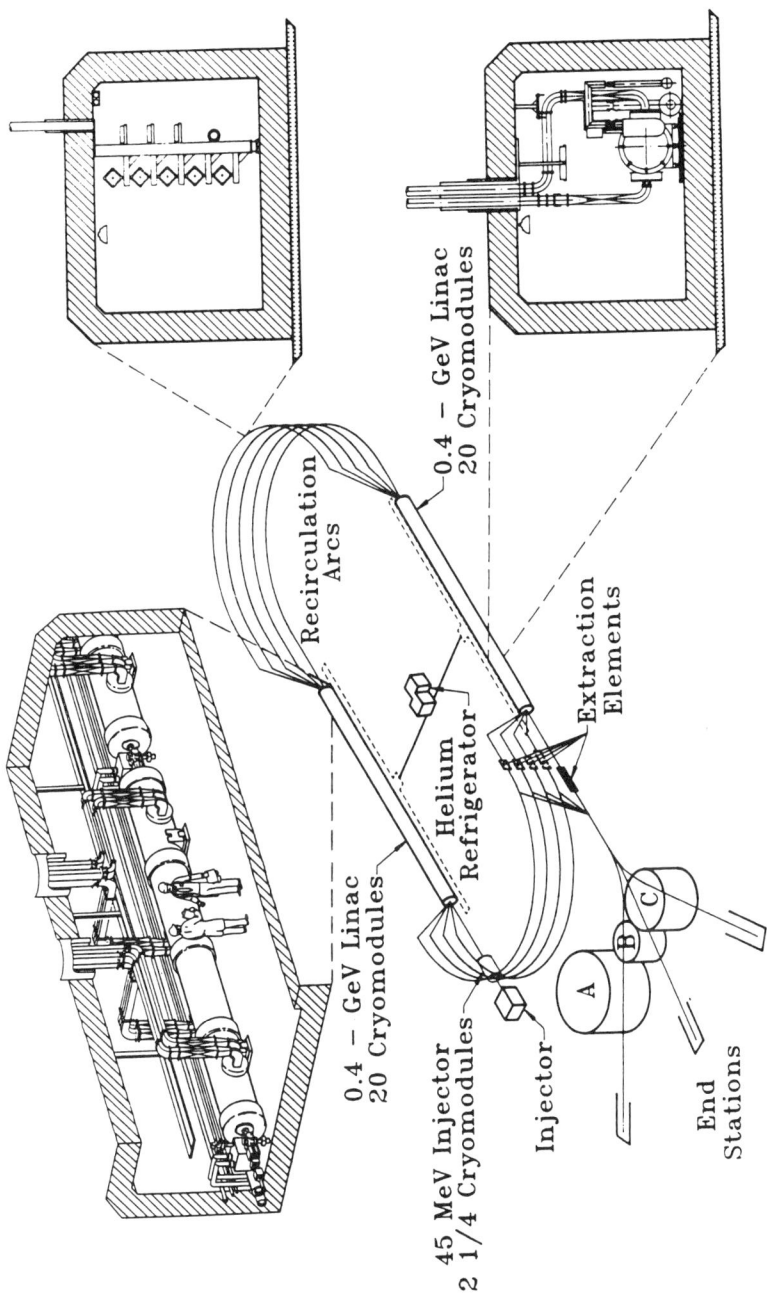

Figure 1. CEBAF machine configuration.

periodic orbit corrections, it is necessary to track beam position through the entire five pass orbit.[2] In the CEBAF arcs, this can be accomplished with standard position monitoring techniques. In the linacs, the problem is complicated by the fact that five different energy beams following five different paths are present simultaneously in the same beampipe. In order to distinguish between passes in the linacs, it is necessary to mark or modulate the beam in a controlled manner so that the beam current for each pass in the linacs possesses unique characteristics.

In order to disturb experimental processes in the end stations as little as possible, it is necessary to modulate the beam current only when an actual orbit correction is being performed. Therefore, the CEBAF BPM system actually consists of two subsystems, one for the arc regions and one for the linacs. The arc system consists of pickups and receivers which operate at the fundamental RF frequency of 1.497 GHz. These BPMs monitor beam position in each of the five arcs continuously. Because beam optics in the arc regions are more critical than linac optics, a drift requiring orbit correction is most pronounced in the arcs. In the event that a drift severe enough to require a global orbit correction is detected in the arcs, the beam modulation and linac BPM system can be activated. With beam position information for the entire machine available for each pass, an orbit correction can then be performed.

It is presently estimated that 400–600 BPMs are required for the CEBAF arcs. In addition, there are 25 BPMs in each linac. Because of the large number of monitors involved, low cost designs for pickups and receivers are of primary concern. It is also a requirement to measure beam position to an accuracy of .1 mm with intensities as low as 1 μA. Therefore, it is important that the pickups are highly sensitive and that the receivers exhibit low signal-to-noise characteristics. Features common to both the linac and arc systems for achieving these goals include simple high impedance – low cost stripline pickups and synchronous heterodyne receivers.

Quite fortunately, the larger of the two BPM systems, that of the arcs, is also the simpler. In this system, the pickups consist of open-ended thin-wire striplines[3] tuned to the fundamental RF frequency of 1.497 GHz. At this frequency, the magnitude of the beam current is $2I_{av}$ and can range from 2 μA to 400 μA peak. The arc monitor receivers synchronously demodulate the 1.497 GHz signals from the pickups, in two stages, to DC. The standard difference-over-sum technique is then used to determine beam position. Because bandwidth requirements in the arcs are essentially zero, extremely high accuracy with very low beam currents can be obtained.

As previously mentioned, the position of each of the five different energy beams present in the linacs must be determined separately. Therefore, at the injector, the beam must be marked or encoded in such a way that each pass in the linacs can be distinguished. As a first step, a 100 MHz, 1 μA amplitude modulation is impressed on the beam at the grid of the injector electron gun.

In addition, by amplitude or phase encoding the 100 MHz carrier properly, the different passes in the linacs can be distinguished by using correlation techniques.

Regardless of how the beam is encoded, the linac pickups are designed to operate on the 100 MHz carrier. Specifically, the linac pickups consist of short-circuited-end thin-wire striplines[4] and are the electromagnetic duals of the arc pickups. At 100 MHz, where these pickups are small compared to a wavelength, they respond as simple inductive loops and are moderately sensitive. To boost the sensitivity, the pickups are externally resonated at 100 MHz with lumped capacitors. Again, the linac receivers employ synchronous demodulation in order to maximize signal to noise and therefore the accuracy of position measurements.

There are, of course, many options available for encoding the beam. Perhaps the simplest and historically the first technique tested at CEBAF consists of pulsing the 100 MHz modulation for a time equal to one recirculation (4.2 μsec) every five recirculation periods (21 μsec). As shown in figure 2, every time a pulse goes by a BPM, the position readings correspond successively to each pass. Therefore by keeping track of the pulse number modulo 5, the position of the beam for each pass is determined separately. This is in a sense a correlation technique, because position readings are correlated with pulse number. The major drawback of this system is that a position measurement for a particular pass can only be performed once every five passes (21 μsec). In addition, in order to meet position measurement accuracy requirements, many such measurements must be performed and averaged, further increasing the time required for a complete position measurement. Therefore, it is desirable to employ a correlation technique that allows continuous signal detection for each pass.

A simple, effective, continuous correlation technique for measuring the position of individual passes in the linacs makes use of the 100 MHz carrier for encoding the beam. In this case (see figure 2), the 100 MHz carrier is turned on continuously but the phase of the carrier is switched in a pseudorandom fashion between 0° and 180° every circulation period (4.2 μsec). Such a sequence has an autocorrelation function that is unity for zero time delay and $1/T$ for all other time shifts, where T is the total length of the sequence. By supplying five replicas of the sequence delayed by $n\tau$; where $n = 0$ to 4 and $\tau = 4.2$ μ sec, to each BPM, the position of each beam pass can be measured continuously but separately by correlating the signals from the BPMs with the time shifted copies of the pseudorandom sequences. In this case, noise and cross coupling between passes decrease with total sequence length.

In both linac BPM systems mentioned above, the amplitude of the modulation remains constant at 1 μA. Because the pickups respond only to the modulation, regardless of the average beam current, the only dynamic range requirement for the linac BPM receivers is that associated with changes in beam position – approximately 4:1. This is a fortunate by-product of the beam modulation technique in general.

In this paper, the CEBAF arc and linac BPM receiver systems are described in detail from an electrical engineering standpoint. The physics of the BPM

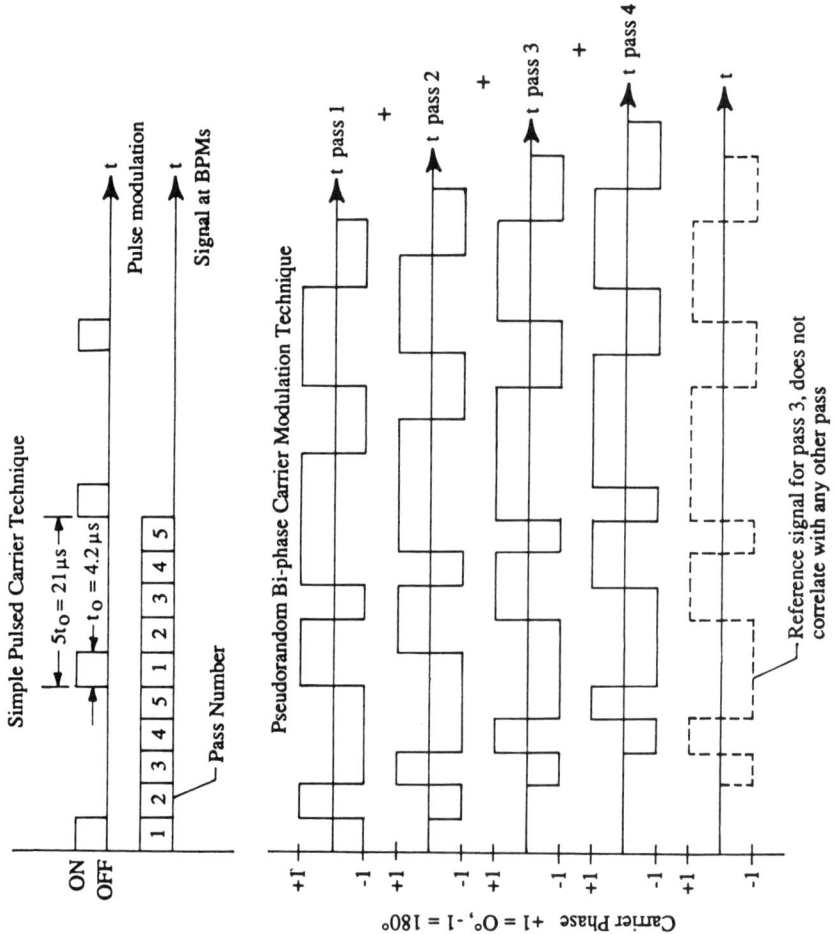

Figure 2. Simple pulse and pseudorandom modulation for spreading linac passes.

pickups themselves are thoroughly described in the references mentioned above and will not be addressed here. Both the simple pulsed linac system (presently being used at CEBAF) and the pseudorandom sequence system (planned upgrade) will be discussed in detail with experimental results given.

100 MHz LINAC RECEIVERS (SIMPLE PULSED VERSION)

The first linac BPM system developed at CEBAF (presently used in the north linac) uses the simple pulsed 100 MHz carrier technique. A system diagram of the pulsed 100 MHz beam position monitor electronics is shown in figure 3. For optimum S/N, the system uses synchronous detection[5] to convert the 4.2 microsecond current modulation bursts detected by the four BPM pickups into pulses that are read with an analog-to-digital converter. The basic system components are: the matching network and tunnel line driver board, the synchronous detector board, and the integrate-and-dump and microprocessor board. Each of these components is described below.

A schematic for the tunnel line driver board is shown in figure 4. The matching networks resonate the BPM pickups and match them to 50 ohms, at 100 MHz. The matching networks consist simply of a capacitor for resonating the pickup and an impedance transformer. The networks, located right on the vacuum feedthroughs of the beam position monitor, are connected to the tunnel line driver board by several feet of 50 ohm cable. The matching networks are necessary for several reasons. The first and most obvious is that the input impedance of the amplifiers used on the tunnel line driver board is 50 ohms and a perfect match will couple the maximum amount of power from the loops into the amplifiers. This increases the signal-to-noise ratio of the system. In addition, a matched system allows the tunnel line driver board, which contains radiation sensitive active components, to be placed several feet away from the beam pipe. Lastly, a front end shunt capacitance in the networks resonates the BPM pickups for higher sensitivity.

The tunnel line driver board is used to amplify (20 dB) the 100 MHz signal detected by the loop prior to transport from the accelerator tunnel to the above-ground service buildings. A line driver is necessary because the distance between the monitor and the detector electronics in the service buildings can be several hundred feet. The use of a front end line driver and gain stage allows the signal to be transported many hundreds of feet without a serious degradation in the system noise figure. The L-C networks on the outputs of each channel of the board match the 50 ohm amplifiers to the 75 ohm RG-6/U cable that is used to transport the signal from the tunnel to the service buildings. It should be pointed out that RG-6/U cable is used for these long runs because it is considerably lower in cost than most 50 Ω cable.

A closer examination of the schematic shown in figure 4 reveals that there are two diode switches that are used to switch the autocalibration signals into

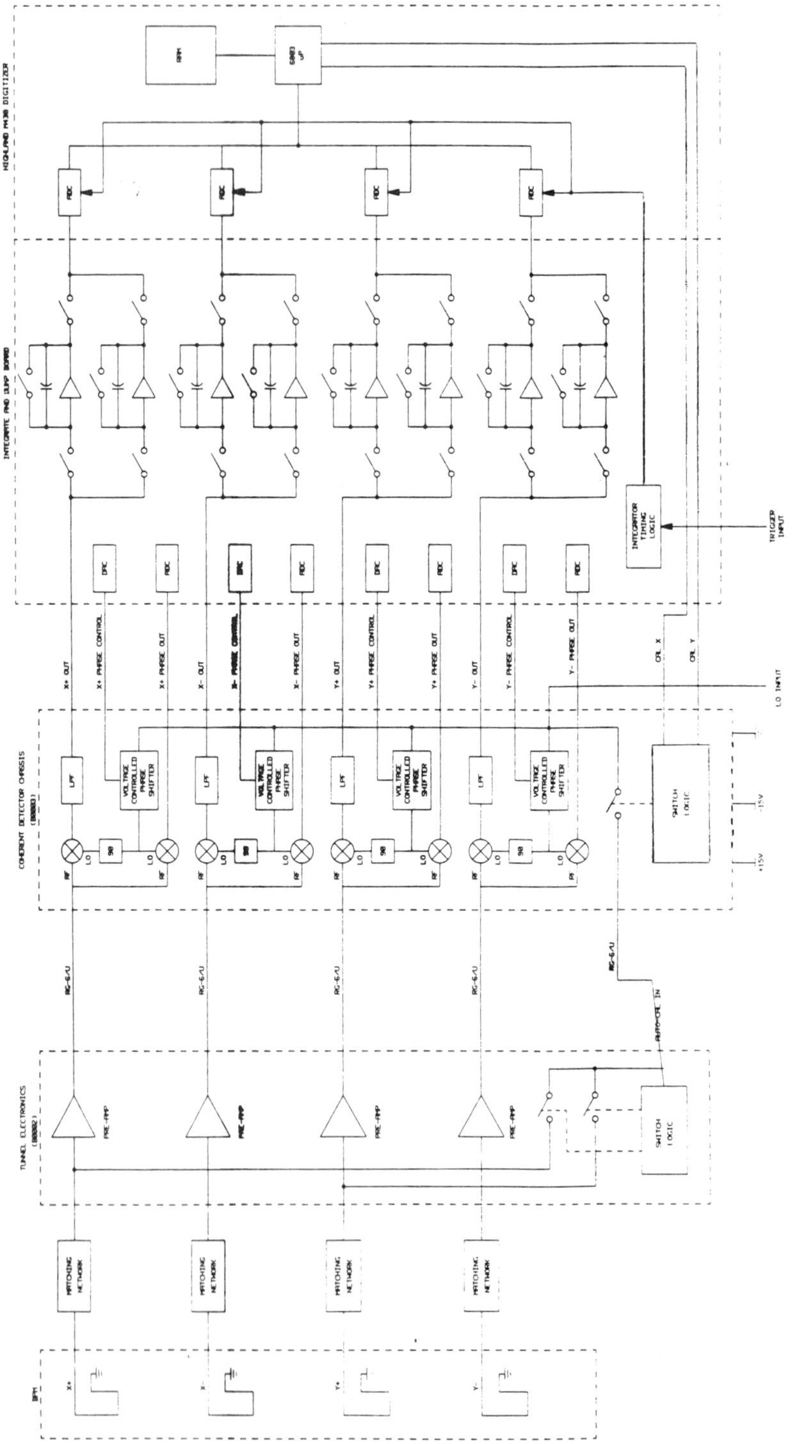

54 Beam Position Monitors at CEBAF

Figure 3. 100 MHz pulsed receiver system diagram.

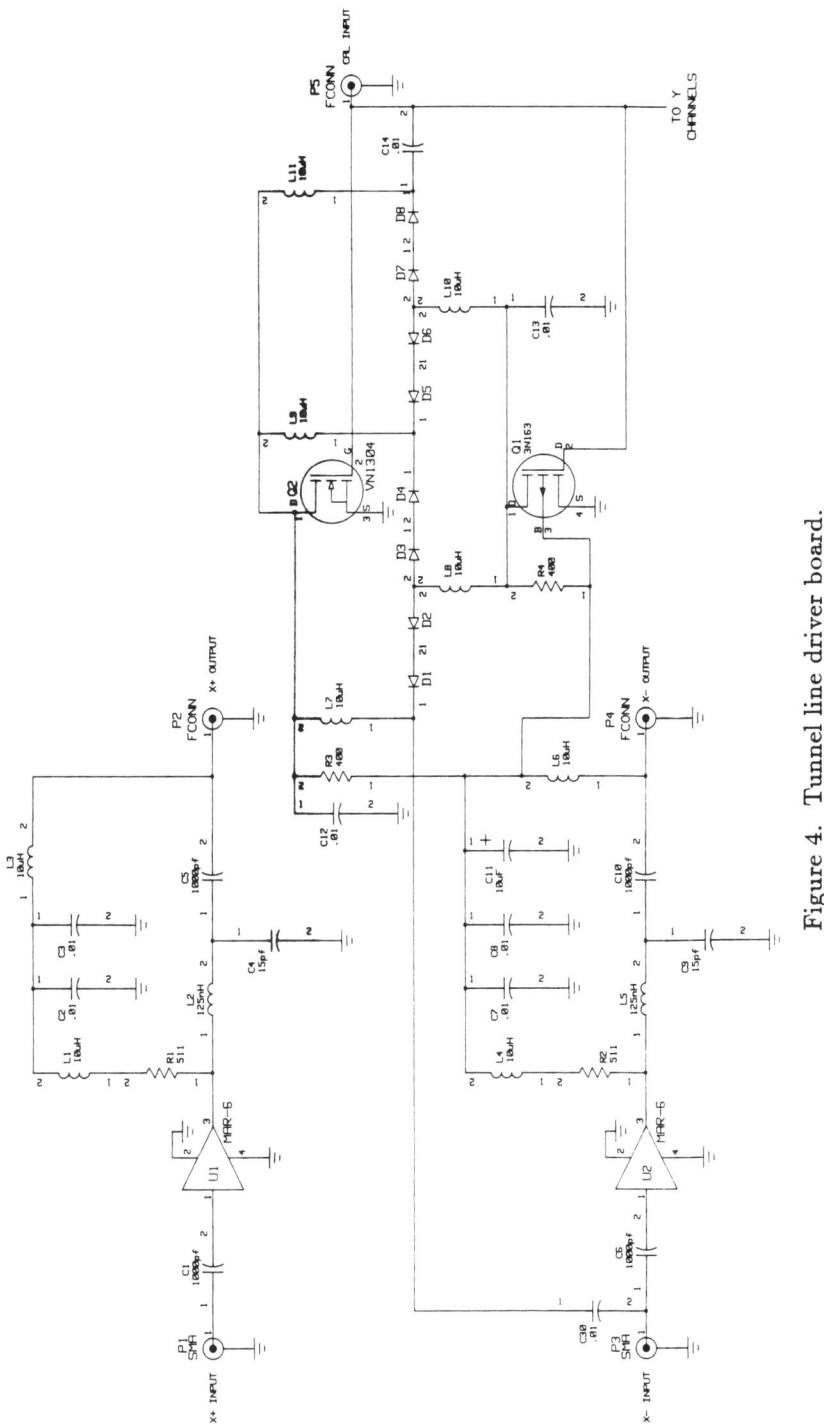

Figure 4. Tunnel line driver board.

the BPM. The autocalibration function of the system is discussed later in this paper. Another unique feature of the tunnel line driver board is that the power for the active components is supplied to the board from the coherent detector board through the output coax cables connecting the two. This dual usage of the cables reduces the number of cables that are required to be pulled through the already crowded penetrations that connect the service buildings to the tunnel.

A schematic of the coherent (synchronous) detector board is shown in figure 5. There are five basic parts to the board: RF amplifiers, voltage-controlled phase shifters for the local oscillator signals, quadrature phase detectors, down conversion mixers, and baseband amplifiers. The board is located in the aboveground service building and is used to convert the 100 MHz signal to baseband so it can be read with an analog-to-digital converter. Signals from the tunnel board are amplified by two stages of Mini-Circuits MAR-6 amplifiers. The gain of each stage is approximately 20 dB. The amplified signals are then down converted to baseband using an SBL-1 Mini-Circuits mixer. The local oscillator port of the mixer is driven by a Plessey SL952 limiting amplifier. Baseband signals from the IF port of the mixer are then amplified and passed to the Integrate-and-Dump Board. The gain of the baseband amplifiers is adjusted such that the total gain for the board from input to output is 75 dB.

The local oscillator signal used for the synchronous detector is derived from the same master oscillator that impresses the 100 MHz modulation on the beam at the injector gun. The signal is routed to each of the coherent detector boards via RG-6/U cable and coaxial cable couplers similar to those used in cable television systems. The local oscillator signal for each of the four channels on the board is phase shifted using the voltage-controlled varactor phase shift networks in figure 5. Four separate phase shifters were used in order to account for and eliminate phase differences between each of the four signals coming from each BPM. These channel-to-channel phase errors are due to matching network mismatches, differences in cable lengths, and differences in amplifiers. The varactor networks are capable of providing a full 360 degrees of phase shift at 100 MHz for a 1 to 10 volt control signal input. Each channel of the coherent detector board also uses a Motorola MC1496 multiplier as a phase detector. The phase detector output is passed to the Integrate-and-Dump Board and is used by the CEBAF control system to phase lock each of the channels. The phase lock algorithm is discussed later in this paper.

The schematic representation for the Integrate-and-Dump Board is shown in figure 3. The board resides in a CAMAC crate slot next to the microprocessor board and has several functions: integration of the baseband signal from the coherent detector board over the 4.2 microsecond modulation burst, analog-to-digital conversion of the phase detector signals for each channel, digital to analog conversion of the phase adjust signals for each channel, CAMAC interface to the control system computer for these signals, and timing circuitry for the integrators and analog-to-digital converters on the microprocessor board.

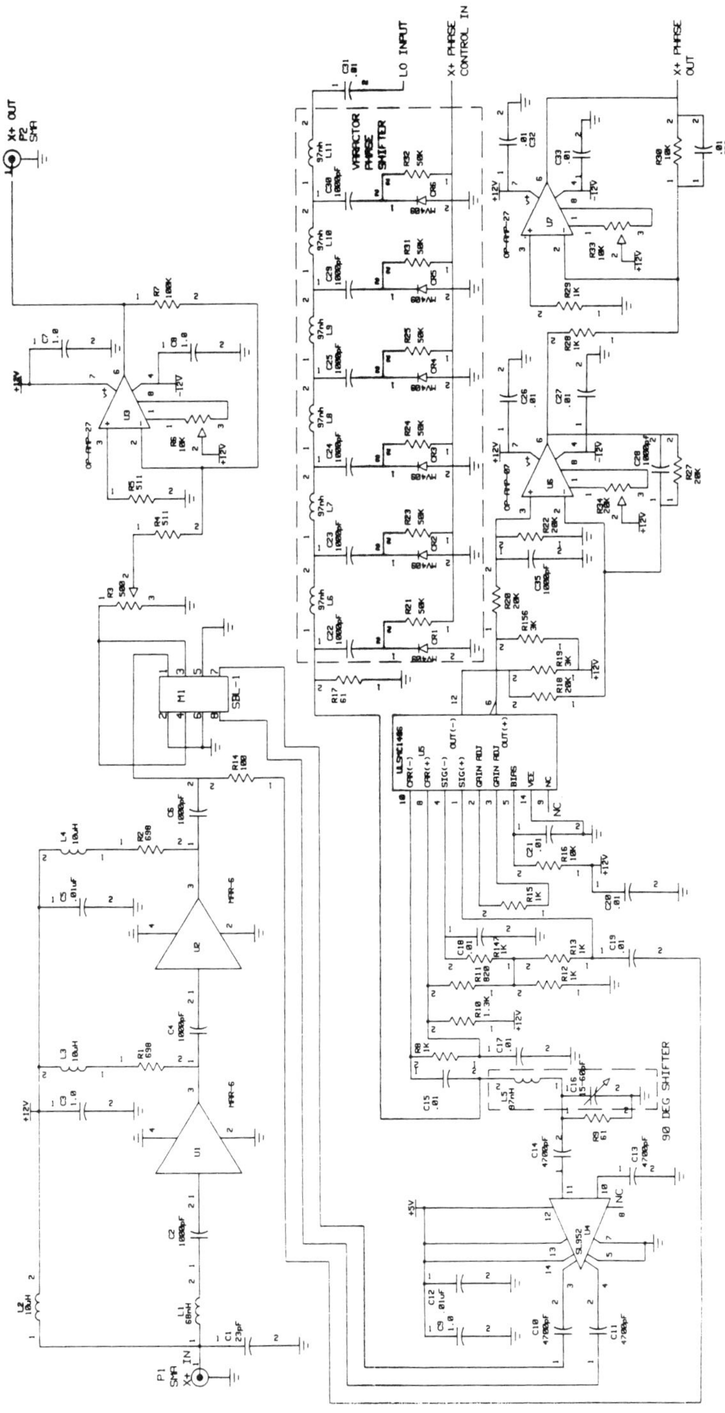

Figure 5. Coherent detector board.

58 Beam Position Monitors at CEBAF

Each channel of the coherent detector board feeds dual integrators on the Integrate-and-Dump Board. The integrate-and-dump technique was used because it provides an optimal filter for a pulse such as that produced during a modulation burst.[6,7] The dual integrators allow the board to acquire each of the five passes every time the current modulation is injected into the accelerator. A timing diagram for the integration of each pass is shown in figure 6. As can be seen, integrator number 1 integrates the first burst. At the end of 4.2 microseconds the integration is held and the analog-to-digital converter on the microprocessor board is triggered. Meanwhile integrator number 2 is integrating the second burst. Once the ADC has finished with burst 1, the integrator is dumped and readied to integrate burst 3. This sequence continues until all five bursts have been integrated and digitized.

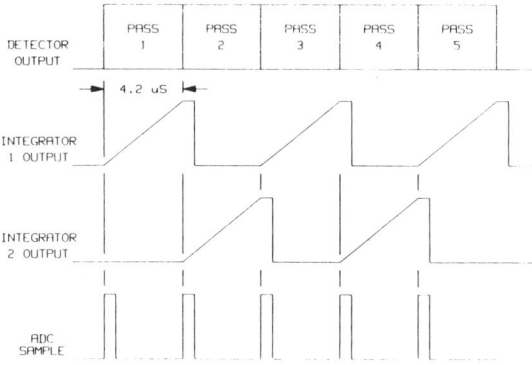

Figure 6. 100 MHz BPM Integrate-and-Dump Board timing diagram.

The microprocessor board used for the 100 MHz BPMs is a Highland Technologies model M430 eight channel digitizer module. It is a single width CAMAC module that uses a Motorola 6803 microprocessor and eight fast analog-to-digital converters to digitize the signals fed to it from the Integrate-and-Dump Board. The module can store up to 256 data points for each of the five passes. Once the data are stored in buffer memory, the microprocessor can perform various filtering and averaging operations and calculate the position of each of the five beam passes. The microprocessor also controls the autocalibration of each monitor channel and uses the offset and relative gain data acquired from the calibration for the beam position calculation. The actual signal processing and calibration algorithms are described below.

The 100 MHz BPM receiver has a built-in autocalibration function. The autocalibration function is used to check the offset and relative gain of each of the four channels and is controlled by the microprocessor module. The autocalibration sequence is as follows (refer to figure 3):

1. The CEBAF control system commands the microprocessor to perform a system calibration. The command is issued over the CAMAC bus.

2. The microprocessor module then reads the offset voltage of each of the four BPM channels. This reading is taken in the absence of beam modulation and with no signal applied to the loops. The offset voltage for each channel is stored in memory.
3. The microprocessor then sets a bit indicating that a portion of the local oscillator signal should be split off at each coherent detector and sent to the tunnel board. If the X channels are being calibrated, the diode switch connected to the Y^- channel of the tunnel board is closed and the signal is sent to the Y^- loop. The switch is controlled by a DC bias applied to the autocalibration signal line. Each of the X loops of the BPM picks up this signal. The control computer then performs the phase lock algorithm described below. After the system has been locked, the microprocessor reads the voltage of the X channels with the ADCs and calculates the relative gain of each channel. Again, this process is performed in the absence of beam modulation.
4. The sequence is repeated for the Y channels. The relative gains for each channel are then stored in memory.

The type of calibration described above checks the gain and offset of all components of each channel and is performed each time a position measurement is to be made or when the system accuracy is in question. Inherent in the calibration technique used is the assumption that all gain and offset drifts are slow and negligible on the time scale of the measurement. This is a reasonable assumption because most gain and offset drifts are induced by changes in ambient temperature which are indeed small on the time scale of a few seconds.

The Highland Model M430 Digitizer module is capable of performing various signal processing algorithms and position calculations. At this time the only signal filtering that is used is an averaging algorithm. The control computer instructs the module how many measurements of each of the signals are to be averaged. The maximum number of averages is presently set at 256. Each of the signal averaging calculations is performed to 16 bit accuracy. In the future, the microprocessor will perform the correlation function that will be used in conjunction with the pseudorandom modulation technique described elsewhere in this paper.

The module also performs the actual position calculation for each of the five beam passes. The position calculation is performed to 16 bit accuracy and is given by equations 1 and 2.

$$X_{pos} = K_{BPM} \frac{K_{x+}(X^+ - X^+_{off}) - K_{x-}(X^- - X^-_{off})}{K_{x+}(X^+ - X^+_{off}) + K_{x-}(X^- - X^-_{off})} \quad (1)$$

$$Y_{pos} = K_{BPM} \frac{K_{y+}(Y^+ - Y^+_{off}) - K_{y-}(Y^- - Y^-_{off})}{K_{y+}(Y^+ - Y^+_{off}) + K_{y-}(Y^- - Y^-_{off})} \quad (2)$$

60 Beam Position Monitors at CEBAF

where $K_{x+}, K_{x-}, K_{y+}, K_{y-}$ are the relative gain coefficients of each channel, determined during auto-cal.

$X^+_{off}, X^-_{off}, Y^+_{off}, Y^-_{off}$ are the offsets of each channel, determined during auto-cal.

$X^+, X^-, Y^+, Y^-,$ are the processed values for each channel and pass.

K_{BPM} is the constant relating difference-over-sum voltage ratio to beam position for the BPM.

The CEBAF 100 MHz BPM electronics are phase locked to the beam modulation and the autocalibration signal by the control computer. A flowchart for the algorithm is shown in figure 7. Phase locking of the system is performed at system startup, during autocalibration and whenever it is determined that the system phase has drifted. The basic procedure is as follows:

1. The modulation or autocalibration signal is turned on, and the phase detector output for each channel is read by the control computer through CAMAC.
2. The actual phase difference between the local oscillator and the signal is calculated.
3. The phase of the local oscillator is adjusted as per the algorithm using a digital-to-analog converter and the varactor phase shifter for each channel.
4. This process is repeated until the phase difference is less than 2 degrees. The DAC voltage is then held constant until a new phase adjustment is necessary.

The phase locking technique described above, like the autocalibration algorithm, relies upon the assumption that drifts in phase are slow and negligible during the time period of a measurement or series of measurements.

The 100 MHz system described above was tested using the CEBAF injector during July of 1990. The tests included tests of the actual system components and the phase-lock and autocalibration algorithms. In addition, this system is installed and will be used in the Front End Test and commissioning of the north linac.

The performance of the BPM was examined to the extent possible by comparing position readings from the BPM with those determined with a standard phosphorescent view screen located next to the BPM in the beamline. The view screen, normally used for rough visual feedback when steering and focussing the beam, is not absolutely calibrated with respect to the center of the beamline. In addition, persistence and blooming, which enlarge the apparent spot size of the beam, further degrade the accuracy of beam position measurements with the view screen. These effects, combined with position readings obtained from a TV monitor, make the view screen method of measuring beam position approximate at best. However, it was still of qualitative importance to compare BPM readings with beam position as measured with the view screens. The re-

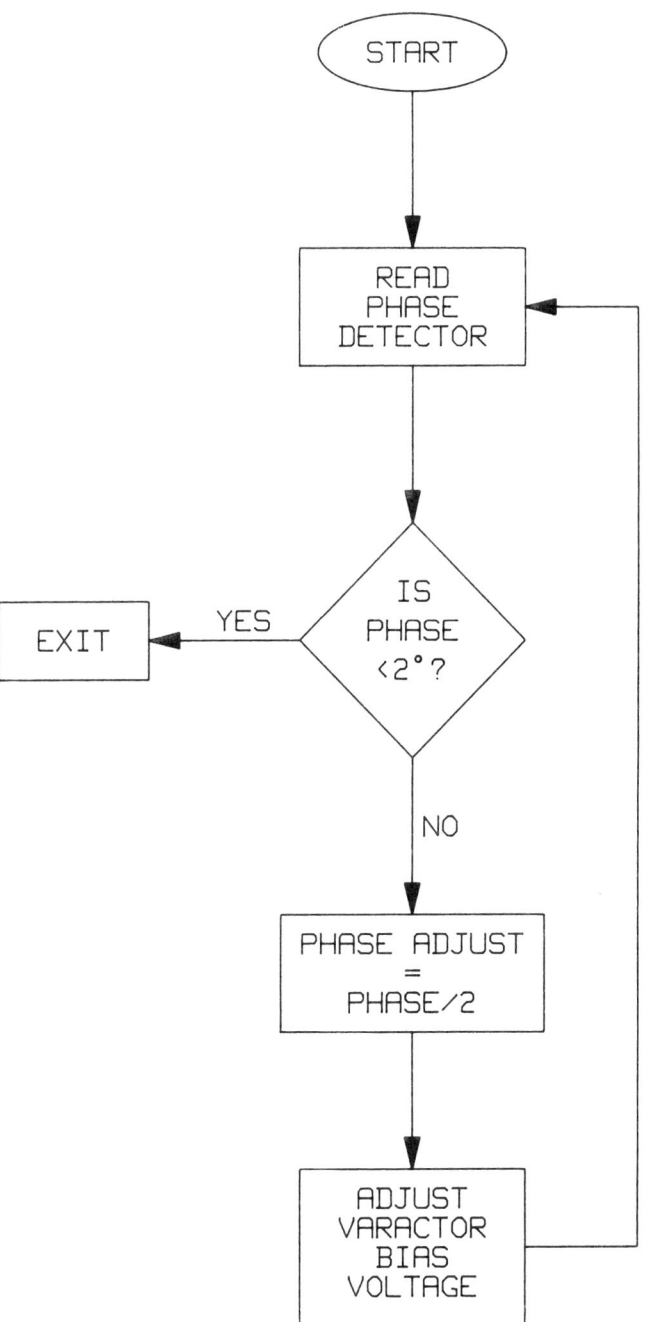

Figure 7. Phase lock algorithm.

sults of such a comparison appear in figure 10. Here, the beam was positioned with x and y steering coils about a square grid in 4 mm increments as determined by view screen readings (dots). The associated BPM readings for each position are indicated by crosses in figure 8. As shown, qualitative agreement exists between the two devices. Note: Due to beam steering difficulties data could not be obtained for positions (−4, 4), (4, 4), and (4, −4).

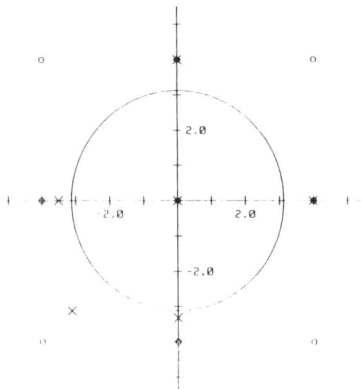

Figure 8. 100 MHz BPM injector test results
X = BPM measured position. O = Viewer measured position.
Large circle is beam spot size on viewer.

A system diagram of the 1497 MHz arc monitor electronics is shown in figure 9. The system uses a down converter and phase locked loop to detect the amplitude of each of the four monitor signals. The basic system components are the tunnel board which down converts the 1497 MHz signal to 1 MHz and the detector board which contains the phase locked loops and analog-to-digital converters. The control computer uses the ADC data to calculate the beam position.

A schematic of the tunnel board is shown in figure 10. The purpose of the board is to down convert the 1497 MHz beam signal to 1 MHz prior to transport to the detector board. The two Mini-circuits MAR-6 amplifiers on the front end of each channel amplify the 1497 MHz signal approximately 17 dB prior to down conversion. The 1496 MHz local oscillator is a Wilmanco model 4S-U-1496/PC. Following down conversion, the 1 MHz signal is amplified and sent to the detector board via RG-6/U cable.

The tunnel board is used to synthesize and inject the calibration signal during autocalibration. The 20 dB microstrip couplers and RF switches are used to inject the signal on the appropriate wire. The autocalibration algorithm is described later in this paper.

A block diagram representation of the detector board is shown in figure 9. The electronics are housed in a single width CAMAC module located in the above-ground service buildings. The main purpose of the board is to detect

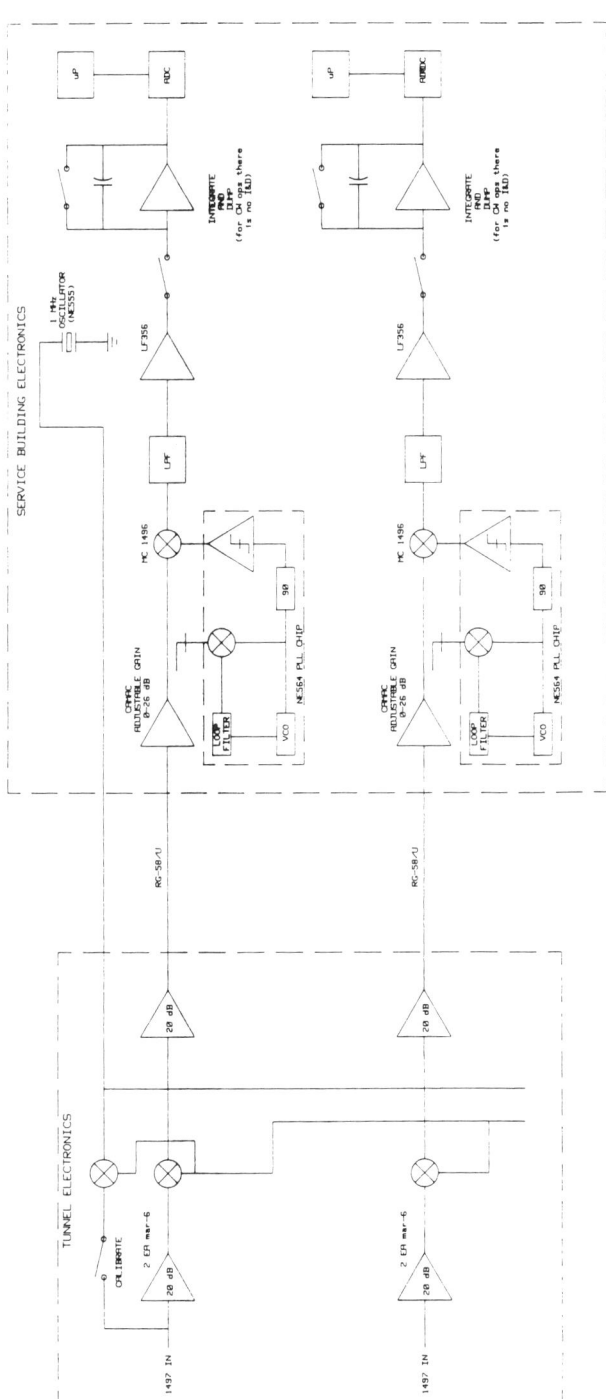

Figure 9. 1497 MHz receiver system diagram.

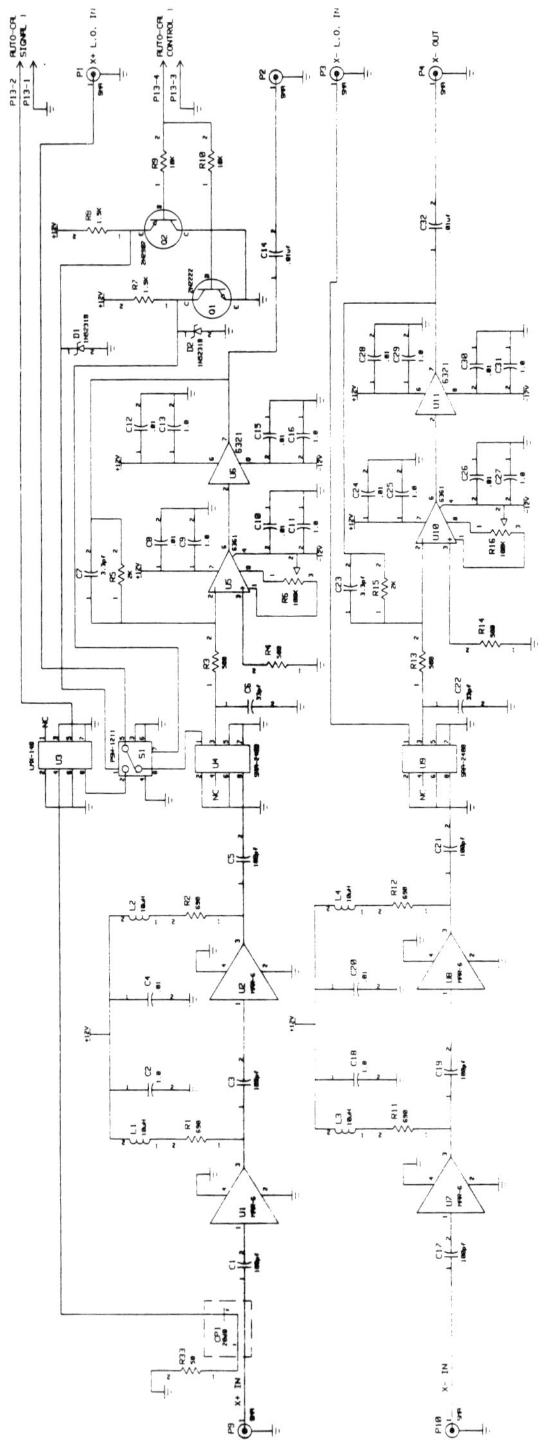

64 Beam Position Monitors at CEBAF

Figure 10. 1497 MHz tunnel electronics.

the amplitude of each of the four BPM signals and provide the CAMAC interface for the system. The front end amplifier for each channel is a Motorola MC1490 programmable-gain amplifier. The amplifier gains are controlled with a quad DAC. The 1497 MHz BPM detector board requires a programmable or adjustable gain because, unlike the 100 MHz BPM, the 1497 MHz BPM must operate over a variety of beam currents (from 1 to 200 microamps). The control computer adjusts the gain of the amplifier according to the accelerator beam current.

The heart of the detector board is the Signetics NE564 phase locked loop chip. The NE564 locks to the incoming 1 MHz signal at the output of each of the programmable gain amplifiers. The lock time for the NE564 is less than 50 microseconds. The VCO output signal from the NE564 and the 1 MHz signal from each programmable amplifier are then multiplied using a Motorola MC1496 multiplier. The output of the multipliers are low-pass filtered and sampled using ADCs.

The control computer reads the amplitude of each of the BPM signals via the ADCs and CAMAC. The module can also be programmed from CAMAC for single or continuous ADC scanning. In the single-scan mode the trigger for the ADC is provided from an external source via a front panel connector. Autocalibration functions, which are described below, are also controlled by the control computer via CAMAC.

The following discussion refers to circuits shown in figures 9 and 10. The 1497 MHz BPM electronics are calibrated in much the same way as the 100 MHz BPM. A major difference is that the beam must be turned off for 1497 MHz autocalibration so as not to interfere with the injected signals. A signal is injected onto a BPM pickup and is read back through the channels in the opposite BPM plane. The computer can then calculate the relative gain and offset of each channel and use these coefficients in the position calculations of equations 1 and 2. There are only a few differences in the actual implementation.

The 1497 MHz BPM system does not have a local oscillator signal that is distributed around the accelerator; therefore the 1497 MHz calibration signal must be synthesized. A 1 MHz oscillator is located on the detector board. This signal is sent to the tunnel board via RG-6/U cable. On the tunnel board this signal is mixed with the 1496 MHz local oscillator signal. The resulting sum frequency (1497 MHz) signal is then coupled into the pickup via a 20 dB microstrip coupler. The RF switches on the tunnel board which direct the local oscillator signal to the appropriate mixers are controlled by signals from the detector board. It should also be noted that the amplitude of the 1 MHz oscillator signal on the detector board is adjustable from CAMAC. This is done in an attempt to simulate the various beam currents over which the BPM electronics are required to operate.

The 1497 MHz detector and tunnel amplifier boards were tested during July of 1990 on the CEBAF injector. The on-line autocalibration function was not ready at the time, although the system was calibrated by injecting a 1497 MHz signal from a signal generator into a BPM pickup and recording the offset and relative gain of each channel. The tests were conducted in the same manner as described in the 100 MHz test results section of this paper, and for the same reasons the results can only be used to qualitatively verify operation. Figure 11 is a plot of BPM position readings versus a 4 mm square grid measured on the view screen. As shown, qualitative agreement does exist between the BPM readings and the view screen positions.

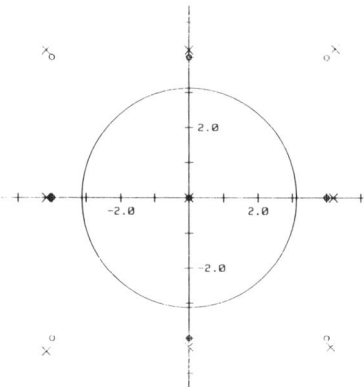

Figure 11. 1497 MHz injector test results
X = BPM measured position. O = Viewer measured position.
Large circle is beam spot size on viewer.

100 MHz BI-PHASE PSEUDORANDOM MODULATION TECHNIQUE FOR LINACS

As described in previous sections, one way to measure each beam individually is to modulate the beam for a period of less than or equal to the revolution time, 4.2 μsec. The signal will appear sequentially as the modulated pulse passes a BPM on each of the passes through the linac. The first signal corresponds to the lowest energy pass, the second to the next higher energy pass, and so on. Although this method provides a straightforward measurement of the position of the beam in each pass, it has the disadvantage that each measurement for each pass can only be made once every five revolution periods (21 μsec).

Recently another solution to the problem of individual beam position measurement has been devised.[8] In this method, the 100 MHz carrier is turned on continuously, but the carrier phase is switched from 0° to 180° every revolution period, τ, in such a way that each pass can be distinguished with no interference from other passes.

The separate identification of each beam can be illustrated for the simple case of two passes. Let us assume a modulation amplitude

$$f(t) = A \sin(\omega t) x(t/\tau), \qquad (3)$$

where ω is the 100 MHz modulation frequency, τ is the beam circulation period in the machine (4.2 μs for CEBAF), and $x(t/\tau)$ is a function with values ± 1, changing at integer values of its argument. Consider the sequence

$$x = 1, 1, -1, -1, 1, 1-1, -1, \ldots \tag{4}$$

where ± 1 corresponds to a $0°$ and $180°$ phase shift of the carrier. If the signal is mixed with itself, the resulting product is

$$f(t)f(t) = A^2 \sin^2(\omega t), \tag{5}$$

which represents the detected signal for the first pass. The time average is nonvanishing and is equal to $A^2/2$. If $f(t)$ is delayed by τ (second pass) and then mixed with the undelayed signal, the product is:

$$f(t)f(t-\tau) = A^2 \sin^2(\omega t) x(t/\tau) x(t/\tau - 1). \tag{6}$$

The time average is then

$$\frac{1}{T}\int_0^T f(t)f(t-\tau)\,dt = \frac{A^2}{T}\int_0^T \sin^2(\omega t)\,dt\,(1 - 1 + 1 - 1\ldots) \tag{7}$$
$$\approx 0.$$

Thus, the correlation function in this case picks out the signal from the first pass and suppresses that from the second. To suppress the first and observe the second, the signal is mixed with the delayed sequence $f(t-\tau)$.

The sequence $x(t/\tau)$ that has been given for illustrative purposes is not sufficient for three or more passes, but it is easy to extend the argument. It is necessary only to generate a sequence with values ± 1 such that the autocorrelation function with delay will vanish. Such sequences are already well known in coding theories for communications systems where they are referred to as pseudorandom sequences.

The correlation function $R(x, y; n)$ is defined as

$$R(x, y; n) = \frac{1}{T}\int_0^T x(t/\tau) y(t/\tau + n)\,dt, \tag{8}$$

where $x(t/\tau)$, $y(t/\tau)$ are pulse sequences with values ± 1, T is the total time of the sequence, and $n\tau$ is the time delay. We are particularly interested in the autocorrelation, $x(t/\tau) = y(t/\tau)$, which should satisfy the following orthogonality condition in order for the sequence to act as a filter:

$$R(x, x; n) = \frac{1}{T}\int_0^T x(t/\tau) x(t/\tau + n)\,dt = 1 \text{ if } n = 0, \tag{9}$$
$$= 0 \text{ if } n \neq 0.$$

Perfect sequences,[9] which satisfy equation 9 exactly, are hard to find and implement. Shift register sequences, however, which will satisfy equation 9 to within an arbitrarily small residual error determined by the length of the sequence, can be easily implemented by shift registers with appropriate feedback connections. Programmable array logic (PAL) devices can produce shift register sequences in a single device that needs only a clock pulse. A length-63 PRS generator is indicated in figure 12.

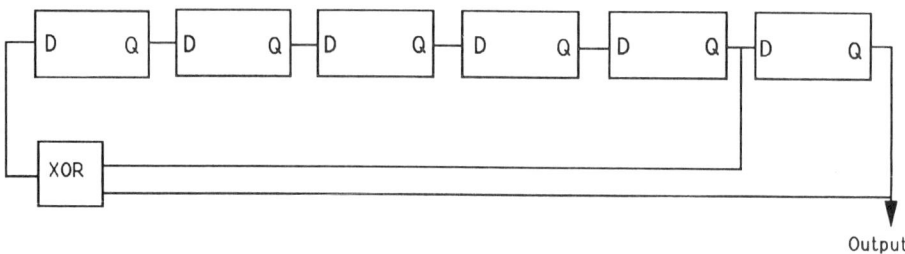

Figure 12. Shift register pseudorandom sequence generator (length 63).

The so-called maximal-length shift register sequence has nearly perfect autocorrelation characteristics. It is referred to as pseudorandom because it has several randomness characteristics[10]:
1. the numbers of $+1$ and -1 pulses are approximately equal, with a maximum difference of 1,
2. the number of runs of length l is proportional to 2^{-l},
3. the autocorrelation function has only 2 values.

The first of these characteristics is convenient because it means that an electronic implementation will not have to compensate for large shifts in the voltage or current levels introduced by the PRS modulation. The second implies that the levels will not shift appreciably within the time span of the sequence; for example, the longest possible run for a PRS of length 1023 is 10 consecutive $+1$s or -1s, and this longest run will occur only once. The third is the required autocorrelation condition. The autocorrelation values for maximal-length sequences are

$$R = \frac{1}{N} \sum_{r=0}^{N} x(r)x(r+n) = 1, \ n = 0,$$
$$= -\frac{1}{N}, \ n \neq 0.$$
(10)

For sequences of reasonable length, the deviation from perfect autocorrelation is insignificant.

For a PRS that is generated by a shift register of n stages, the maximum length is $2^n - 1$. Note that the orthogonality condition is satisfied only for a complete sequence; a sum of less than a complete sequence will have spurious peaks. It is not usually difficult to use a complete sequence. The characteristics of a three-stage shift register that generates a length-7 pseudorandom sequence are illustrated in figure 13.

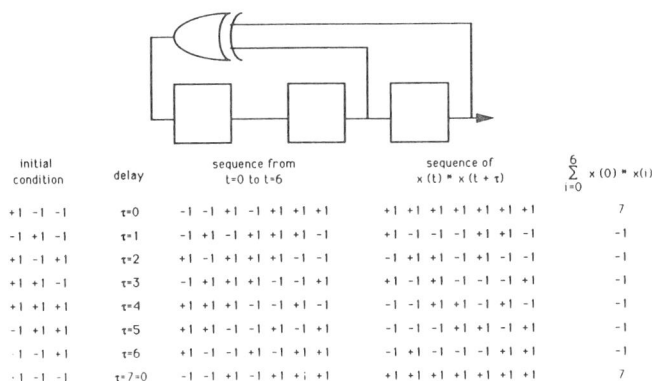

Figure 13. Orthogonality of shifted length-7 pseudorandom sequence.

Shift registers also provide another benefit in that they are easily expanded. Addition of another register stage will double the length of the maximal sequence, so with small expenditure one can select a sequence length that will give the processing time required to process the signal and average out the interference.

The block diagram for the pseudorandom modulation and detection is shown in figure 14. The beam is modulated with the frequency ω and in addition by a pseudorandom sequence $Ax(t)$,

$$I = I_0 + Ax(t/\tau) \sin(\omega t), \qquad (11)$$

where $x(t/\tau)$ is either $+1$ or -1. For five beam passes through the detector, the output signal at a BPM pickup is:

$$S_d = \sum_{r=0}^{4} \left(k_r Ax(t/\tau - r) \sin(\omega t - r\tau) \right) + \text{noise}, \qquad (12)$$

where k_r is a constant for each pass which depends upon its position. Because noise is generally quite system-specific and cannot be easily handled analytically,[11] we will henceforth neglect its effect; as usual, however, noise decreases with sequence length and will determine the processing time needed for a given signal/noise ratio. When the detector signal is mixed with the input delayed by j circulation periods, the output is

$$S = \sum_{r=0}^{4} k_r A^2 x(t/\tau - r)\, x(t/\tau - j) \sin^2(\omega t). \tag{13}$$

Integration over m complete sequences will remove the high-frequency time-varying terms, so the signal average will be

$$\overline{S} = \left[\frac{1}{2} k_j A^2 - \frac{1}{2N} \sum_{r \neq j} k_r A^2\right] \left[1 - \frac{\sin 2\omega T}{2\omega T}\right], \tag{14}$$

where T is equal to $mN\tau$. It is seen that the last term in the second brackets is $O(1/T)$ and the signal from the uncorrelated passes is down by a factor of $1/N$.

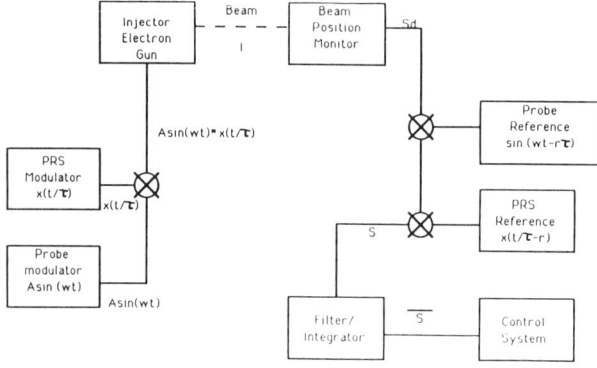

Figure 14. Beam position monitor processing.

A preliminary evaluation of a PRS generator developed at CEBAF has been performed (figure 15). It is a good simulation of the interaction between a beam position monitor and two beam passes, one delayed relative to the other. Our data show the orthogonality characteristic of the PRS autocorrelation function. The results (figures 16–18) show great promise for arriving at a firm design that will be able to distinguish individual beams in the accelerator. In particular, figure 16 shows that two uncorrelated signals have little influence upon one another's signals; this corresponds to two beams' simultaneous signals on one pickup of one beam position monitor. Figure 17 shows the orthogonality of an uncorrelated beam to a particular PRS delay; all uncorrelated PRS delay values show this same resistance to interference. Finally, figure 18 shows the output of the integrator as the PRS sequence is stepped through the entire sequence of delays with both channels at full amplitude. The two correlation peaks are located at the delays at which the PRS sequence is exactly correlated with the two channel sequences.

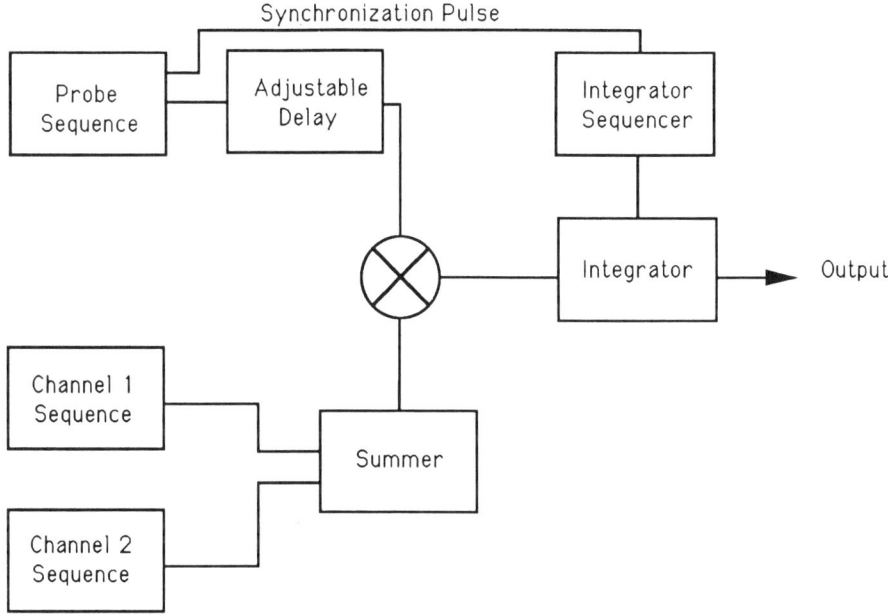

Figure 15. Pseudorandom sequence evaluator.

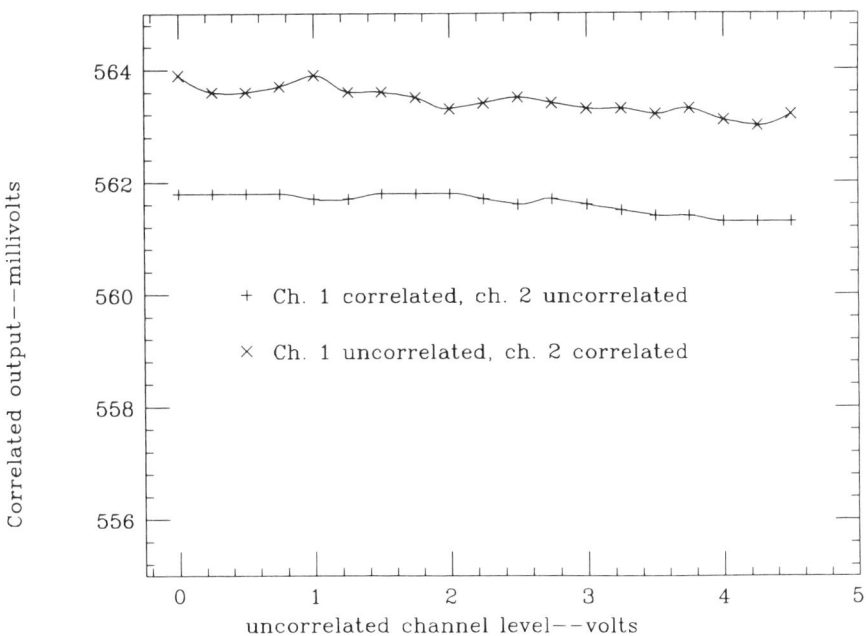

Figure 16. Correlated output vs. uncorrelated signal level.

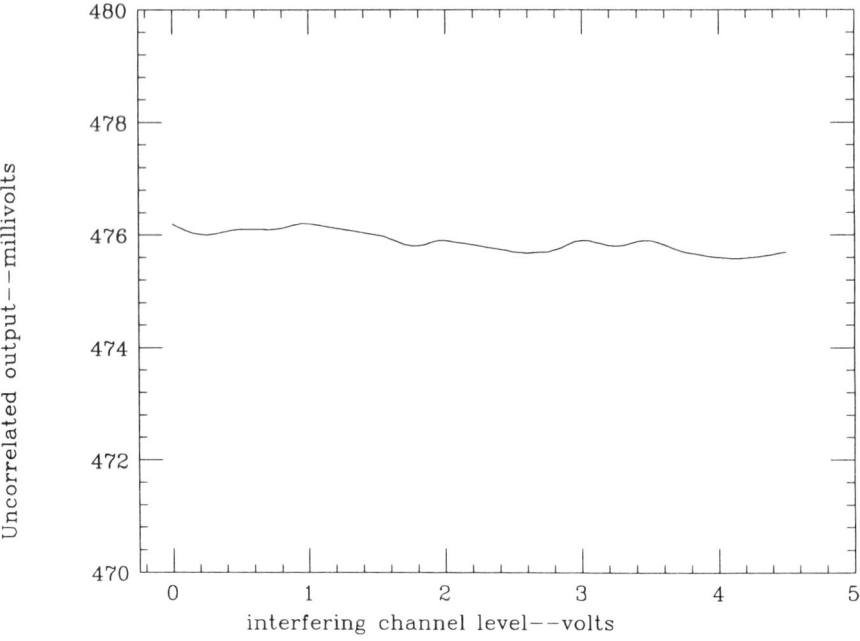

Figure 17. Uncorrelated output vs. interfering signal level.

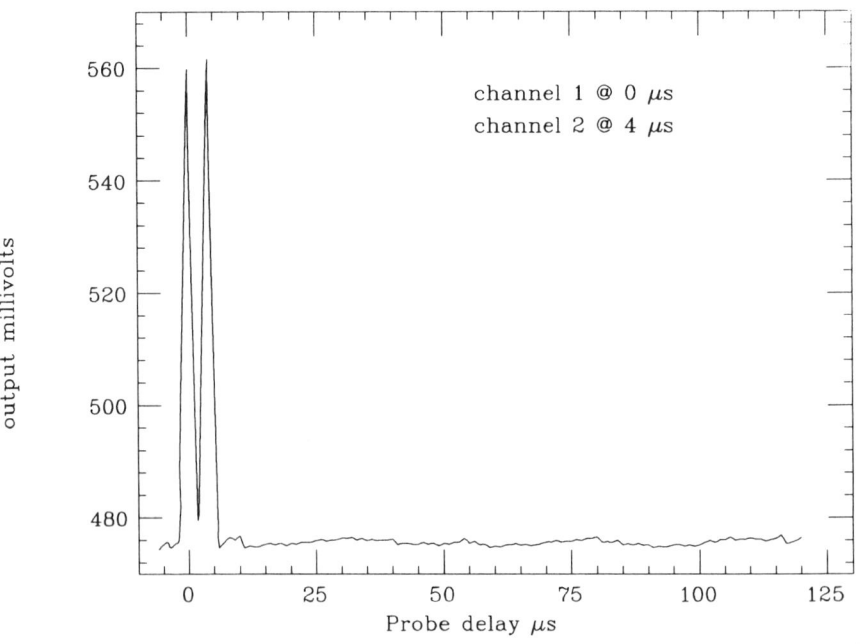

Figure 18. Autocorrelation for two channels.

IMPLEMENTATION OF PRS AT CEBAF

The pseudorandom sequence BPM system that is described in the previous sections of this paper will be implemented and tested during the CEBAF Front End Tests scheduled to begin in November 1990. A block diagram of the system is shown in figure 19. Comparison with figure 14 will reveal that the actual PRS correlation is performed in software within the microprocessor. There are several advantages to software processing of the PRS:

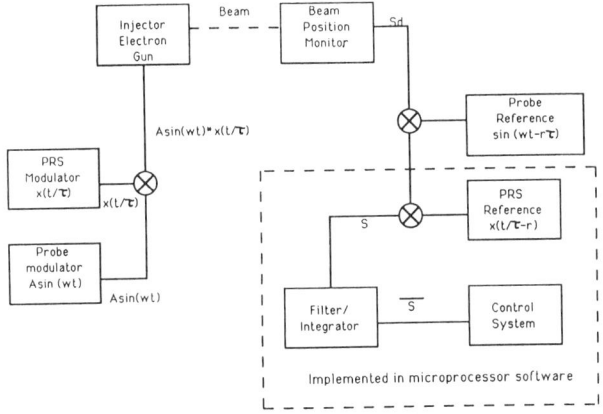

Figure 19. Integration of PRS into existing line receivers.

1. No additional hardware is required over and above that already designed and developed for the CEBAF Linac BPMs.
2. The microprocessor is not subject to the drifts and offsets that are inherent in hardware-based systems.
3. The microprocessor generates the PRS internally and therefore does not require that the sequence be piped around the accelerator, or be generated locally in hardware.

The only changes that will be required in the present system are the reprogramming of the microprocessor to handle PRS correlation instead of signal averaging, reprogramming of the PLDs that control the integrate-and-dump timing, and the addition of a PRS modulator to the injector grid modulation system.

The CEBAF construction schedule does not call for multiple-pass operation of the accelerator for several years. This means that a full functionality test of the PRS BPM system will not be possible during the Front End Test, but multiple-pass operation can be simulated in two ways. The first way is to time shift the PRS modulation at the injector and verify that each BPM must also time shift its internal sequence to locate the beam within the BPM. The second way is to add several time shifted copies of the PRS together prior to the injector grid modulator. Each BPM should then time shift its internal sequence accordingly and verify that the position of the beam is the same for each of the significant copies of the PRS. Both of these tests are planned for the Front End Test.

REFERENCES

1. B. Hartline, "CEBAF Progress Report", Proceedings of the 1990 Linear Accelerator Conference, Albuquerque, NM, September 1990.
2. A. Barry, B. Bowling, J. Kewisch and J. Tang, "Orbit Correction Techniques for a Multipass Linac", Proceedings of the 1990 Linear Accelerator Conference, Albuquerque, NM, September 1990.
3. W. Barry, "General Analysis of Thin-Wire Pickups for High Frequency Beam Position Monitors", CEBAF, September 1990 (To be published).
4. W. Barry, "Inductive Megahertz Beam Position Monitors for CEBAF", CEBAF-PR-89-003.
5. A. B. Carlson, Communication Systems, An Introduction to Signals and Noise in Electrical Communication, McGraw-Hill Inc., New York, 1986.
6. W. P. Robins, Phase Noise in Signal Sources, IEE Telecommunications Series, 1982.
7. M. Schwartz, Information, Transmission, Modulation and Noise, McGraw-Hill Inc., New York, 1980.
8. W. C. Barry, J. W. Heefner, G. S. Jones, J. E. Perry, and R. Rossmanith, "Beam Position Measurement in the CEBAF Recirculating Linacs by Use of Pseudorandom Pulse Sequences", Proceedings of the 1990 European Particle Accelerator Conference, Nice, France, June 1990.
9. H. D. Lüke, "Sequences and Arrays with Perfect Periodic Autocorrelation", IEEE Trans. on Aerospace and Electronics Systems, May 1988.
10. S. W. Golomb, Shift Register Sequences, Holden-Day Inc., San Francisco, 1967.
11. G. Krafft and J. Bisognano, "Fluctuations in the Results of Finite Time Averages for Signals of Arbitrary Spectrum", CEBAF TN 90-240.

Panel Discussion on
Beam Position Monitors Around the Labs

Scott Miller, Moderator
The Superconducting Super Collider Laboratory, Dallas, TX 75237
Jay Heefner
The Continuous Electron Beam Accelerator Facility, Newport News, VA 23606
Thomas J. Shea
Brookhaven National Laboratory, Upton, NY 11973
Robert C. Webber
The Superconducting Super Collider Laboratory, Dallas, TX 75237
Robert E. Shafer
Los Alomos National Laboratory, Los Alomos, NM 87454
Jean Borer
The European Center for Nuclear Research (CERN), Geneva 23, Switzerland

[Editor's note. This is the transcribed text of the panel discussion which occurred at 3:00 on Monday October 1, 1990 at Fermilab. It has been edited for clarity.]

Scott Miller: The panel discussion this afternoon is about beam position monitors and their use at different laboratories. The panelists today are: Bob Webber with the SSC, Tom Shea with BNL, Robert Shafer with LANL, Jay Heefner from CEBAF and Jean Borer from CERN. I am going to let things open up right now by having panelist introduce themselves. Jay?

Jay Heefner: My name is Jay Heefner and I work at CEBAF. I am an electrical engineer in the Instrumentation and Control Group. Since CEBAF is a new and relatively small laboratory, compared to Lawrence Livermore where I used to work, we have a wide variety of job assignments. One of the things I do is beam position monitors. We also work with everything from the PLCs that controls the safety system to harps and viewers and beam loss monitors. We have a very small crew, so I do not pretend to be an expert on BPMs.

Tom Shea: I am Tom Shea from Brookhaven National Laboratory. I have been hired by the RHIC project to work on their instrumentation effort. RHIC is the Relativistic Heavy Ion Collider at Brookhaven. It features a pair of 3.8-kilometers circumference superconducting storage rings, 100 GeV on 100 GeV per nucleon gold on gold. The beam position monitoring there will be approximately 250 stripline positioning monitors per ring. So I am considering to the economies of scale in both the production of the position monitors themselves and in the readout of the 500 channels. The BPMs, incidentally, have to operate at 4 degrees kelvin.

The other thing I am involved with is the Booster project at Brookhaven. I am developing the electronics for the single pass BPMs in the transfer line from the 200 MeV Linac which is almost identical to the Fermilab Linac. The peak current during this H- beam pulse is about 15 to 25 milliampere average current modulated at 200 MHz.

Bob Webber: I am Bob Webber, presently with SSC Laboratory. I have worked there for three months. A large part of my work has involved helping organize this workshop. I worked for at Fermilab for eighteen years up until June 1990. At Fermilab I was involved in the beam position systems for the Tevatron Project and for the Anti-Proton Source. I am in charge of the Hardware Section of the Instrumentation and Diagnostic Group now at the SSC Laboratory.

Bob Shafer: I am Bob Shafer, I have been in beam position monitoring off and on for over 30 years. I started at a small linac in Livermore in the late 50's and then I decided to go into high energy physics and get my Ph.D. at Berkeley, so I have worked around the LBL machines. Then after a short stay at Cambridge Electron Accelerator near Harvard, I went to Fermilab and was a high energy physicist here for many years, until Helen Edwards decided I should become the beam instrumentation man on the Tevatron. So I became an accelerator physicist overnight I guess, at Helen's demand. So I built the beam diagnostic systems on the Tevatron with a great deal of help from Bob Webber and Steve Jachim, and several others in the audience. Four years ago I moved to Los Alomos and I am working with the beam instrumentation people there. We are working on high frequency pickups, 425 MHz and 850 MHz, on proton beams, and we are also getting information such as synchronous phase out of the detectors as well as beam position. Synchronous phase information is used for making energy measurements by time of flight for drift-tube Linac commissioning measurements. So, I am still in instrumentation and I have worked on a wide variety of systems.

Jean Borer : My name is Jean Borer, from CERN, as you know from this morning. I have been in this business almost as long as my neighbor. I was hired hired in 1965 to work on the ISR Project in CERN. I was responsible for the beam position monitoring system using large electrostatic monitors. We later adapted that system to handle antiprotons when the ISR was adapted to handle those particles. We also experimented for the first time in changing from an analog processing system with the ADC in the vicinity of the control system to one with the microprocessor and the analog processing very close to the analog instrumentation.

We gained experience in single shot acquisition because that system had to catch THE first ejection of antiprotons. This was possibly the first time in history that a nondestructive system like the Beam Orbit Measurement system was so fundamentally relied upon. It had to catch the low level signal from the first antiproton injection into the ISR. It did work, thanks to the autotriggering and to a lot of testing and simulation without beam beforehand. We really made sure that when that rare injection of antiprotons came we would catch it.

Then we began something very different, the Beam Orbit Measurement system, or BOM, which I described this morning. Long machine, striplines, buttons, much higher frequencies were all new to us. Now we are being pushed forward to the next generation in the LEP tunnel, the LHC. So now our group is the combination or the SPS Instrumentation Group and the LEP Instrumentation Group. We, as the instrumentation group of those two accelerators, are going to build LHC instrumentation. Of course, the instrumentation for the LHC is still very much in the planning stages. In the SPS part of our group, we have also experts in coupling loop or directional couplers used as monitors. So we have experience with the full range of beam position monitors.

Scott Miller: Mr. Borer, since you mentioned the strip lines and the buttons, what are the relative pros and cons? In your talk you talked about button electrodes and their use at CERN. Are you happy with button electrodes? If you had to do it over again would you do it differently?

Jean Borer: The buttons work well. They have the same transverse non-linear characteristics as the striplines. The advantage of a stripline is its lower impedance and higher power taken out of the beam which leads to a better signal. The buttons have the advantage needing only one feedthrough, if it is not used as either short circuited loop, which is not a direction coupler anymore. But if you really need the directivity of a directional coupler, to separate electron from positrons or protons from antiprotons revolving in different directions, then a directional coupler is good. But it is difficult to be the directivity above 40 dB, which may be critical for separation of particles. Also, striplines take up much more space in a machine. Even in long machines, like I said this morning, you often have no space for monitors because most of the space has to be for distributed magnet fields. Buttons are good, too. They are especially useful when combating the synchrotron radiation problem because you can build them at a tilted position, but then you need to ruse a computer to rebuild the X and Y information.

Linearity Issues

I feel that the advantage for low frequency, enclosing electrodes is their linearity. But linearity may not be such a problem. On most machines you only want to center the beam. So a nonlinear device may not be critical. In LEP, our linearity has been acceptable for reduce orbit distortion to about the 0.5 millimeter RMS level.

But now we enter another world, the world of optic optimization. If you want to locally measure the magnitude of the betatron oscillation, it is necessary to have a linear detector. I have had no time this morning to speak of this aspect. Once you have stored the data, there are a lot of things you can do with it: average normalization; frequency analysis; transfer functions; the integral part of Q, if you have a single shot orbit analyzed from pickup to pickup; the real number of oscillation, which often is difficult in a machine. Moreover, you can, if you take pickup to pickup with a permanent excitation, take the magnitude and the phase of this excited transverse motion in order to know locally the betatron phase advance and betatron magnitude. With that you can check if you have a problem in your optics locally. So that is where we are now. It is only data processing.

Button Pickups

Bob Shafer: I would like to say a few things about striplines and buttons. One gets the same signal power from a strip line and or from a button if they have the same surface area exposed to the beam. Buttons are used extensively around electron machines. In the past, they have been used in the time domain where the very high frequency component of the signal is important because the beam bunches are so short. The LEP application is the first exception to this. For longer bunches, typical of a proton machine, where there is no, or very little, high frequency response, they are really not as good as striplines. This is exhibited for example in PETRA, which is a ring built for electrons which they now plan to use as an injector for the HERA proton ring. The button electrodes, which are ideal for electrons, have poor response for the PETRA proton current signals.

Jean mentioned that the stripline electrodes have two external connections. I have seen situations where people have tried to put a back-termination resistor on a stripline inside a vacuum system and regretted it because they could not get access to it when they needed to repair it. On the other hand, it turns out that you can short one end of a strip line, make it bidirectional and still have the other good characteristics of a stripline, including a very smooth Z/n function, or the beam coupling impedance as it is sometimes called. It is not necessarily resonant as you might have on pickups where you are not impedance matched and connected at an end.

I should mention that as soon as you get to the point where you can get a standing wave on a button, that is, when the half wave length is nominally the button diameter, then you start getting into the regime where you can have non-smooth Z/n. So even button pickups can be resonant at certain frequencies. In fact, strip lines can be resonant, too. When the perimeter of a strip line is of the order of a wave length then you can get resonances there. We think we observed those here at Fermilab on the stochastic cooling electrodes in the p-bar rings.

Shorted, Terminated or Open Striplines?

Scott Miller: Mr. Heefner, your system uses both shorted loops and open loops. The open loops are unique in our community. Do you have any comments about the relative choices that were made?

Jay Heefner: The short circuited loop is the original design, optimized to work at 100 MHz. It turns out that it is just an inductive pickup, a one turn transformer. We picked this design because it is cheap and easy to manufacture. At the time we thought we would have these types of devices throughout the machine; the original number of monitors was around 700 so we have since gone to open circuited devices for the majority of the monitors. They are the electromagnetic dual of the short circuited device for the high frequency pickups. The choices were the same. It was mostly cost, ease of manufacturability and the amount of space that was allotted to us to put our beam position monitors in. It is on the order of 10 cm or something like that, I cannot remember the exact lengths. There were a lot of other considerations, which had nothing to do with sensitivity requirements or optimization; they were mechanical and cost constraints.

Bob Shafer: I would like to make a comment on that. It turns out that the last thing to be thought of and the lowest priority in any accelerator is the diagnostics. When we built the diagnostics for the Tevatron after the magnets and the correctional elements had been designed. Then it was proclaimed that we had 20 cm for the diagnostics. That was all.

Tom Shea: The is precisely what has happened in RHIC, so the lessons have not been learned.

Jay Heefner: The same thing happens at CEBAF.

Scott Miller: The SSC has not been built yet and we already do not have enough space.

RHIC's monitors

Mr. Shea, what are the special BPM requirements you have at RHIC? How much is the dynamic range, for instance, an issue?

Tom Shea: Dynamic range is, magically, not too much of a concern. We can actually let in a full bunch and, I am assured by Magnet Division, that if I let it crash into the wall, it is not going to hurt anything. That, of course, remains to be seen. We can limit the intensity by letting in a single bunch, whereas a full storage ring full of beam would be 57 bunches. So the constraint on the diagnostics, and particularly the beam position monitors, is that you be able to look at a the first turn of a single bunch. Even though we are looking at storing protons through gold, the dynamic range is limited because we expect a top intensity for protons of 1011 protons per bunch. The fully stripped gold charge state 79, and we end up with a top intensity of 10^9 particles per bunch. That turns out to be only about a factor of two apart in charge. Additionally, after acceleration in the storage mode, the bunch lengths for protons and gold are similar: They are both are nanosecond long bunches.

Not only is the dynamic range rather small under normal operating conditions, but the frequency response of pickups can be very similar for proton though gold. Of course this is all "in theory;" probably we will end up using the lower part of the dynamic range they provide. But once again, we do not foresee any scenario where we can get ourselves in trouble by requiring dynamic range.

This is far different from the situation in our new booster ring. It will have a huge dynamic range. If anybody wants to learn about the problems associated with that they can talk to Dick Witkover or Dominick Ciardullo. They have made quite a bit of progress, particularly in development of the front-end electronics to make a position monitoring system that operates over a great dynamic range in a small ring.

I am also concerned with the coupling impedance of the monitors in RHIC. Because it is a storage ring, we have chosen a directional coupler style pickup. However, I mentioned before that these things have to be cold. So it has been decided that we will short one end of the pickup. We will get the same frequency response. We do not need the directionality and therefore we save on trying to have a cold termination. We save the heat load which the extra back terminations would cause by terminating them externally. We will have one cable coming out per plate and there will be a 20 cm long strip line that is shorted on one end. The entire thing will be impedance

Bob Shafer: I would like to make a comment on that. It turns out that the last thing to be thought of and the lowest priority in any accelerator is the diagnostics. When we built the diagnostics for the Tevatron after the magnets and the correctional elements had been designed. Then it was proclaimed that we had 20 cm for the diagnostics. That was all.

Tom Shea: The is precisely what has happened in RHIC, so the lessons have not been learned.

Jay Heefner: The same thing happens at CEBAF.

Scott Miller: The SSC has not been built yet and we already do not have enough space.

RHIC's monitors

Mr. Shea, what are the special BPM requirements you have at RHIC? How much is the dynamic range, for instance, an issue?

Tom Shea: Dynamic range is, magically, not too much of a concern. We can actually let in a full bunch and, I am assured by Magnet Division, that if I let it crash into the wall, it is not going to hurt anything. That, of course, remains to be seen. We can limit the intensity by letting in a single bunch, whereas a full storage ring full of beam would be 57 bunches. So the constraint on the diagnostics, and particularly the beam position monitors, is that you be able to look at a the first turn of a single bunch. Even though we are looking at storing protons through gold, the dynamic range is limited because we expect a top intensity for protons of 10^{11} protons per bunch. The fully stripped gold charge state 79, and we end up with a top intensity of 10^9 particles per bunch. That turns out to be only about a factor of two apart in charge. Additionally, after acceleration in the storage mode, the bunch lengths for protons and gold are similar: They are both are nanosecond long bunches.

Not only is the dynamic range rather small under normal operating conditions, but the frequency response of pickups can be very similar for proton though gold. Of course this is all "in theory;" probably we will end up using the lower part of the dynamic range they provide. But once again, we do not foresee any scenario where we can get ourselves in trouble by requiring dynamic range.

This is far different from the situation in our new booster ring. It will have a huge dynamic range. If anybody wants to learn about the problems associated with that they can talk to Dick Witkover or Dominick Ciardullo. They have made quite a bit of progress, particularly in development of the front-end electronics to make a position monitoring system that operates over a great dynamic range in a small ring.

I am also concerned with the coupling impedance of the monitors in RHIC. Because it is a storage ring, we have chosen a directional coupler style pickup. However, I mentioned before that these things have to be cold. So it has been decided that we will short one end of the pickup. We will get the same frequency response. We do not need the directionality and therefore we save on trying to have a cold termination. We save the heat load which the extra back terminations would cause by terminating them externally. We will have one cable coming out per plate and there will be a 20 cm long strip line that is shorted on one end. The entire thing will be impedance the SSC are both superconducting machines which are able to take less of a hit from the beam. I believe that in this situation, you really do require a beam loss monitoring system.

RF Signals vs. Processed Signals for Long Runs

One-sixth of the SSC monitors will be instrumented with an AM-to-PM type processing. The other five-sixths will have simple diode detectors connected to the outputs of the beam position pickups. This would have to be located physically near the pickups so that dispersion in long cable runs, up to 270 meters, is minimized. This would reduce the peak pulse height at a detector far away. Some problems we need to face there, aside from the obvious one of having electronics right near the magnets, is feeding calibration signals into the front-end electronics which are far away from any source of power and control.

Another problem, which perhaps Jean Borer can comment on, is this. When one does a peak detection process near the pickup and then tries to transmit that signal back to some electronics a long distance away, then you have changed the frequency spectrum of the signal. In the ultimate case you take an AC signal and turn it into a DC signal much like a power supply rectifier does. So you end up transmitting the signal over a long cable. I believe the important part of the spectrum is now base-band sort of frequencies. That means this long signal transmission path needs to deal with problems of pickup and interference from relatively low frequency sources like power supply harmonics and things like that. How did you deal with this problem?

Jean Borer: I cannot agree more. We avoided a DC connection and we always have galvanic separation. What we try to do is have the monitor and the front-end electronic grounded at the vacuum ground. Once the signal has been prepared, we use some kind of narrow band, one could say transmission broad-band in a certain sense but without DC and without very high frequencies. We then use the familiar baseline restitution procedure to get back the absolute information. In fact we use double-sampling flash-ADC: the first one in front of a pulse for which amplitude is the useful information and then the second one of the peak value. Those two values are stored in the system. In doing so and in having no galvanic connections, we have not suffered any coupling problems.

We have had problems locally in racks due to another subject you mentioned, the test generators. In order to feed those, we need relatively high voltage, something like 200 volts. This goes through a multiplicity of multicore cables. Just the fact that you switch on and off those high voltage power supplies will capacitively couple them to the rest of the system. So we had to do a quick transformation on those switching boards to have them switch slowly. But it was a small problem which was quickly solved.

Importance of Simulations

I think it is absolutely essential to have a means of simulating a system without the presence of beam, both during the installation period and between accelerator running periods. We have built small test generators for LEP which are able to produce pulses close to the monitor. So now, depending on the type of monitor, you can inject a signal into the monitor through a capacitive coupling. We also use simple linear directional couplers to inject a signal into the signal path. It is only a simple avalanche transistor plus a piece of a stripline to define the lengths of the pulse. There was one small trick: we used those avalanche current sources in the middle of a dividing network to ensure, with resistors, that we have identical outputs at both ends of the system.

We always do a calibration, pair by pair, either diagonally or horizontally/vertically with the signals to know the zero balance.

Another test we perform is to use the attenuation network to inject a signal "in the middle" for the center, or to inject on two other points which give defined ratio of output. This defined ratio is used to calibrate the transverse sensitivity, or scaling factor, of the system. Those test monitors are not too complicated. A stripline monitor would have the advantage for the injection of the signal to use it directly as a coupler for the signal output. It is not essential, use but one can do it in another manner.

Bob Shafer: One of the things Bob alluded to but did not say much about is the political connection. His diagnostic systems are going to cost something like $50 to $80 million. The $50 million I think does not cable up half the pickups in the main rings, among other things.

Political connections are as follows: It is important, because of the industrial partnership liaison between the SSC and the US industries which are looking for a piece of action, that we select industrial partners who are in the congressional district of Representatives on the appropriate panels back in Washington D.C. It is unfortunate, but it is the kind of thing we have to consider now.

Jean Borer: I can say the BOM system costs around 10 million Swiss Francs. Just for the BOM including cables monitors and electronic materials, no CERN labor.

Signal Processing Techniques

Scott Miller: We have seen a variety of signal processing techniques, such as phase conversion, synchronous detection, diode detection, and high speed sampling. Is there any direction that the signal processing is headed in? One of the things I noticed here with Heefner's paper was a lot more electronics than I would have normally have expected in a radiation field. Any comments by the group in general?

Jay Heefner: I do not think there are that many electronics in the radiation area. There are just a few amplifiers, and some mixers in the case of each system.

Scott Miller: Well, active devices in general I consider bad things.

Jay Heefner: Yes, that is true. We have used them during the injector test and we are now making estimates of radiation levels for shielding purposes. We did not have much choice because of the signal to noise levels and the noise figures associated with transporting such a signal over several hundred feet of coax up to a safe environment. It would have been virtually impossible to achieve the goals that we set.

Scott Miller: Any other comments?

Compensating for Response Losses

Jean Borer: I have not mentioned the cable lengths. The maximum lengths we have for the narrow band system is 800 meters. The cable lengths go from about 30 meters to 800 meters. The cables join in the central crates where we have compensating networks for cable losses and cable transfer functions for each individual pickup so that finally you have about the same response after these compensations as you have at the detector. The range of presetting or trimming for each is in the range of 1 dB.

Bob Webber: So you attenuate the signals from the nearby pickups.

Jean Borer: Exactly.

Now about radiation. The synchrotron radiation problem scared us very much in LEP. We tested all of our equipment with cobalt source so that we think it should survive for about 10 year. This is about 10^9 rad.

We tried to use passive components in the front-end electronics, like hybrids, filters or ringing filter, which are built on a ceramic plate and are made with thick silk screening deposition. The rest of the circuit is made of only fast amplifiers, like you find in the comparators for limiting function, and an exclusive-OR gate which makes the phase detection. You have only three dual-inline components as active component plus a regulator, for the local power supply, because of the droop along the feeding line.

All that is included in a small aluminum box. We put about 4 cm of lead around it and the whole thing is on the floor below some concrete under the magnets. We hope with that it will stand it. That is the precaution we have taken.

Bob Webber: What happens when you have a water leak?

Jean Borer: We have 2 cm feet on the boxes.

Klaus Unser, in the Audience: I would like to make a proposal which is a bit controversial. Why can't we suppress all analog signal cables and do all the electronic processing right on the spot? We know there is a radiation problem. But it can be relatively easily solved by making the electronics very small, in a form of a cylinder which we put into a bore in the tunnel wall two to three meters deep. Because the electrons is very close, we can simplify part of this electronics with the use of microelectronics techniques. There is a cable connection which we still need. They can be very thin cables. We can do multiplexing right on the spot and we only have to transmit the pictchel signals.

Bob Shafer: LEP foresaw the water problem and they have tilted the tunnel for that reason, 1 degree, so all the water goes to one end.

Jean Borer: Yeah, but we have had modules with heavy leaks under JURA and I found my monitors with a lot of dried mud around it and the connector just above it.

The Voice of Experience

Scott Miller: One more question for Bob Shafer. What lessons have you learned, positive and negative, from your "decades of experience" with BPMs?

Bob Shafer: Make absolutely sure that what you have designed is going to work. Very often you do not have a chance to fix it after you have installed it. Furthermore almost every BPM system is going to have an Achilles heel, so you better be looking for it. We had one at Tevatron called Martinsitic Transformations in the connector because we never thought that the connectors at 4 degrees Kelvin might have different characteristics than those at room temperature. Look very, very carefully at every component you design to make absolutely sure.

Communications Among Personnel

I am aware of some really well engineered beam position systems built by really excellent engineers which did not do the physics measurements required for that particular machine. The physicist was just not communicating with the engineer. The engineer went off by himself and designed a beautiful system which really did not make the measurements the physicist needed. We have a ring at Los Alomos where the injected beam has a very well-defined 200 MHz modulation on it. It gets captured into a bucket at 2.8 MHz RF frequency. The beam position system operated at 200 MHz. The 200 MHz signal disappeared after about 10 microseconds but the accumulation time was 800 μs. Therefore there was no monitor on this machine that worked after 10 microseconds, even though the beam is in there for 800 μs. So you really have to be very careful to make sure that what you are designing and building meets the overall requirement of that machine. It requires very good communication between the physicist and engineer. It is not only the physicists job to tell the engineer what he needs but it is also the engineers job to make absolutely sure that he communicates well with the physicist.

Comment from Audience: Bob, you bring up a good point there that the physicist has to think well in advance of the problems he does not want to deal with because if you build it to only see closed orbit, you are not going to see oscillation or other facts which may eventually be the limitations.

Bob Shafer: That's right. For example you would like to be able to see a betatron oscillation on a single turn like the people do here at Fermilab. Just look at the orbit distortions of a single bunch as it goes around the machine rather than looking at closed orbit only. So the engineer really should be a well-trained accelerator physicist as well. It is a big load to put on engineers, but I think the engineer has an obligation to learn a lot about accelerators.

Jean Borer: I agree fully with what you said. We have, I would say, good connection between the builders of the instrumentation and equipment and the other teams. You always find in CERN teams of machine specialists, theoreticians, as well as experimentalists in the control room during an operation period. I appreciate their advice very much. I was initially a specialist in construction, but I have involved in many experiments in the control room. We have discussed how could we change the processing or how could we find this response for question about beam dynamics or the optics in bringing new equipment in. I learned my profession like that. Later on when I went to the LEP program, I figured out the possibilities of what one could do with a system once I knew how many bunches were in the machine had the other general characteristics of LEP. From this I discussed with my boss, Claud Bovet, and we offered a proposal to the machine physicist and we had feedback. Slowly, by successive iteration, we defined what the BOM system should do.

Relations with the Controls Specialists

But the really difficult part has been communication with the controls specialist ton define what the application program should do. While we were dealing with that, we did not even know what the control system structure would be. So we did not know, if you do a call to an instrument, how long it would take to get the answer back. All these problems reappeared during the commissioning of the BOM system. It is distributed around the machine and it relies on all the facilities of the control system.

The control system for us is a data transmission system. We also had the requirements of future operation people who wanted to have things presented in certain manner. The most difficult part was to convince computer specialists that they were constructing an instrument and not a new software system they could present in the world of computing. And now we are redoing it. We have removed all those many layers of computers so we can go back to much more direct connection using OS9 and Unix. (There is no Unix system for VME yet.) OS9 and Unix are connected through an ethernet bridge and the token ring system. What we lack in LEP right now is the speed and the reliability of the control system. Otherwise you cannot dream of any control, surveillance of beam loss or motion of the beam in intersection points to monitor these things. These things are impossible to do with the present control system.

So I think the designer of a tool called the Beam Orbit Measurement system has to have the power to decide on some technical solutions used by his colleagues of the control world. Otherwise it does not work.

Scott Miller: Yes, I will use this opportunity for a brief advertisement. The panel discussion tomorrow is the interface between the controls group and the instrumentation engineer and I imagine there will be some interesting discussions coming out of that also.

Interfacing to the Timing System

One other system that one interfaces with is the timing system. One is required to get timing signals for a variety of reasons. How do your different systems interface with the timing systems? Some people do beam synchronization, some use absolute pulses. What is considered better? A lot of people seem to lean very favorably toward the beam triggered type of sensor.

Tom Shea: I do not think that there is a need to actually trigger an acquisition with an external clock other than to select the bunch that you are interested in. You should really self-trigger on the bunch, use that for your acquisition on that particular bunch. But leave open the capability particularly with all the advances in digital electronics, to have very flexible timing. For example you might want to switch on the fly from measuring closed orbit to having turn by turn measurements of a single bunch. In RHIC that might be important for machine studies because each bunch can come from a different AGS and booster cycle. So you might want to do turn-by-turn measurements on one selected bunch. Another thing you might want to do is something that actually Steve Kramer has been talking about: looking for a coupled bunch modes by taking a snap shot of the ring. Here, every position monitor is possibly looking at a different bunch but as close in time as possible to reconstruct, given enough monitors, what the oscillations look like all the way around the ring during a short period of time.

So you want a lot of flexibility in the timing system, but given a nanosecond bunch length, you do not want to require a part of a nanosecond for the resolution of your clock all around the ring.

Bob Shafer: We thought a lot about this before we implemented the beam position systems here at the Tevatron. What we finally did was develop a system called Tevatron clock, TCLK, which in essence is a clock ticking at a 10 MHz rate. It has a di-phase code on it for sending information and timing information as well. But in addition we used, at the multinanosecond level, beam synchronization inside the beam position system to locally get the exact timing signal for gates. In circular accelerators, except for interbunch type instabilities which was just mentioned, beam motion and phase oscillations are very slow. It is synchrotron oscillations take hundreds of turns to go through a phase of synchrotron oscillation. This means that you can make a measurement at one point in the ring and expect that the phase has changed a very small amount by the time the bunch comes around again. Therefore you do not need to measure that globally around the ring. You just need to measure it in one or two places. So it does not need to be included in a global beam position system, or a beam orbit monitor. My feeling is that using local beam synchronization is adequate for the global system with, of course, timing signals.

Bob Webber: I would like to comment on the problems that we had with the beam synchronization in the Tevatron. That system worked fine while we were running the collider in the fixed target mode. But once we started to run in the collider mode, where we had isolated bunches for which we wanted to measure the position, we learned that there are some important dynamic characteristics in the electronics, both of the AM-to-PM circuitry and of things related to beam intensity. In the single bunch mode, the time window in which you had to sample the position signal in order to get the best information was a fairly narrow range: 50 nanoseconds. One of the difficulties there was to be able to always stay within that window. The beam synchronization/self-triggering scheme operated off a threshold that was set ahead of time. Once the beam intensity signal passed that threshold, it would delay a certain amount of time and then do the sample on the position signal.

We learned that there were problems with this scheme. The beam intensity would vary, so the threshold of the system was quite important. We were not quite certain if we were sampling the position signal at the proper time for this beam intensity, for instance. That is not to say that the beam self triggering was not the best choice for the Tevatron. It likely was the best choice, but when we began operating in the collider mode we started asking questions that basically boiled down to how bad would the problems have been had we chosen an external timing signal to try to control this. The answer is not known yet.

LEP Timing

Jean Borer: In case of LEP, we are following two directions at the same time. One is timing, which determines the value of a measure signal like sample-and-holds, peak-

and-holds or integrate-and-hold circuits, which may be sensitive to time. In that case, if possible, the best solution is self-triggering. I have used it for the antiproton acquisition circuitry at ISR and now for LEP. It works perfectly reliably and it also works on simulated pulse signals. So you are always in the same situation, if it is beam or if it is simulation. The acquisition through a flash ADC of any kind and data transfer to a buffer memory should be beam triggered.

The second situation is the problem of having autonomous timing stations around a large machine. As soon as you want to do single-turn measurements or to make sure that you follow the same bunch around the machine you need some means to do it. In such a case, a less precise system can be used, where the window is broad relative to the signal to catch. For this, we have invented a Beam Synchronous Timing system in which we stick a label to each data acquisition. It's resolution is 400 nanoseconds, from a locally created fast clock, vernier between turn clock pulses. This time label is very useful to distinguish between the bunch and parasitic signals, as I said this morning. This transmission of a turn number and the recording of a time sticker with the data is not difficult.

Jay Heefner: On the CEBAF system we use an external trigger generated from the same trigger that is used on the injector. The major reason is that both the single pulse measurement and the pseudo-random sequence measurement require reasonably precise timing. With the signal-to-noise levels that we have out of our detector on a single pulse, self triggering would not meet those requirements, so we have no choice.

Scott Miller: Are there any general questions or BPM questions from the audience?

More on Dumping the Beam

Comment from the Audience: I have a question regarding future machines. What are the requirements for a beam loss monitors in future machines? More specifically, how will the decision be made to dump the beam? I understand that any future machines with very high currents could be destructive and there must be some means of dumping the beams and how would a decision be reached to dump the beam.

Bob Shafer: The dumping of the beam itself is done, in a case the Tevatron and the SSC, by an abort magnet that extracts the beam from the machine in one turn and then an abort beam dump which absorbs the full energy of the beam. The abort is triggered by a variety of systems which can include either electronic malfunction or, in the case of beam loss monitors, the indication that a beam loss is occurring. One of the issues at the SSC is whether or not they can abort one beam without having to abort the other because beam loss monitors are not very directionally sensitive.

Comment from the Audience: But do think that these monitors are reliable enough to make the decision to dump the beam?

Bob Webber: I think they have to be.

WIRE SCANNER SYSTEMS FOR BEAM SIZE AND EMITTANCE MEASUREMENTS AT SLC

M. C. Ross
Stanford Linear Accelerator Center, Stanford University, Stanford Ca. 94309

ABSTRACT

The SLC wire scanner beam profile monitors provide accurate measurements for use in beam feedback systems and automated beam optimization procedures. Beam profile measurements can be performed throughout much of the SLC with no interruption to normal machine operation and no adverse impact on interaction region detector backgrounds. In this paper we describe the design, construction, performance and uses of SLC wire scanners with particular emphasis on the integration of the scanner with the accelerator controls. In keeping with the workshop nature of this meeting the difficulties encountered in the design and commissioning process will also be described.*

INTRODUCTION

For several years steady progress has been made on the Stanford Linear Collider (SLC) project. The performance of this first linear collider improves following the installation and commissioning of new diagnostic and controls systems. Measurements of beam size and associated optical parameters are key to SLC and progress has been hampered by the inability to measure beam size accurately in a rapid, non-invasive manner.

In this paper we describe the design, construction, commissioning and ultimate uses of wire scanners in the SLC, focusing on the linac and upstream systems scanners. These are more or less conventional devices and many of the issues encountered in their construction and commissioning are not special to the SLC. The small SLC beams and the high degree of active, automatic control required for SLC are perhaps the most significant differences between SLC wire scanners and scanners in use at other accelerators. Of particular interest is the interaction between the wire and the scattered radiation from the wire with the extreme electric field of the beam. As this field reaches the level of several volts/angstrom, as it does easily at the SLC interaction point (and may in upstream parts of SLC), field emission from the wire may occur.

Wire scanners have been in widespread use in many labs and are well developed. References to these systems are included where appropriate.

We begin this paper with a short discussion of the interaction region diagnostics and then give some examples of how wire scanners are used in the SLC. This is followed by the device specifications and a description of its mechanical, controls system and detector design. The results of performance tests are given in the final section.

INTERACTION POINT WIRE SCANNER

In the early stages of machine commissioning a special wire scanner[1] was installed at the collision point. This device, capable of measuring beams with transverse sizes ($\sigma_{(x,y)}$) ~1µm, has been described and we will not go into the details of its construction. Two factors have combined to make it less operationally important that it initially was. The first, and most significant, is that the beam-beam deflection scan, or use of the one beam to probe the other, is more accurate, more appropriate, since it also ensures that the two beams are in direct collision, and much more convenient since it can be done continuously with no impact on the users detector which surrounds the interaction region.[2] The second reason is that, at intensities low compared to typical operating intensities, the wire will not survive a single direct hit from the beam. Use of this wire is limited to intensities of 10^{10} e+/e- per pulse well below the nominal SLC operating intensities of 5×10^{10} e+/e- per pulse. This prediction has been tested when a 4 micron wire was broken by the beam at low repetition rate.

The interaction point (IP) wire scanner provided the information necessary to perform the final optical corrections to each of the beams and bring them into collision. Since the threshold for strong beam-beam deflections is now easily passed, it is only used as a backup diagnostic.

ACCELERATOR CONTROL FUNCTIONS

A key feature of SLC operation is the degree of high level active control required to keep it optimized. In sharp contrast to colliding beam storage rings, active beam parameter control is vital. The high level of demand takes the instrument out of the category of a device primarily used for machine development or failure diagnosis purposes and elevates it to an online device.

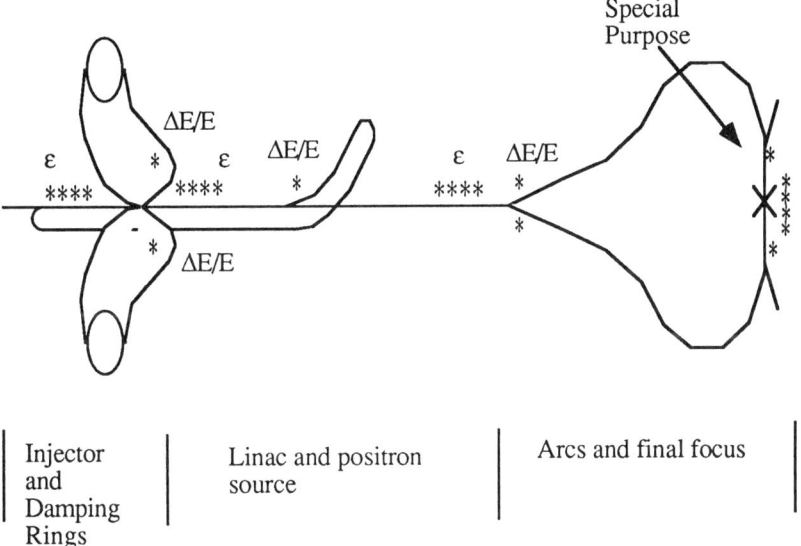

Figure 1. <u>Schematic diagram of the SLC showing wire scanner locations and related functions</u>. Groups of four scanners are installed at the end of the injector and at the beginning and end of the linac to provide emittance information. The final focus region has scanners for the IP, the final triplet and the arc to final focus transition area for emittance, beam tail and beam size monitoring.

Active beam parameter control in the SLC has been done, for the most part, with slow feedback loops operating in the control system central host computer [3].

Wire Scanner Systems at SLC

Beam parameters requiring control include intensity, trajectory, energy, energy spread and transverse beam size. The SLC schematic diagram in figure 1 shows wire scanner locations and their feedback related functions.

Operational feedback requires good knowledge of the transfer function between control parameter and measured quantity. Some knowledge of the stability of the transfer function is also required.

Automated procedures are used in order to accurately and effectively determine the control transfer function. Thus, in practical terms, the step preceding feedback is automated tuning. An example of such a procedure is illustrated in figure 2. The linac output beam emittance can be changed if a large dispersion (energy - transverse position correlation) is present at the linac launch point. In the presence of this correlation the beam emittance measured at the launch with the techniques discussed below will be larger than that of a monoenergetic beam. A sensitive method for controlling the residual dispersion is to map this pseudo-emittance as the focusing magnets that affect it are adjusted. The minimum, usually close to the monoenergetic emittance, is then accepted as the final set point.

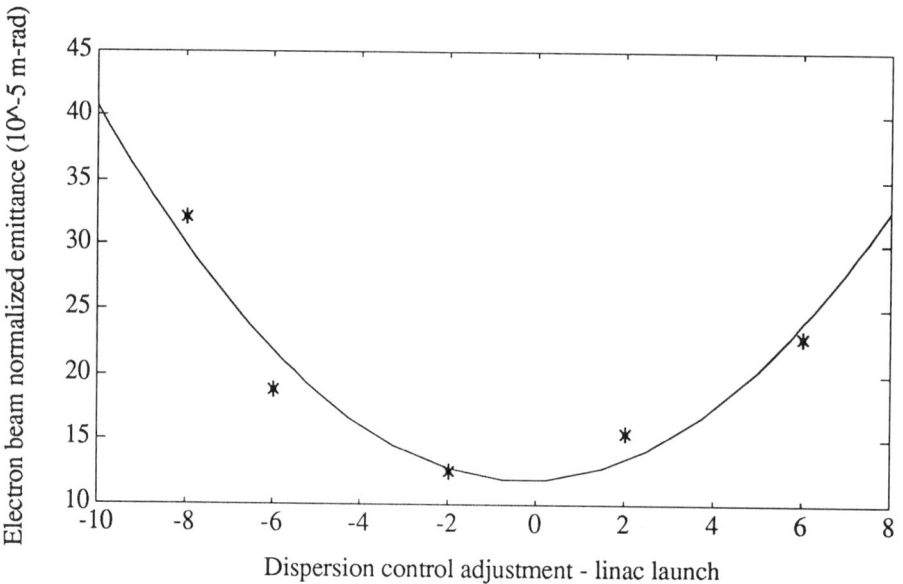

Figure 2. <u>Final linac emittance vs. linac launch dispersion.</u> The units of the dispersion control adjustment are in cm of peak dispersion.

The collider luminosity is given by:

$$L = \frac{f_{rep} \, I^+ \, I^- \, H_D}{4\pi\sigma_x\sigma_y}$$

where f_{rep} is the collider repetition rate, $I^{+/-}$ are the positron and electron intensities, H_D is the disruption enhancement factor and $\sigma(x,y)$ are the interaction region beam sizes. In practice, $\sigma(x,y)$ depend on the emittance because $\sigma(x',y')$, the angular divergence at the collision point, must be kept below the threshold of secondary or off-axis particle generation in order to keep detector backgrounds minimum. Since the collision region optics produce an upright phase space ellipse at the collision point, the spot sizes are determined by the emittance:

$$\varepsilon = \sqrt{\sigma_x^2 \sigma_{x'}^2 - \sigma_{xx'}^4} = \sigma_x \sigma_{x'}$$

The luminosity thus depends directly on $1/\varepsilon_x\varepsilon_y$. Many mechanisms can cause an emittance increase in the SLC. In order to quantify a series of small effects and properly control this blowup, the emittance must be accurately measured in several places and tracked.

The specified damped SLC beam normalized emittance is $\gamma\varepsilon = 3 \times 10^{-5}$ m-rad., where γ is the relativistic normalization factor. Typical SLC damped beam sizes range from 300 μm at 1 GeV to 100 μm at 50 GeV. The scanners must measure emittance with at most 10% systematic error throughout this range. The beam size errors typically enter quadratically into the emittance so that a 5% systematic error on the beam width becomes a 10% error on emittance.

The method used to measure emittance with a phosphor screen profile monitor is to insert the device and digitize the image while varying an upstream quadrupole magnet [4]. This should be done in a nominally dispersion free region. In the thin lens approximation, the square of the beam size plotted against the quadrupole magnet strength fits a parabola

$$\sigma^2 = a(Q-Q_0)^2 + c.$$

where Q is the quadrupole strength and a, Q_0 and c are the fit parameters. The emittance is then

$$\varepsilon = \sqrt{\frac{ac}{R_{12}^4}}.$$

where R_{12} (R_{34}) is the horizontal (vertical) transfer matrix element between the quadrupole and the screen. Systematic errors associated with resolution, i.e. those that add in quadrature to the real beam size, simply offset the parabola, adding to 'c', and enter directly into the emittance estimate.

The above method requires the insertion of a screen profile monitor and adjustments to quadrupole magnets in the beam line. Even if the screen can be made thin enough for the beam to pass through, the changes made to the magnets and the scattering in the screen will spoil it for most downstream users. This problem is most severe at the entrance to the SLC linac where the positrons and electrons destined for collision must pass along with a bunch of electrons destined to produce positrons for use on a subsequent SLC pulse. Any attempt to measure the positron emittance usually results in the removal of the positron production beam from the target thus preventing further production. To circumvent this, the multiwire emittance measuring technique was adopted which does not require changing any magnetic element.[5] A minimum of three wire scanners, placed approximately 45° of betatron phase advance apart, must be used since the incoming phase space orientation is unknown. We have installed the scanners in groups of four in order to overconstrain the result in locations with normal, fixed phase advance per cell and provide measurements in both planes in cases where non-degenerate locations are not available. Given this overconstraint, it is possible to determine the internal consistency of the scanner measurements. This redundancy can be very useful in the presence of badly blown up beam tails or for hardware checks.

Energy spread is critical since the SLC final focus, which contains a strong chromatic correction system, can only properly correct ±0.5% energy spread beams. In addition to poor correction, detector backgrounds may increase significantly outside this limit. The energy spread is controlled by RF phases which in turn are controlled through wide area low level RF systems. The phase stability tolerance on this control is very tight, typically a fraction of an S-band degree. Experience has shown that daily temperature fluctuations can be a few degrees, even after care has been taken to provide temperature stabilization.

The energy spread wire scanners at the entrance to the SLC arcs will, when combined with other monitoring systems, provide the information necessary to track down remaining problems. A straightforward feedback loop will be constructed using these scanners. Since they are downstream of the high power linac beam tail collimators, the scan is invasive and the scanned beam must be disposed of using a nearby pulsed magnet and dump. Typical scans are completed in about 1/2 second so a scan performed every five minutes lowers the full repetition rate luminosity by a negligible 0.2%.

The optimum SLC beam energy distribution is not gaussian but instead has a double peaked structure. By measuring this distribution accurately, we will be able to estimate other beam parameters such as bunch length.

Aside from the actual collision point size, the angular divergence and tilt of the beam in the final focus region are important. The beam size at the entrance to the final focus triplet provides an estimate of the angular divergence at the IP and the beam x-y coupling can be measured using diagonally oriented wires. The multiwire scan performed to derive the skew component of the beam shape is called a skew scan.

The collider beams are not expected to be gaussian beyond several σ. By adjusting the scanner detector gain, accurate measurements of the beam 'tails' are possible. The SLC beams must traverse several background suppression collimators which cannot withstand the high peak or average power (100KW) of the full beam and must be not be allowed to intercept too much of the beam in order to retain their integrity. The beam size, as well as the population in the non-gaussian tails, must be well known in order to maintain reasonable levels of beam loss on these devices.

Closer to the interaction region, where the beam tails should have been removed by the upstream collimator system, the scanner data can be used to show some details of the extremes of the particle distribution in order to check the collimator setup.

PERFORMANCE REQUIREMENTS

Feedback requires a fast, non-invasive (or minimal impact) device which in turn means that the wire, not the beam, must perform the scan. The speed range, vibration and other mechanical specifications can be generated from this and from the expected beam sizes and rates. Table I shows the expected performance of the scanners.

General		
Beam size resolution	measurement stability	<3% σ
Systematic error	most challenging design effort	<3% σ
Emittance error	for linac scanners	10% (for $\gamma\epsilon \approx 3 \times 10^{-5}$ m-rad)

Dynamic range	Intensity range over which the above applies	10^9 - 10^{11} particles/pulse
Vibration	Oscillation of the wire about its expected location	Peak amplitude $< 0.2\sigma$
Mechanical		
Absolute alignment	Wire monitors are not used for absolute position measurements	±0.5mm
Relative positioning	Ability to put the wire back to a known position	20μm
Speed and acceleration	For typical linac beam sizes and repetition rates from 10 to 120Hz.	1cm/s max 0.3mm/s min; 0.2m/s^2 accel.
Other		
Multibunch operation	separate e+/e- bunches which have 60 ns interbunch spacing	<5% signal contamination from nearby beam
Minimal interference with normal operation	Must be able to determine emittance and energy spread at key locations during routine operation without significant interruption	
Radiation resistant	Major maintenance performed annually	10Mrad/year
Lifetime		100,000 cycles/year
Adaptable Geometry	There are often significant space constraints, including nearby high power collimators	
Wire presence test	Wires must be electrically isolated to allow testing	

Table I: <u>Wire scanner performance specifications.</u> The most severe design specification is the 3% σ limit on systematic error. Typical sources of this error come from wire vibration and from detector linearity.

Over 100 phosphor screen beam profile monitors are used in the SLC beamlines. Table II lists the relative drawbacks and advantages of wire scanner profile monitors and phosphor screen profile monitors.

Phosphor Screen	Wire Scanner
-	+
Very fine resolution (<20μm) not possible	Resolution down to few μm has been achieved
Large systematic errors Spatial resolution dominated by optical problems - hard to test	Simpler systematic errors - Detector non-linearity - relatively easy to test
Camera non-linearity and optical aberrations	

Limited dynamic range (from peak to tail)	Wide dynamic range
Phosphor desensitization after prolonged use	
Complex frame grabber data acquisition and background subtraction	Many possible signal detectors - increases reliability and flexibility
Image lag and slow scan speed	Time resolve 60 ns. spaced SLC bunches
Camera and optics radiation sensitivity (<100KRad without complex optics)	Rugged
Invasive	Non-invasive upstream of background suppression collimators. Minimally invasive downstream
+	-
Single pulse capture at low machine repetition rate	Multi-pulse sampling required, difficult to unfold beam jitter
Full two dimensional display	Only projections are available. Hard to get detailed information about x-y coupled non-gaussian beams
	Thin wire fragility
Visual presentation, rich intuitive content, real time display	

Table II: <u>Advantages and disadvantages of wire scanner profile monitors and phosphor screen profile monitors</u>. The non-invasive property of scanner profile monitors is most significant for SLC.

MECHANICAL DESIGN

The mechanical design effort addressed the following problems: 1) wire and wire retention, 2) vibration over the large speed range and 3) positioning errors and position transducers. An overriding concern was radiation damage control.

A schematic diagram of the scanner is shown in figures 3 and 4. Several labs have built scanners of a similar design [6,7,8]. The wire is strung around 1.5mm stainless steel studs set in a 3/16in thick alumina fork in such a way so that it can carry wires of three different orientations across the beam and provide x, y and u (45°) scans. The carriage motion is actuated by a stepping motor through a 2mm pitch ball screw, chosen because of the expected large number of cycles. Some difficulty was experienced obtaining the small pitch, high quality ball screw with no plastic parts. Screws from two vendors are used[9,10]. The wire arm support flange is attached to a crossed roller bearing translation stage[11]. A 125µm thick stainless steel vacuum window opposite the wire allows low energy wide angle scattered radiation to emerge from the vacuum chamber.

Both the cantilever nature of the wire and the stepping motor contribute to wire vibration. We found that the key to careful control of this is adequate testing, both in the lab and in situ. We have used a piezo-electric accelerometer to quantify the motor related system vibration. Figure 5 shows typical accelerometer test results. It is difficult to use the voltage induced by the vibrating wire in a uniform magnetic field for vibration checks, as done with other flying wire scanners[12], in this range of speed and wire loop size because the signal is very small (10^{-7} volts). Further vibration tests will use long range microscopes with video readout.

The wire chosen for the scanner is gold plated tungsten. Tungsten has high strength and its high Z helps contribute to the scattered particle signal. The wire diameter was chosen to be 0.3 σ_{beam}. The wire has an effective 'σ' of r/2 which, when added in quadrature to the beam size, causes a 3% or greater effect when $\sigma_{beam} \leq$ wire diameter. Under normal conditions the wire size can be subtracted in quadrature from the measured size. In the extreme case where the beam is much smaller than the wire, the measured shape has a sharp non-gaussian fall off.

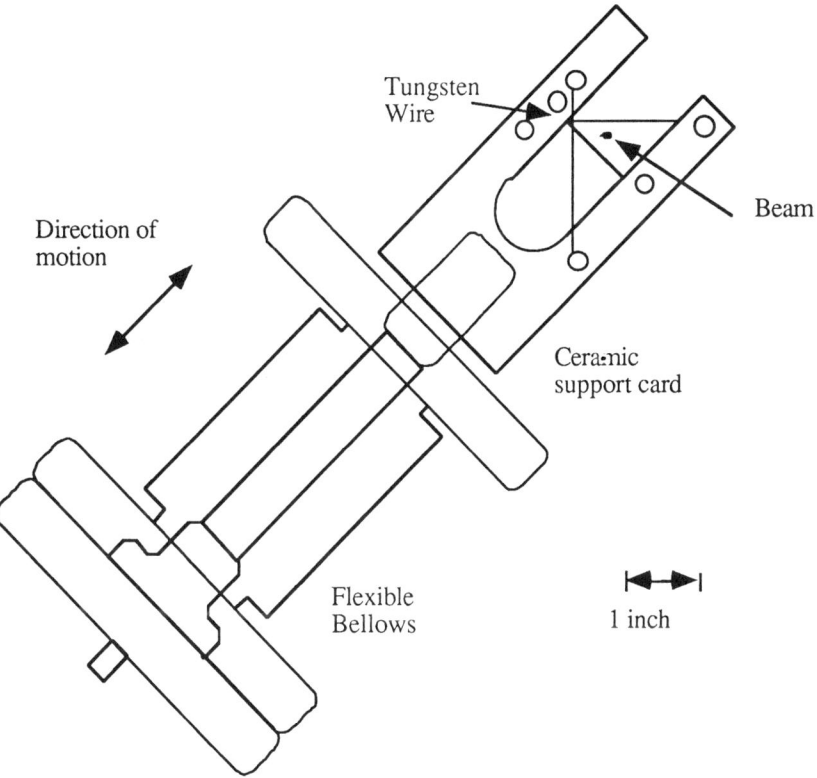

Figure 3. <u>Wire scanner wire support card and flange</u>. Typical scanner maximum travel is 2 in. In the inserted and in the retracted state the beam cannot strike the wires. Electrical connections to the wires are made through ceramic sleeves inside the flexible bellows and studs (not shown) mounted on the ceramic card.

At full SLC currents and rates, the beam can heat the wire substantially reaching a steady state temperature of 1000°C. In the fast scan mode, the wire is subjected to continuous beam for no more than a few seconds in the worst case. However, the wire may be parked in the beam in error or for diagnostic purposes and must be able to withstand continuous beam. By measuring the increase in resistance of the wire assembly an estimate of the wire temperature rise can be made. Tests show good agreement with a calculated rise of about 4° C/pulse.

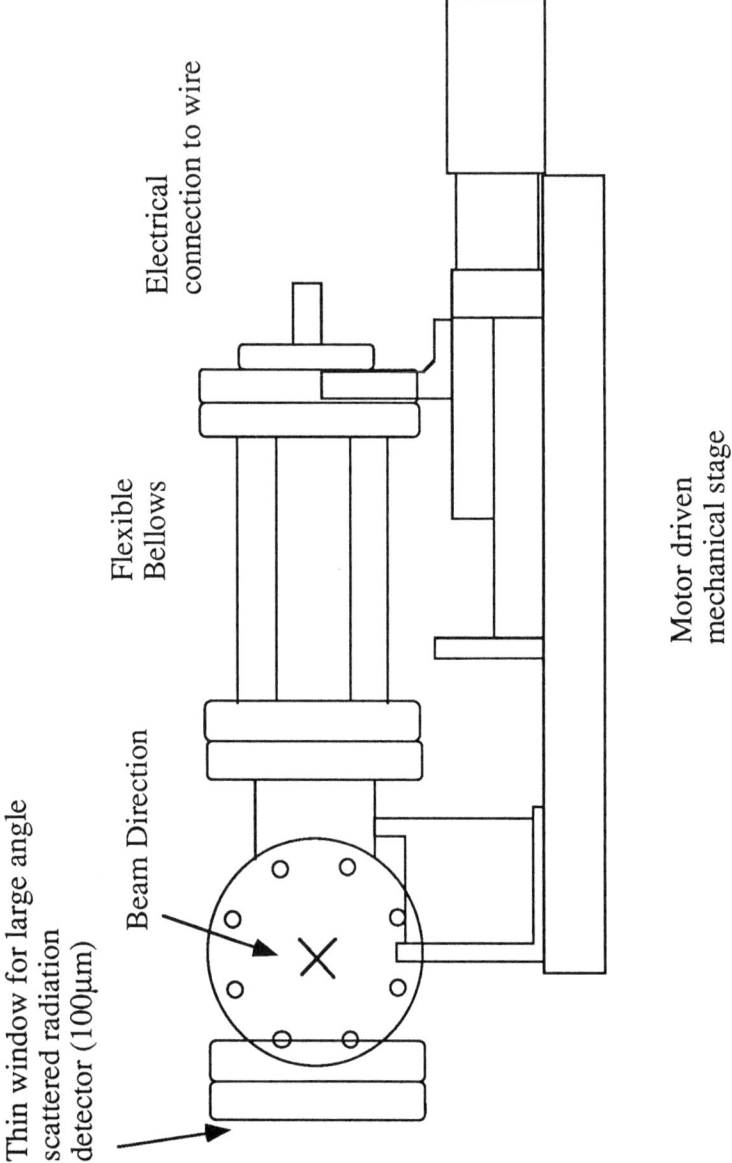

Figure 4. Wire scanner assembly showing translation stage and thin vacuum window.

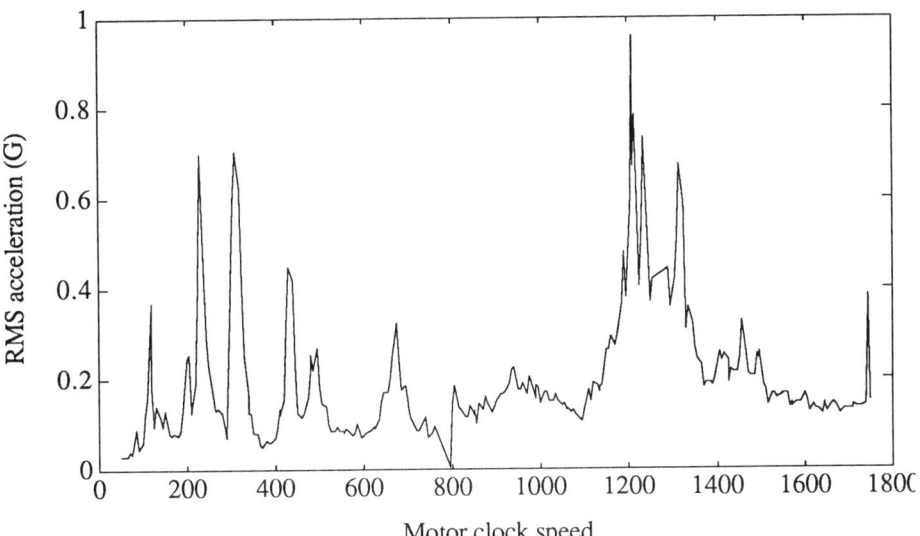

Figure 5. <u>RMS scanner support flange acceleration vs stepping motor clock speed.</u> Typical scanner speeds are 0.5 (5.0) mm/sec at 10 (120) Hz machine repetition rate. This corresponds to 110Hz (1320Hz) with a 1.5mm pitch screw and a 200 steps/rotation stepping motor. The expected peak to peak vibration amplitude can be estimated using $\Delta x = 1.4*\sigma_a/(2\pi f)^2$ where f is the motor clock speed. Thus the maximum allowable peak to peak vibration amplitude of 0.2 σ_{beam} corresponds to about 1G rms at 100 Hz. At higher frequencies the observed amplitudes are small.

A further concern is the possible sag of the wire as it is heated. The wire retention scheme uses small bore ferrules crimped onto the wire and welded to springs to maintain a fixed tension on the wire. The tension is transmitted around the support studs in the ceramic card if the wire is thinner than about 100μm.

The wire position is not encoded on each successive beam pulse, rather the current position of the wire is inferred from a check of the remaining step count. A position measurement using a radiation hard LVDT (linear differential transformer) is done at the limits of the scan to check that the expected position was reached.

CONTROLS DESIGN

Figure 6 shows the controls and data acquisition hardware. A commercial stepping motor controller [13] and LVDT[14] processor are used. The data acquisition sequence is shown schematically in Figure 7. The scan consists of three steps: 1) move from PARK (near, but not in, any of the beams) to the start of scan at maximum speed, 2) scan at the speed corresponding to the desired inter-point spacing and beam rate and 3) return to PARK at full speed. Only brief pauses, to allow the device to come to a complete stop, occur during the scan sequence. During multiwire emittance or skew scans, the wire moves to the next appropriate park at step 3. An important feature of this scheme is the use of machine wide data acquisition codes which coordinate the readback of the scanner step count, the signals from the wire scanner detector and allow a great deal of flexibility in the choice of detector including, for example, the use of detectors several miles away.

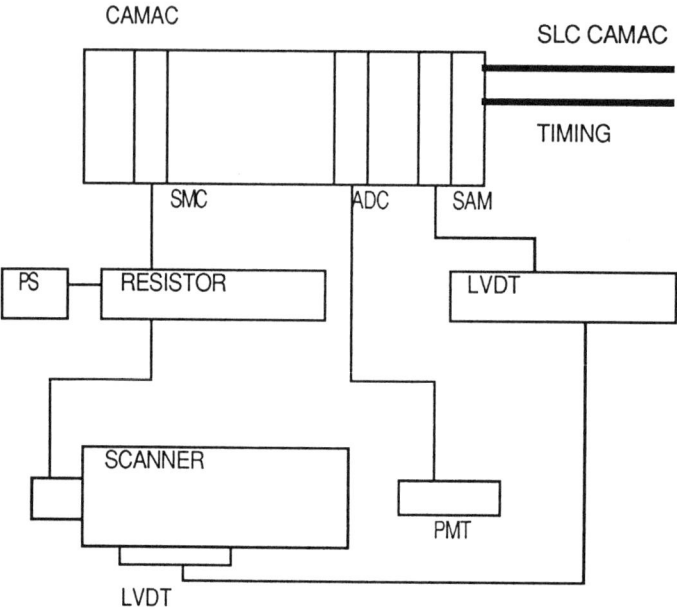

Figure 6. Controls hardware schematic. The SLC CAMAC system houses the stepping motor controller, photomultiplier (PMT) digitizer (ADC) and LVDT signal digitizer (SAM). Full step L/R stepping motor drive is used.

Because the scanners are to be used for feedback, the application software that controls them must have sophisticated exception handling, error logging and status reporting. The control system software built around these devices allows use of the wire scanner at several levels. The lowest is the single scan and associated single pulse detector signal readout. This allows checking the fit quality, scan ranges and other details. At the next level higher, the fit results can be used in the SLC control system correlation plot utility. This extremely powerful tool, allowing the acquisition of scan data with other beam diagnostic data and machine parameters, has been invaluable for commissioning the scanners. All aspects of the gaussian fit to the scan data are available and are automatically acquired as an upstream device setting is controlled in a programmed way. Most automated optimization procedures are built around this facility. The next level of software does multiple scans and accumulates these results in correlation plot utility. This includes four wire emittance scan results and skew scan results. Finally, feedback can perform the scans as a background task and implement the desired corrections. The first feedback loops will control energy spread.

Another background task takes and records data from the multiwire and single wire scans. Figure 8 shows a history of beam size measurement data taken once every 30 minutes at the entrance to the linac.

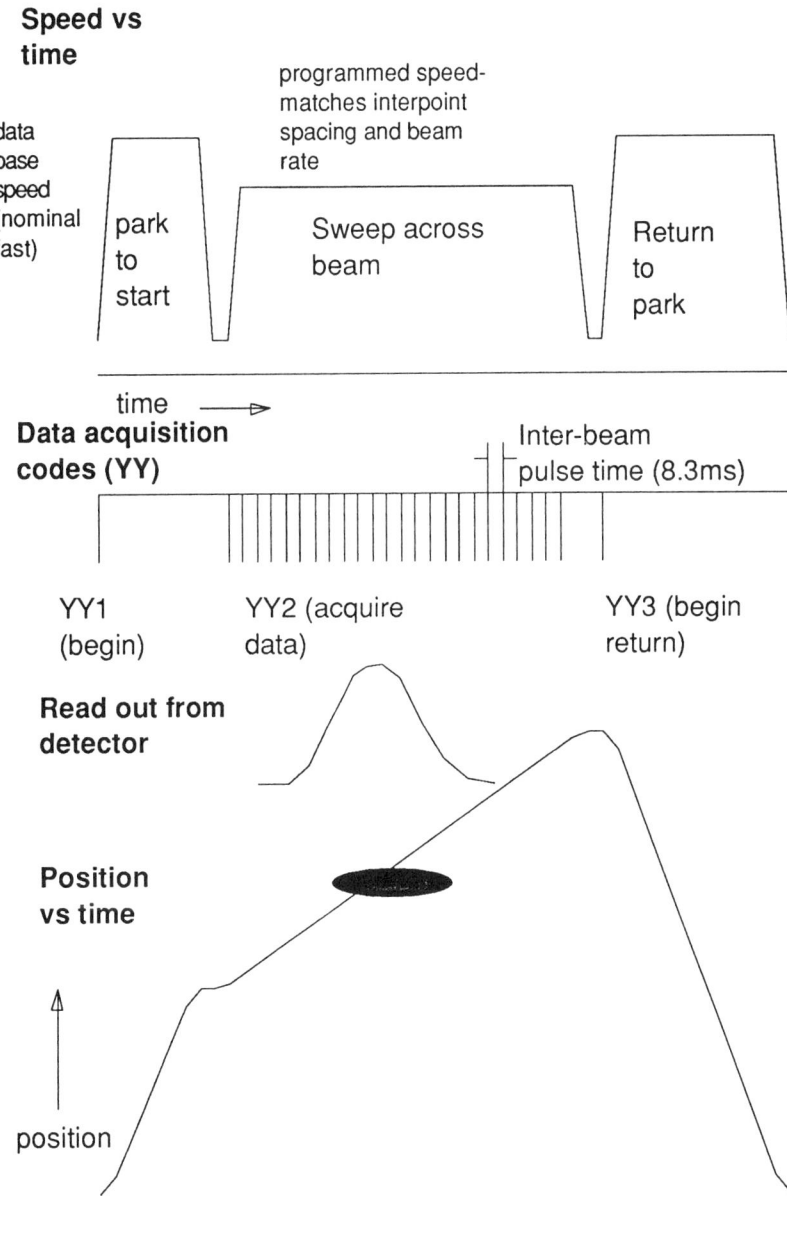

Figure 7. <u>Single beam scan sequence.</u> The speed, machine wide data acquisition codes and wire position are shown vs. time. A programmable acceleration ramp (0.1m/s^2) is used. Controlling codes initiate the start of the procedure and the return to park.

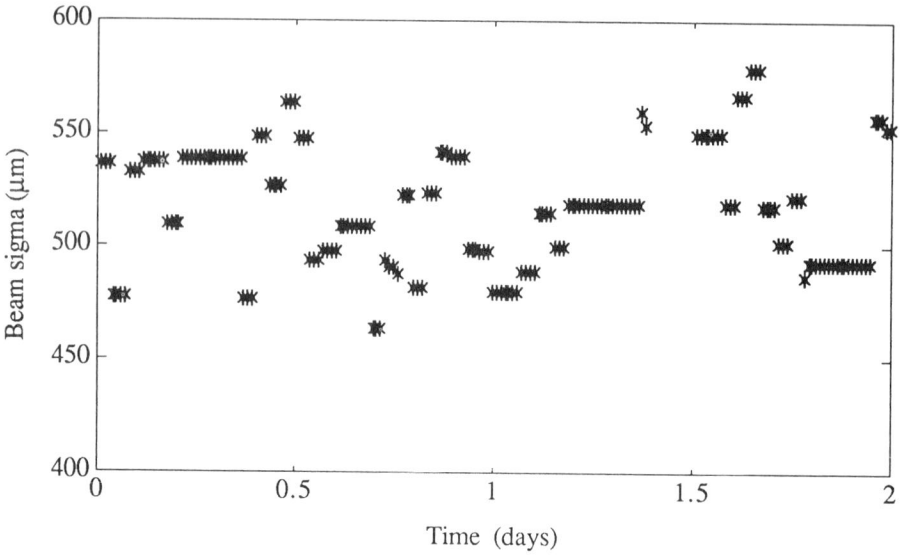

Figure 8. Beam size measurements vs time. Beam size measurements are made once every 20 minutes and recorded by a control system background process

SIGNAL DETECTOR

The purpose of the wire signal detector is to indicate the amount of charge striking the wire. One of the reasons wire scanners were not adopted early in SLC development was the lack of an adequate wire signal source. Secondary emission and forward scattering, used in many wire scanner systems, are often not practical at the SLC. The first because of problems discussed below and the second because of the very different beam line areas in which these devices have to operate. The most difficult of these is just upstream of a high power collimator system. Radiation scattered by the wire in the forward direction is completely overwhelmed by the scattering from the collimator jaws thus making the use of small angle scattered radiation impossible. Furthermore, it may be difficult to place a detector downstream of the wire in a location where the detector acceptance is constant for various beam conditions, such as steering and focusing, at the wire.

When the IP wire scanners were first tested it was found that the secondary emission signal would increase dramatically when either the beam current exceeded about 5×10^9 or the beam size fell below about 10μm. The onset of this dramatic increase is signalled by a very unstable signal. In the SLC linac, where the beam sizes are about 100μm, a beam current of about 2.5×10^{10} is required. Figure 9 shows scans taken just below and above this threshold in the linac. This effect appears to be field emission induced by the field of the beam.which peaks at about 20V/Å.

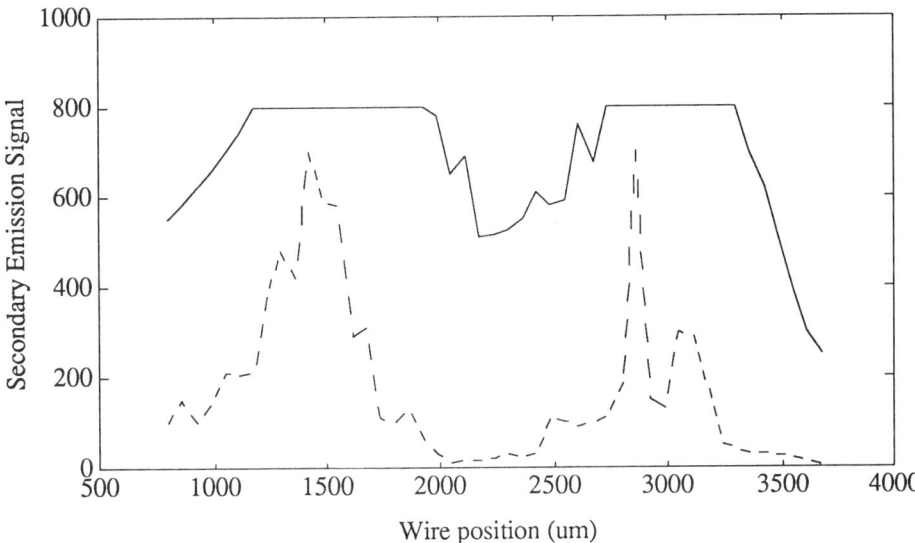

Figure 9. Secondary emission signals from wire scans at 2.5×10^{10} (dashed) and 2.8×10^{10} (solid) e- per pulse showing a large increase in signal strength. This data was taken with a prototype double wire unit.

Because of these problems, tests were made to determine if a significant signal was present at 90° to the beam direction, directly opposite the wire support card. A strong, very low energy signal was seen in a bare photomultiplier (PMT) placed about 30cm from the wire. A thin window is required so that this scattered radiation is not absorbed by the vacuum chamber wall. Possible problems associated with this signal include the interaction of the the electric field of the beam with the scattered radiation. An advantage is that there should be no acceptance related cutoffs. Substantial shielding (±50 radiation lengths) is required in some locations to protect the PMT's from background generated by upstream beam losses. In regions where no collimators follow the scanners, small angle scattering monitors have been placed about 10 m downstream to use for comparison with the PMT.

The detector linearity must be better than a few percent. PMT's of the type used in HEP detectors showed signs of non-linearity in tests using a spark gap and a normalizing PMT with neutral density filters. Because it is somewhat difficult to reproduce the very fast (few ps) pulses seen by the PMT at the wire, these tests may not show time related saturation effects near the PMT photocathode. A moderate gain, excellent linearity tube was chosen[15]. In order to cross check PMT performance a small ion chamber was built and installed downstream of the scanners. The special feature of this device is it's fast signal, required to separate the bunch signals. It uses a 2mm gap, 1mm thick wires to prevent multiplication, and a fast gas (CF_4) to achieve 10 - 20ns pulse width signals.

102 Wire Scanner Systems at SLC

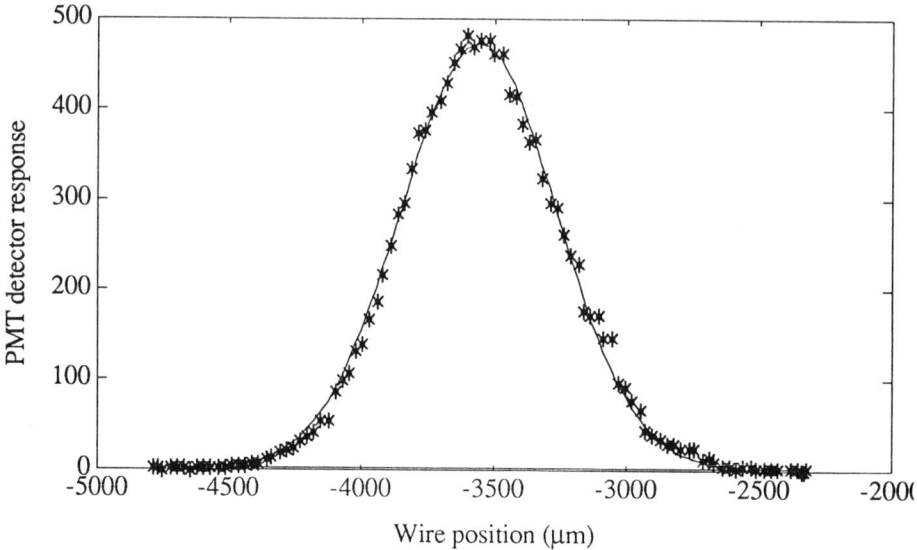

Figure 10. <u>Typical single scan.</u> The data are fit to a gaussian with an offset to allow for backgrounds that may arise from nearby beam losses or microwave breakdown.

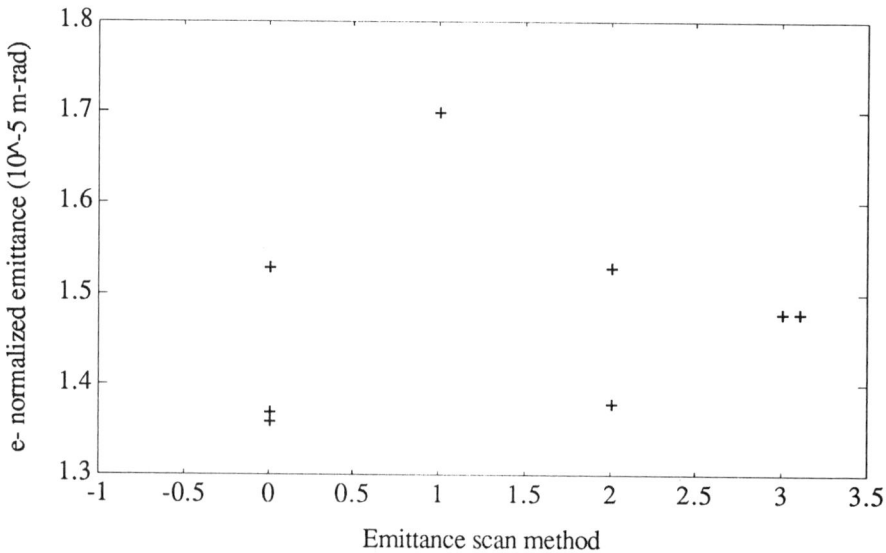

Figure 11. <u>Comparison of emittance measured with the multi wire technique (0) and with quadrupole scans using three different magnet and wire scanner combinations (1-3)</u>. The data agree to ± 0.2 m-rad., consistent with the specified performance.

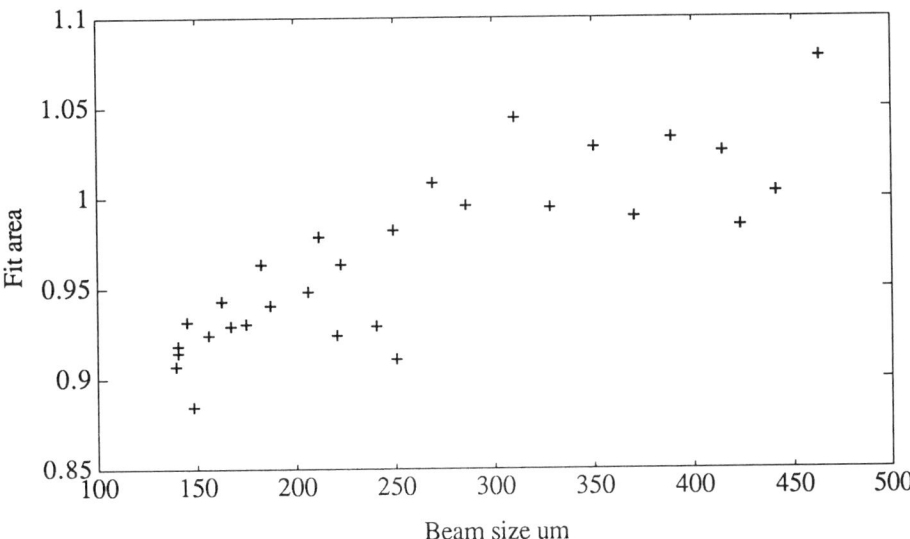

a)

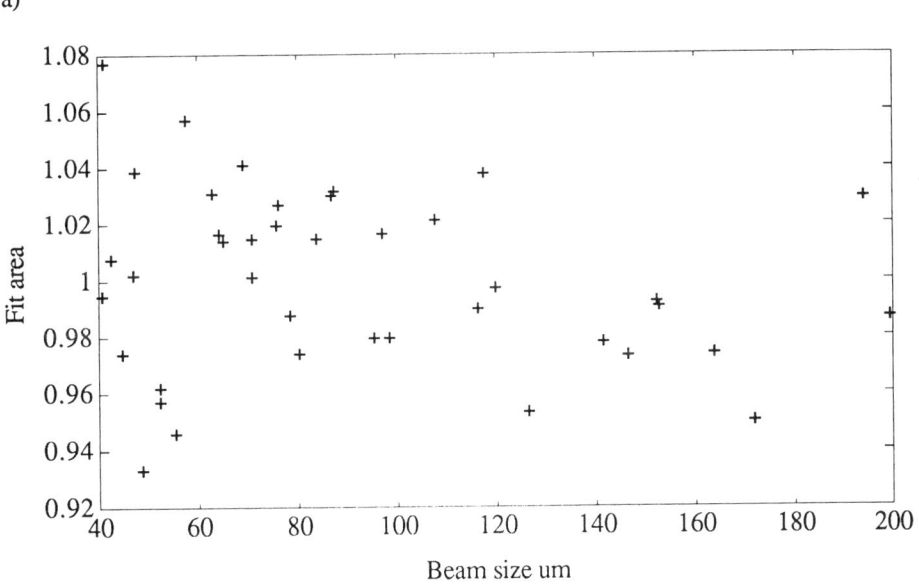

b)

Figure 12. <u>Computed scan area from gaussian fit parameters vs beam size.</u> The beam size was controlled using a nearby quadrupole. The area should depend only on the beam intensity for fixed detector gain. Figure 12 a) shows some saturation effects. The point scatter in these plots is consistent with the estimated error on the fit area.

PERFORMANCE TEST RESULTS

Figure 10 shows a typical single scan. The data are fit to a gaussian with a offset. Work is in progress to determine the functional form of the errors associated with each data point.

Performance tests have focused on understanding systematic errors. These tests fall into two broad categories: 1) saturation tests made by varying beam size and / or intensity and 2) tests made with different detectors sensing scattered radiation from a single wire.

Figure 11 shows the beam emittance measured using quadrupole scans on three different wires and multiwire scans at 1 GeV and 4.5×10^{10} e-/pulse, near the SLC nominal operating intensity. This is a good test of saturation effects since the beam sizes and signal strengths vary considerably over the scan range and from scanner to scanner. These tests were done using the downstream fast ion chamber.

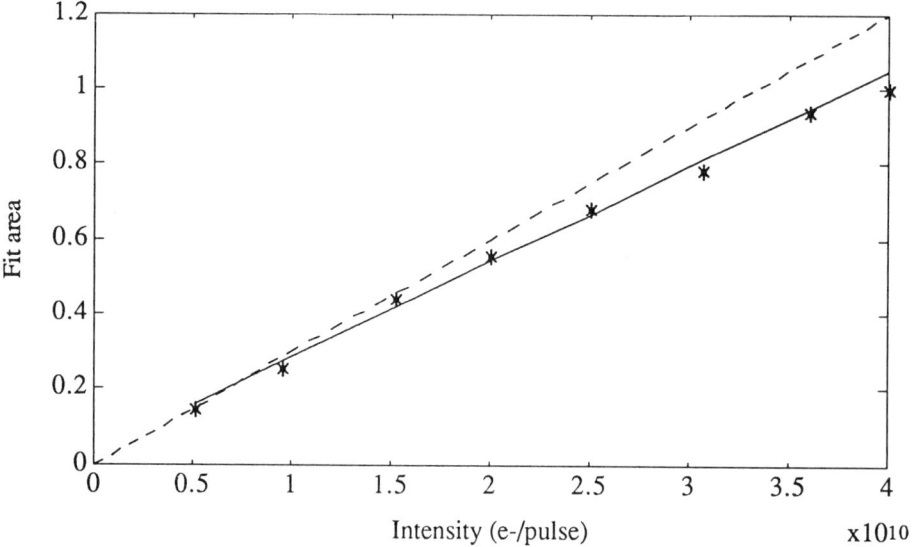

Figure 13. <u>Scan area from gaussian fit parameters vs beam intensity.</u> The solid line is the best fit to the data and the dashed line is a simple extrapolation of a line with zero intercept passing through the first point.

Figure 12 shows the integrated scan area, derived from the fit parameters, vs. beam size. In the presence of saturation the scan area should drop at very small beam sizes, where the density of particles striking the wire is greatest. There is evidence of this in figure 12a. This effect has been difficult to reproduce in a quantified way and work is underway to further our understanding of it. The results from another saturation test result is shown in figure 13. The fitted gaussian area is plotted vs. intensity. The data fall on a good line but the slope is in error by 15% and the intercept is not zero. Both of the above effects give an estimate of our emittance uncertainty of about 20%, two times larger than the specified device accuracy.

Another way of looking at non-linearities associated with possible saturation is illustrated in figure 14. This figure shows a comparison between several detectors detecting radiation scattered from a single wire. Downstream ion chambers give a consistently smaller beam size. This difference, as well as the problems shown in figures 12 and 13 will be improved with the use of a more linear PMT.

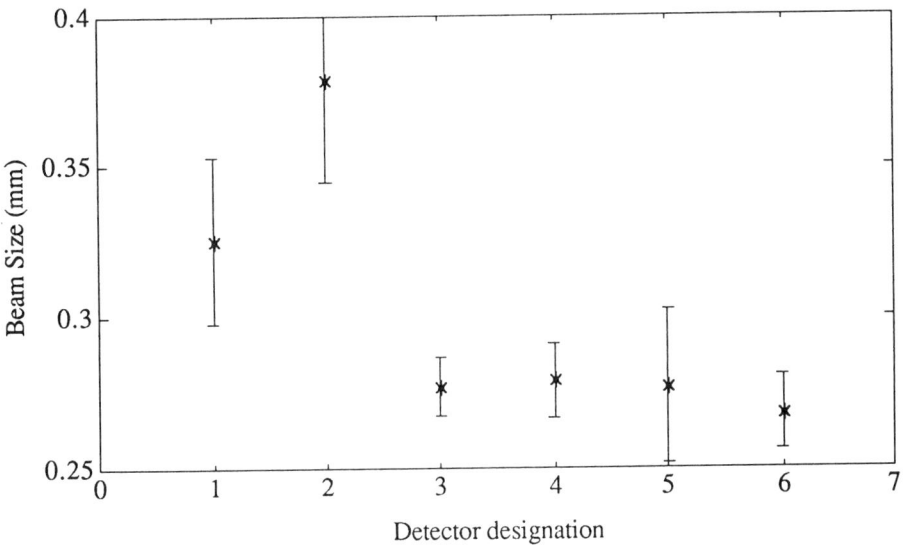

Figure 14. <u>Beam size determined from a single scan using different detectors</u>. The detectors used are as follows: 1) 90° PMT, 2) PMT on downstream wire scanner, 3-6) ion chambers associated with beam line obstruction about 2000m away.

Many other tests, aimed at checking other aspects of scanner performance, have also been done. These include measurements of beam size vs. beam position, wire speed, beam rate, klystron or RF related background and multi-bunch separation. The latter requires the installation of good coaxial cable and careful attention to timing. The achieved bunch to bunch signal contamination is 3%.

CONCLUSIONS

The SLC wire scanners provide beam emittance data that is reliable enough to have already yielded new insights into the performance of the SLC. In the coming months, further tests will be performed to determine the root causes of the remaining non-linearity. Vibration tests and improvements are also planned. By early 1991, 24 scanners will be in use. Future linear colliders will have a tighter emittance budget and will require improved resolution scanners. Piezo-electric motors, with their very small step size and ultra-high vacuum compatibility, may prove to be an appropriate technology for future wire scanners. Two such scanners are in process for SLC.

ACKNOWLEDGEMENTS

I would like to acknowledge the efforts of J. Seeman and the linac group for help in commissioning the scanners, E. Bong, E. Reuter and the mechanical design and engineering group for mechanical design, A. Tilghman for controls, N. Phinney, L. Hendrickson and L. Sanchez for software and C. Field, K. Bouldin and C. Young for PMT testing and data analysis. Final thanks go to D. McCormick for help in every aspect of this work.

*This work supported by DOE contract DE-AC03-76SF00515

---

[1] R. Fulton et. al., ' A High Resolution Wire Scanner for Micron Size Profile Measurements at the SLC', Nucl. Instr. Meth. A274:37, 1989.

[2] W. Koska et. al.,'Beam-Beam Deflection as a Tuning Tool at the SLAC Linear Collider', Nucl. Instr. Meth. A286:32, 1990.

[3] K. Thompson et. al., 'Feedback Systems in the SLC', Proceedings of the 1987 IEEE Particle Accelerator Conference, p748, 1987.

[4] M. C. Ross et. al.,'Automated Emittance Measurements in the SLC', Proceedings of the 1987 IEEE Particle Accelerator Conference, p725, 1987.

[5] K.D. Jacobs et.al., 'Emittance Measurements at the Bates MIT Linac', Proceedings of the 1989 IEEE Particle Accelerator Conference, p1526, 1989.

[6] R. Jung and R. J. Colchester, ' Development of Beam Profile and Fast Position Monitors for the LEP Injector Linacs', IEEE Trans. Nucl. Sci. NS32-5:1917, 1985.

[7] R.I. Cutler et. al., 'Performance of Wire Scanner Beam Profile Monitors to Determine the Emittance and Position of High Power CW Electron Beams of the NBS-Los Alamos Racetrack Microtron', Proceedings of the 1987 IEEE Particle Accelerator Conference, p625, 1987.

[8] K.D. Jacobs et.al., 'The Beam Profile Measurement System at the Bates Linac', Proceedings of the 1989 IEEE Particle Accelerator Conference, p1523, 1989.

[9] Warner, Beaver Precision Products, Troy Mi., U. S. A.

[10] THK Co. LTD, Elk Grove, Ill.

[11] Micro-Slides Inc. Deer Park, N. Y.

[12] J. Bosser, et.al., 'Transverse Emittance Measurement with a Rapid Wire Scanner at the CERN SPS', Nucl. Instr. Meth. A235:475, 1985.

[13] Joerger, East Northport, N. Y.

[14] Daytronic Corporation, Miamisburg Oh., U. S. A.

[15] R580, Hamamatsu Corporation, Bridgewater, N. J., U. S. A.

ION PROFILE MONITORS USING MICROCHANNEL PLATES

John Krider
Fermi National Accelerator Laboratory,
Batavia, Illinois, 60510, USA

ABSTRACT

The presentation on ion profile monitors using microchannel plates consists of two principal parts. The first part describes the design and performance of a profile monitor that was installed in the Debuncher Ring of the Antiproton Source at Fermilab. The information presented relies to a large extent on the contents of a previously published article describing that project [1]. The second part of the presentation describes work in progress on a turn-by-turn profile monitor for the Fermilab Booster. A written description of that work is included elsewhere in these proceedings [2].

DESIGN DIFFERENCES BETWEEN DEBUNCHER AND BOOSTER MONITORS

The main design differences between the Booster monitor and the Debuncher monitor stem from: (1) the fact that the Booster beam current density is 4-5 orders of magnitude greater and (2) the desire to observe Booster characteristics on a turn-by-turn basis (1.6 us revolution period). The Debuncher monitor was designed primarily to observe betatron stochastic cooling, which has a time constant on the order of one second. The first point leads to concerns about beam space charge distorting the ion drift field and about saturation effects, particularly in the microchannel plates. The second point requires that the Booster detector and profile acquisition system operate at roughly a megaHertz, in contrast to 100 Hz for Debuncher monitor. The Booster monitor initially will use the detector assembly (drift gap electrodes, microchannel plates and anode strip printed circuit board) from the Debuncher. However, different readout electronics have been prepared, including transimpedance amplifiers to convert the detector current to voltage and fast CAMAC digitizers with memory. These issues are addressed in more detail in the accompanying article describing the Booster monitor.

REFERENCES

[1] John Krider, Nucl. Instr. and Meth. A278 (1989) 660.
[2] J. B. Rosenzweig, V. Bharadwaj, J. Lackey and P. Zhou, published in these proceedings.

Schottky Signal Monitoring at Fermilab

D. W. Peterson
Fermi National Accelerator Laboratory
Batavia, Illinois 60510

Abstract

Schottky signals arising from the independent motions of particles in an accelerator are extremely useful for observing properties of the beam. This paper will outline the use of Schottky signals at Fermilab. Specific examples of longitudinal and transverse detectors will be shown as well as methods of amplifying and distributing the signals. Various types of signal monitoring equipment will be discussed with some examples of custom built equipment. Measurements of emittance, tune and other beam parameters will be shown.

Introduction

Schottky signals are named for Walter Schottky. Born in Zurich Switzerland in 1886, he received numerous doctorates and worked as a physicist. Among his many accomplishments are his invention of the screen grid tube in 1915 and his discovery of the random thermion emission in vacuum tubes, now known as the Schottky effect. [1]

The signals we observe in accelerators are due to the random distributions of particles throughout the machine. Perhaps an easy way to visualize this is to consider a few individual particles coasting in a circular machine. A pickup electrode would see a pulse (doublet) each time a particle passes. Each particle is free (within some limits) to move independently of the others. The time required for the particle to orbit the machine is the revolution period. At Fermilab this is 21 microseconds in the Tevatron and 1.6 microseconds in the Antiproton Accumulator. The spacings between particles, their individual trajectories and even their revolution periods are all somewhat different. This ensemble of pulses at the detector forms the Schottky signal.

Recall that repetitive pulses in the time domain result in repeated spectral lines in the frequency domain. The general trend of the particles is to return each revolution period. This gives a spectrum of signals all related to the revolution frequency. Ones ability to observe signals in various frequency ranges depends upon the type of pickup electrode chosen.

Schottky Pickups

A wide variety of frequency ranges and types of Schottky pickups are represented at Fermilab. Table 1 lists the presently installed systems. Detailed descriptions of some of the detectors not covered in this presentation may be found in the references.

The proliferation of Schottky detectors in the Accumulator is due to the desire to observe small amounts of injected antiprotons. Beam currents on the order of tens of nanoamperes can be measured.

Table 1.
Schottky Monitors at Fermilab

Accelerator	Type	Number	Frequency
Tevatron	Transverse	4	21.4 MHz [2]
Tevatron	Longitudinal	1	DC - 50 MHz [3]
Tevatron	Transverse*	2	2 GHz [4]
Debuncher	Transverse	2	79 MHz
Debuncher	Longitudinal	1	79 MHz
Accumulator	Transverse	2	240 kHz
Accumulator	Longitudinal	1	23 MHz
Accumulator	Longitudinal	1	DC - 50 MHz
Accumulator	Transverse	2	79 MHz
Accumulator	Longitudinal	1	79 MHz
Accumulator	Transverse	2	200-400 MHz
Accumulator	Transverse	2	.5-1 GHz
Accumulator	Transverse	2	1-2 GHz

* An interesting use of transverse modes in a cavity.

Longitudinal Pickups

The basic longitudinal pickup is simply an open ended beam pipe cut to 1/4 wavelength of the frequency of interest. Adjustable tuners (either capacitive or mechanical) are used to set the exact frequency of resonance. Signal output can be from a direct tap, an inductive loop or a capacitive pickup. The quality factor is given by

$$Q = \text{Center Frequency} / \text{Bandwidth} \qquad (1)$$

The longitudinal pickup shown in figure 1 is used in the Fermilab Antiproton Source. A 79 Mhz unit is installed in the Debuncher. 79 MHZ and 23 MHz units are installed in the Accumulator. Figure 2 shows the frequency response of the Accumulator 79 MHz unit.

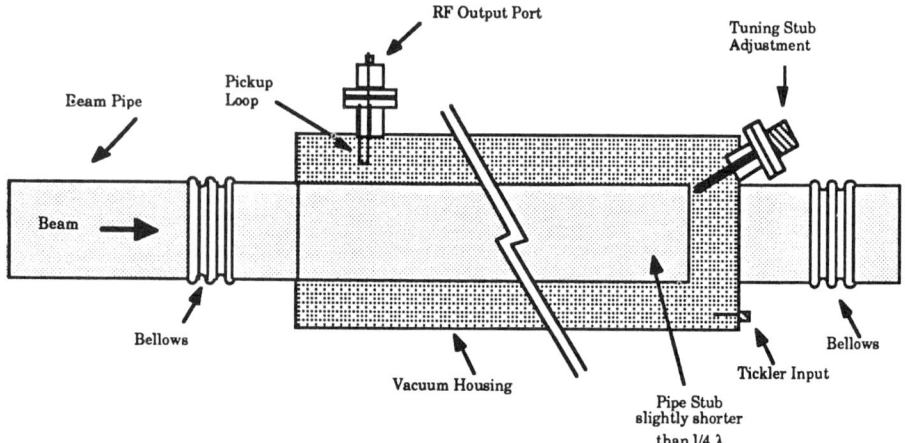

Figure 1.
A Longitudinal Schottky Pickup

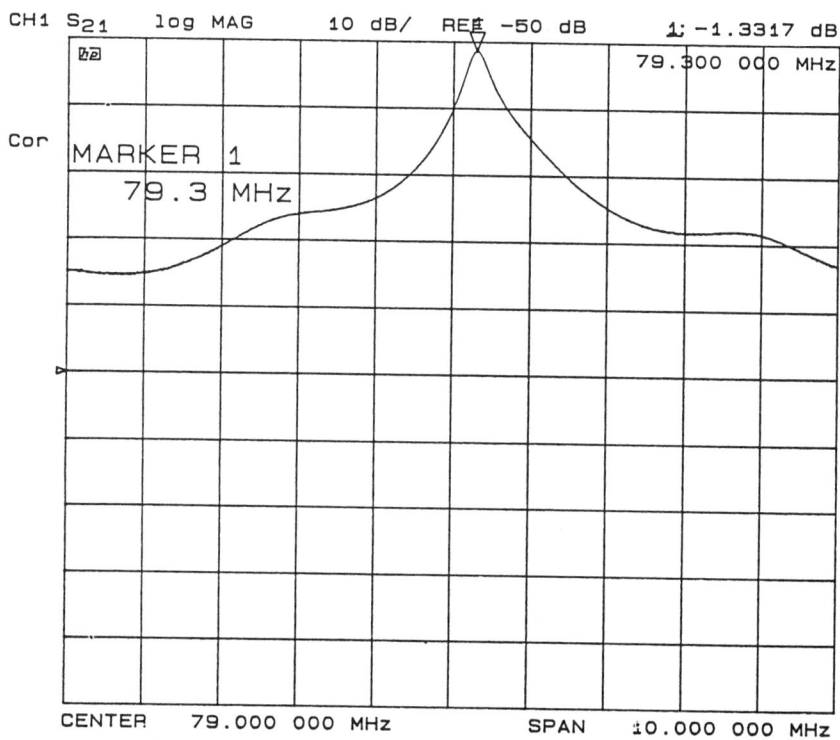

Figure 2.
Accumulator Longitudinal Schottky
Frequency Response

The pickup loop size and placement affect the Q, which in the 79 MHz pickup is presently about 300. A high Q detector is useful for situations with poor signal-to-noise ratio such as observing very small amounts of beam or trying to look at Schottky signals between very strong coherent signals in a bunched beam. The 23 MHz Accumulator pickup has high enough Q that it must be de-Qed when large stacks are in the Accumulator to prevent longitudinal instabilities.

The tickler is a small wire probe placed near the free end of the resonant pipe. In this unit the coupling is about -60 dB relative to a direct connection at 79 MHz. The tickler allows verification of detector operation without the need for beam.

One might wonder why 79 MHz was chosen for the majority of the Schottky detectors in the P-Bar source. The physical size of a 79 MHz detector is roughly 1 meter. This makes the detectors easy to build and install. 79 MHz falls exactly between the first and second harmonics of the 53 MHz RF systems and therefore the Schottky signals are not severely affected by coherent signals when the RF is on. 79 MHz is high enough in frequency that the signals are also not severely affected by the 1.26 MHz RF systems. Components for signal switching, amplification, detection, etc. are readily available for 79 MHz.

Transverse Pickups

The typical transverse pickup consists of two electrodes on opposite sides of the beam pipe. The differential output signal gives a measure of the particle displacement. Differentially driven ticklers provide a means of calibrating the system.

The transverse pickup shown in figure 3 consists of two plates spaced appropriately from the outer pipe to provide a 50 Ohm impedance. This pickup is used at 79 MHz but the plates are shorter than 1/4 wavelength so external matching networks are used to set the resonant frequency. Figure 4 shows the frequency response of the 79 MHz unit.

The same size pickup electrodes are used in low frequency (240 kHz) transverse pickups in the Accumulator. These units require large amounts of external inductance to resonate at the desire frequency. High losses in the inductors decrease the signal to noise ratio in this system. One can also set the wrong resonant frequency and still increase the Q high enough to give a nice looking but completely meaningless "signal". Frequency responses for various settings of Q and center frequency are shown in figure 5.

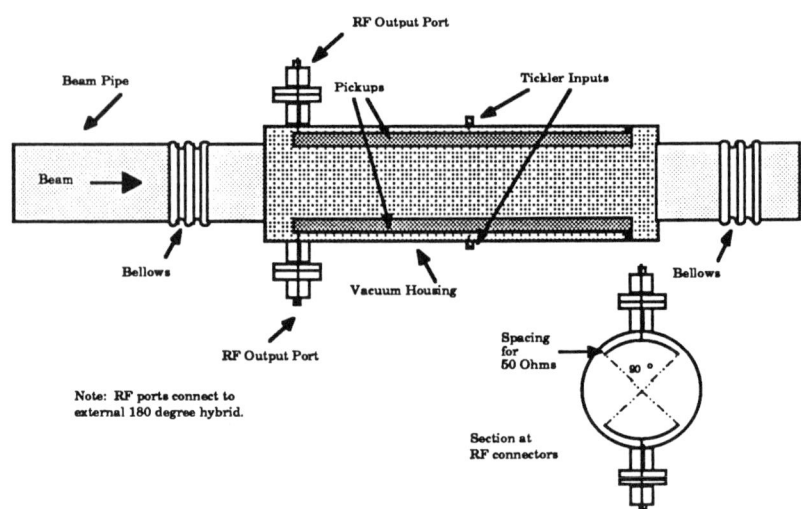

Figure 3.
A Transverse Schottky Pickup

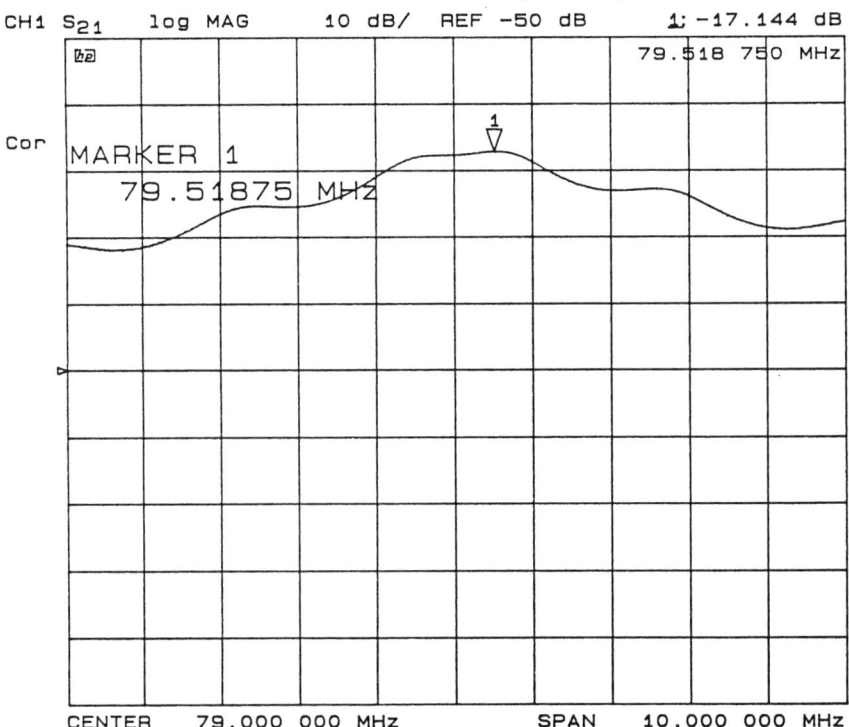

Figure 4.
Accumulator Transverse Schottky
Frequency Response

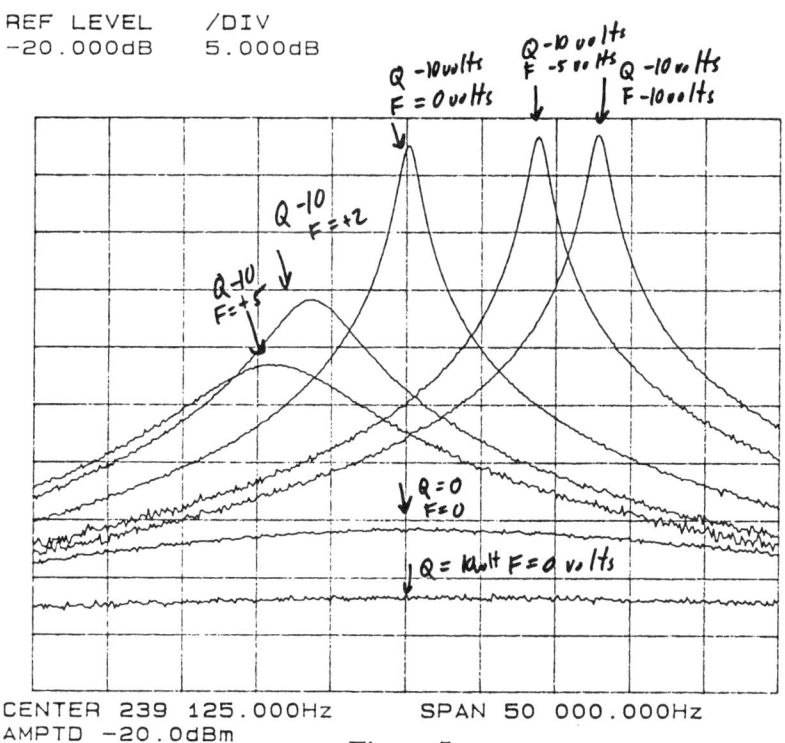

Figure 5.
Accumulator 240 kHz Schottky
A Variety of Q and Center Frequency Settings

A Signal Distribution Multiplexer

Schottky signals from the pickups may require some amplification and filtering. The signals are extremely useful for many different types of measurements and will generally need to be distributed to a variety of users. Figure 6 shows a block diagram of the P-Bar Schottky Multiplexer. The multiplexer has a design frequency range of 0.5 to 500 MHz.

It consists of 8 input amplifier modules, each with a directional coupler for injecting test signals, a low noise amplifier with 30 dB of gain, a PIN attenuator, another 30 dB of gain and an 8 way splitter. Each output module consists of an 8 position switch.

The user can either locally or remotely select any of the 8 input channels to be connected at the desired output channel. The use of the 8 way splitter to distribute each of the input channels allows multiple output channels to be connected to a particular input without signal degradation. High quality double shielded coaxial cables are used to prevent cross-talk in the multiplexer. The PIN attenuators are not used since their attenuation settings are not reproducible to the degree desired for good calibrated measurements of Schottky signal power.

During normal stacking operation four channels are dedicated to stacking rate and bunch rotation monitors and the other four are available for diagnostic measurements.

Signal Monitors

Commercial Equipment

There are a wide variety of devices available for monitoring Schottky signals. Commercially available devices include spectrum analyzers, dynamic signal analyzers, communications receivers and oscilloscopes.

<u>Spectrum Analyzers</u> : Spectrum Analyzers are the most popular devices for Schottky monitoring. The Hewlett Packard 8566B and 8568B are the workhorses of the P-Bar monitoring system. [5] Two units are installed for dedicated monitoring of bunch rotation and stacking while three others are available for general measurements. All units receive their signals through the multiplexer mentioned above. Well calibrated measurements of signal amplitude and frequency are easily done. Advanced features allow remote control and screen hardcopy output. Data can be dumped to the control system host computer in various formats for further analysis. Figure 7 shows the output of a console program which gives revolution frequency, momentum spread, tune, emittance and chromaticity measurements from the spectrum analyzer data. Disadvantages of the spectrum analyzers are their high cost ($36,000 to $59,000) and their popularity.

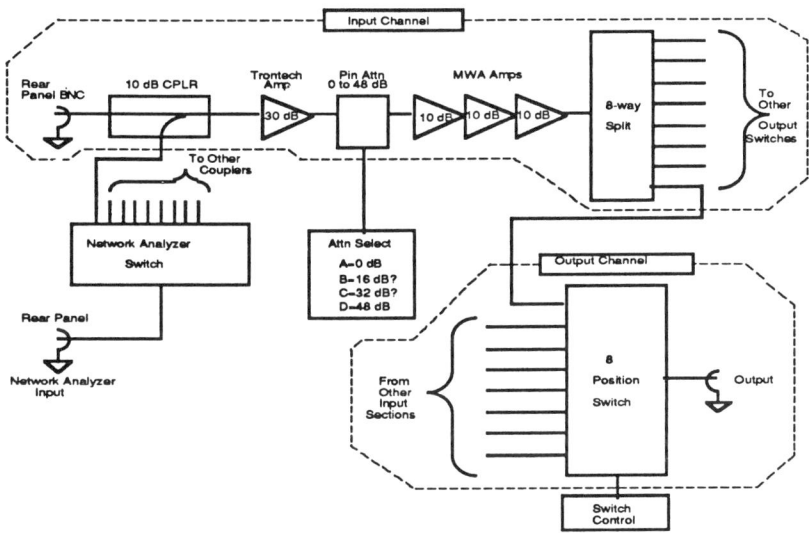

Figure 6.
Diagnostic Multiplexer

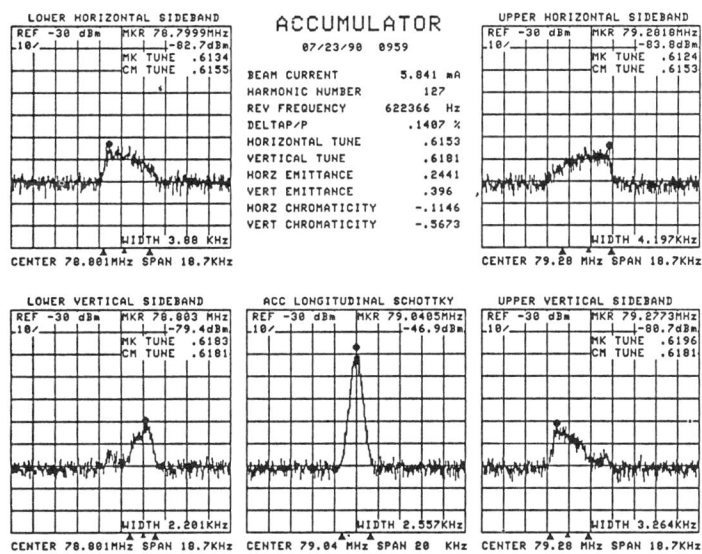

Figure 7.
Console Program using Data from Spectrum Analyzers

Dynamic Signal Analyzers : The HP 3562A Dynamic Signal Analyzer (DSA) is a time domain machine which does a Fourier transform to give frequency information. [6] In the P-Bar source it is used in connection with an external microprocessor to give continuous information about injected beam currents, bunch rotation efficiency and Debuncher to Accumulator transfer efficiency. Longitudinal signals from the multiplexer are mixed down to 20 kHz and distributed in parallel to two DSAs. The microprocessor controls the signal generators used for down conversion and the DSAs over a GPIB bus. Since the signal processing time is slightly longer than the P-Bar Source cycle time the DSAs alternate acquisition to provide continuous information for each injection cycle. Figure 8 shows the console page output for this system.

The DSA can also be used for precise measurements of synchrotron frequency for energy calibration during E760 runs. The DSA has some impressive signal analysis functions and at $20,000 is somewhat less expensive that a spectrum analyzer. It only covers up to 100 kHz so most Schottky signals must be mixed down to low frequencies for this device.

Receivers : Continuous transverse emittance monitoring in the P-Bar Source is done using relatively inexpensive ($500) communications receivers. Two receivers are used for emittance monitoring. One is connected to the Accumulator Horizontal Schottky output and the other is connected to the Vertical. They are manually tuned to the appropriate transverse sideband. The Yaesu FRG-9600 covers 60 to 905 MHz and has a crude computer interface to allow it to be remotely programmed. [7] The output of the internal automatic gain control (AGC) loop is brought out to a filter and amplifier and digitized. The log linearity of the AGC for this receiver is typically +/- 2 dB. One problem as shown in figure 9 is the center of the passband is not very flat so the apparent amplitude of the signal varies by about 0.5 dB depending upon its position in the passband.

There are plans to use ICOM R7000 communications receivers [8] for Schottky monitoring in the Tevatron. The R7000 covers a range of 20 to 2000 MHz. Special intermediate frequency (IF) filters are available which provide much improved passband response. It also has a much better computer interface which, in addition to providing remote setting, allows one to read back the mode and frequency settings. The R7000 with the special IF filter costs $1500. Figures 10 and 11 show the log linearity and passband response of the receiver.

```
P38   NEW FFT STUFF                    SET      D/A     A/D   Eng-U  ♦COPIES♦
-<FTP>+  *SS♦  X-A/D   X=TIME        Y=D:DPFTER,A IBEAM ,A EMITV ,A IB
COMMAND  ----  Eng-U   I= 0         I=-.2      , 0      , 0      , 0
-<23>+   One+  AUTO    F= 600       F= .2      , 8      , 40     , 1200
   MULT:1        PAT'S OTHER PLAY PAGE
  D:FFTINI       FFT Basic Status/Control                                ..
 -D:FFTLOF       FFT Deb Local Osc Fre   74.91478    74.91478    MHz
 -D:FFTWID       FFT Debuncher Width     .825        .825        KHz
 -D:FFTGN        FFT Debuncher Gain      .63         .63
 -D:FFTOFF       FFT Debuncher Offset    .173        .173
 -A:FFTLOF       FFT Accum Local Osc F   79.1995     79.1995     MHz
 -A:FFTWID       FFT Accumulator Width   3.4         3.4         KHz
 -A:FFTGN        FFT Accumulator Gain    494         494
 -A:FFTOFF       FFT Accumulator Offse   .00105      .00105
 -D:FFTSPN       FFT Box Span            10          10          KHz
 -D:FFTAVG       FFT Number of Averages  5           5           avgs
 -D:FFTMSK       FFT Pulse Enable Mask   255         255
  D:FFTTOT       Total Debuncher Curre               49.73661    nA
  D:FFTCEN       Deb Current Central                 51.95828    nA
  D:FFTEFF       Bunch Rotation Eff                  104.4669    %
  A:FFTTOT       Total Accum Current                 40.57658    nA
  A:FFTCEN       Accum Current Central               19.97581    nA
  A:FFTEFF       Deb to Acc Efficiency               36.06221    %
  D:FFTBPC       FFT Beam Pulse Countr               1
 -D:FFT0         FFT RESET TIME          0         * 0           USEC ...-
 -D:FFT1         FFT TRIGGER #1          200000    * 200000      USEC ...-
 -D:FFT2         FFT TRIGGER #2          200000    * 250000      USEC *.L-
 -D:FFT3         FFT TRIGGER #3          200000    * 250000      USEC *.L-
```

Figure 8.
Sample Output from Dynamic Signal Analyzer Program

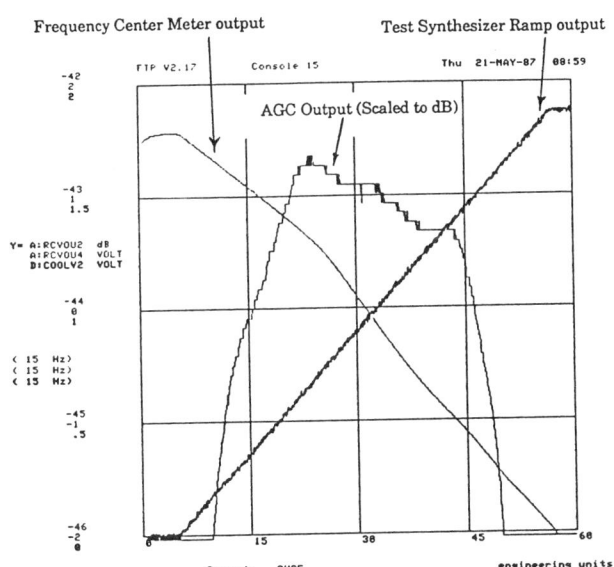

Figure 9.
Passband Response of FRG-9600 Receiver

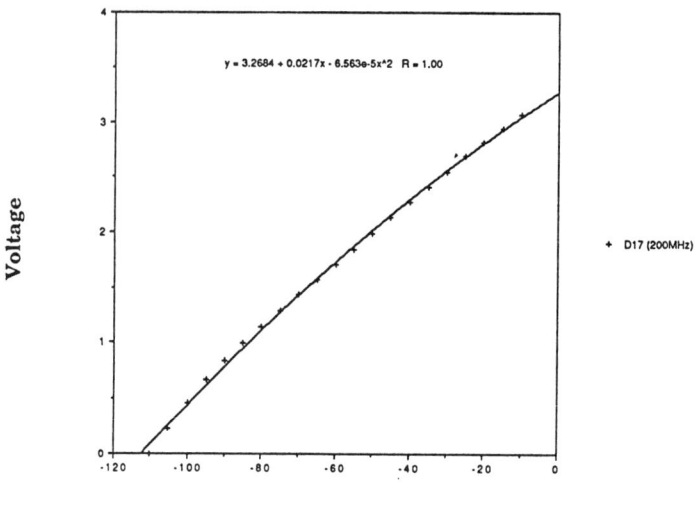

Figure 10.
Detector Calibration for ICOM R7000 Receiver

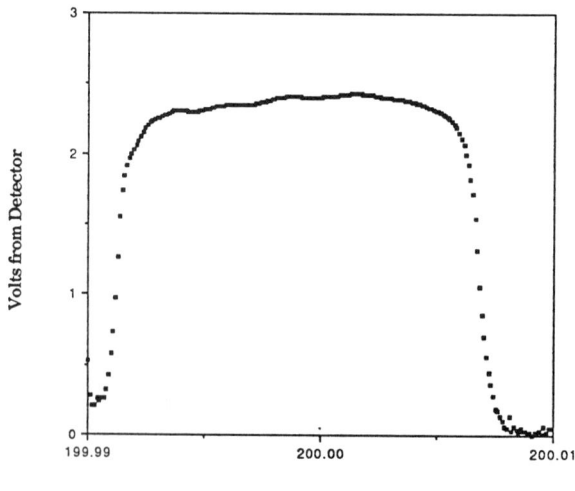

Figure 11.
ICOM R7000 Bandpass Response
(Receiver is set to 200 MHz)

Custom Equipment

Some Schottky monitors need to be custom built due to their specialized nature. A couple of devices covered here are the Tracking Emittance Monitor and the Tune Tracker.

The Tracking Emittance Monitor : The Yaesu receivers have served well for emittance monitoring in the Accumulator during normal stacking operations but the E760 experiment requires decelerating the Antiproton beam to medium energies.[9] The revolution frequency drops during deceleration and so the transverse sideband would move out of the passband of the receiver. One could simply retune the receiver during the deceleration and this was done for a while. Eventually an automatic system was desired with improved log linearity and passband flatness. Knowing (or more likely praying) that the tune in the Accumulator does not change dramatically (< 0.06) during deceleration and that for a given tune the transverse sidebands maintain a particular relationship to the revolution frequency, a tracking emittance monitor was devised.

The tracking emittance monitor shown in figure 12 consists of a tracking synthesizer section and an IF section. To provide a simple formula for tracking and the ability to use a low cost log detector an IF frequency of 0 Hz was chosen. This is not a problem for Schottky monitoring since the negative and positive frequency domains contain essentially noise and do not need to be separated.

The tracking synthesizer consists of a 32 bit Direct Digital Synthesizer (DDS),[10,11] a multiplier and some filtering. The tremendous advantage of using a DDS over a phase locked loop (PLL) is the extremely fine frequency resolution (1 part in 2^{32}). The disadvantage is the output is always less than the clock frequency and so a multiplier (perhaps a PLL) is needed. The IF section consists of input buffer, a mixer, a Butterworth active low-pass filter (20 kHz), and an Analog Devices AD536 RMS to DC converter as the logarithmic amplifier.[12] The 20 kHz low-pass filter provides for a total IF bandwidth of 40 kHz which is narrow enough to not see energy from the revolution harmonics but wide enough to allow the tune to vary within reasonable limits without requiring retuning of the local oscillator. Figure 13 shows the log linearity of the IF section. Figure 14 shows a plot of the residuals from the log calibration.

It is interesting to note that the 0.5 dB steps near the center of the response are actually due to errors in the 30 dB attenuator in the test generator rather than any problem in the detector circuit. Figure 15 shows the ripple in the IF response near 0 Hz due to the AC coupling. Figure 16 shows the overall IF response.

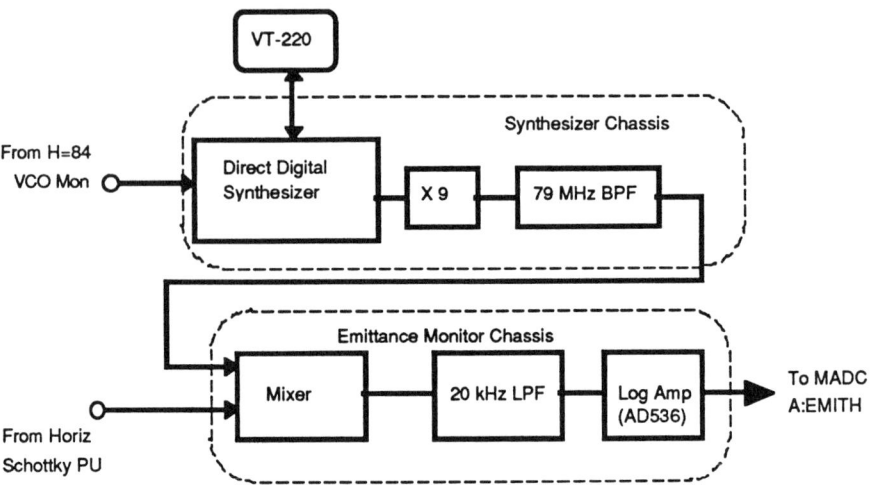

Figure 12.
Tracking Emittance Monitor Block Diagram

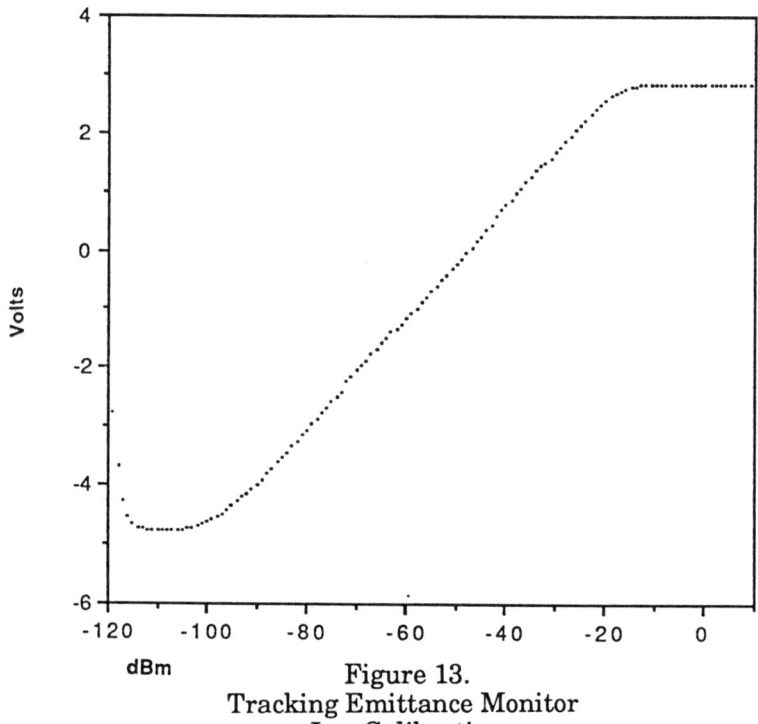

Figure 13.
Tracking Emittance Monitor
Log Calibration

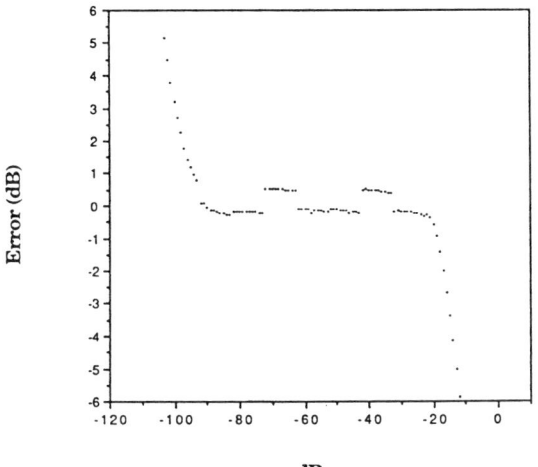

Figure 14.
Tracking Emittance Monitor
Log Calibration Errors

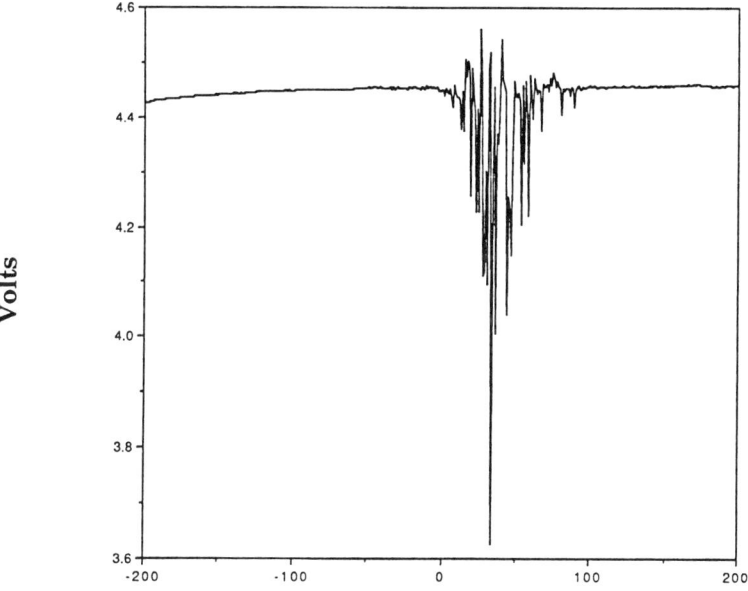

Offset (Hz)

Figure 15.
Tracking Emittance Monitor
Log Detector Response at Low Frequencies

The key to this tracking scheme is the DDS is used as an arbitrary ratio generator. The DDS output frequency can be expressed as

$$F_{out} = (F_{clock} * N)/2^{32} \qquad (2)$$

where F_{clock} is the input reference frequency and N is any integer between 0 and 2^{31}. One can see that virtually any ratio of input to output frequencies can be created. If the input frequency is a harmonic of the beam revolution frequency then the output frequency will also track as a ratio of the revolution frequency.

In the P-Bar Accumulator the deceleration is locked to a synthesizer generating the 84th harmonic of the revolution frequency. The fractional tune is roughly 0.611 and the transverse Schottky operates near the 126th harmonic. This means the tune line of interest is at H=126.611. The DDS clock input is the H=84 signal from the RF system. The DDS is programmed to generate an output which is 126.611/(84*9) of the clock frequency. When multiplied by 9 this output gives the correct local oscillator frequency for mixing the tune line of interest to near 0 Hz.

The use of a DDS as a tracking signal generator for diagnostics is also useful for accelerators (such as the Fermilab Booster) which accelerate (i.e. change frequency) very quickly. One can use the DDS and a mixer to convert signals to a stable frequency domain.

The Tune Tracker : A rather elegant method for tracking the tune in the Accumulator involves using a PLL to act as a frequency to voltage converter.[13,14] The reference input of the PLL is fed from the 240 kHz transverse Schottky and the loop is locked to the sideband. The tuning voltage for the PLL voltage controlled oscillator (VCO) is brought out and digitized. The host computer can then calculate the tune from the VCO voltage scaling factor and the revolution frequency. A block diagram of the system is shown in figure 17.

Data Analysis

Various beam parameters can be calculated from information provided by the Schottky signals. These calculations can be done by the host computer or even by the internal processor in some of the more sophisticated instruments. The following is a summary of some commonly used parameters;

Beam Current

The area of the longitudinal signal is proportional to the number of particles. The power detected by a Schottky pickup can be calibrated to give a measure of the number of particles in a machine. This is useful

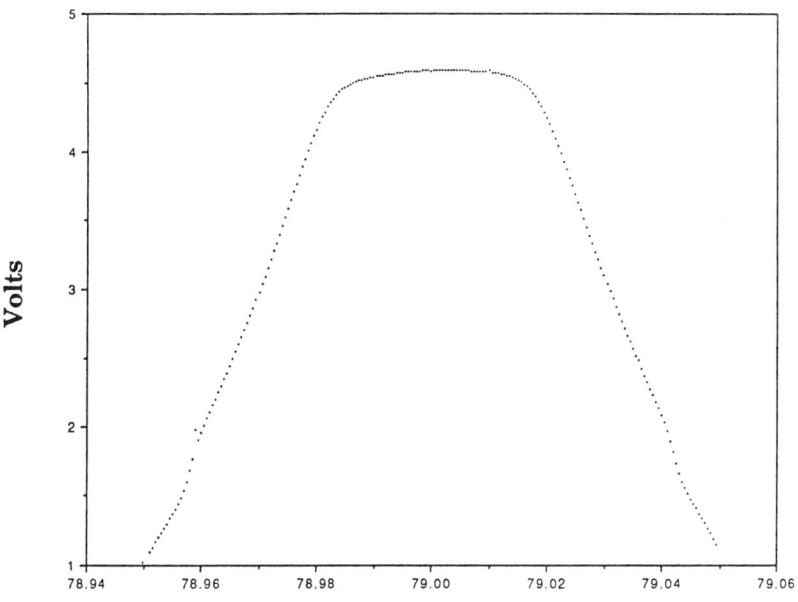

Freq

Figure 16.
Tracking Emittance Monitor
Overall Frequency Response
(Tuned to 79 MHz)

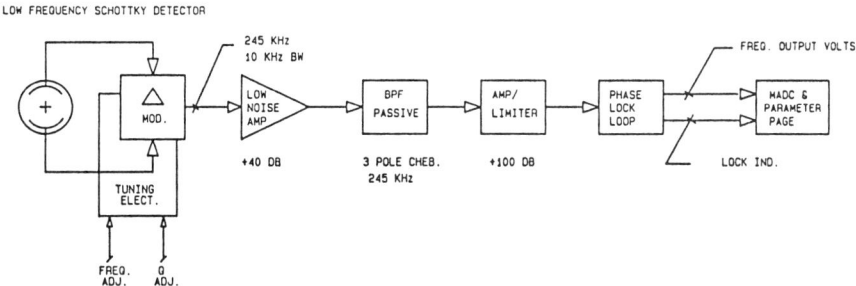

ACCUMULATOR TUNE MEASUREMENT SYSTEM

Figure 17.
Tune Tracker Block Diagram

for small amounts of beam which may not be measurable by other means.

Emittance

The frequency width of the longitudinal signal and the power in the transverse sidebands are measures of emittance. Figure 18 shows display parameters from the transverse emittance monitors. The monitors read out in both dBm and units of Emittance-Milliamps (EmA). The EmA value is then divided by the beam current in milliamps to give emittance. Figure 19 shows a typical Beam Lifetime display with transverse emittance traces along the bottom of the display.

Tune

The transverse sideband frequencies relative to the revolution harmonics gives the fractional tune. Using a calibrated instrument, such as the tune tracker discussed above, one can get good measurements of the tune very quickly.

Chromaticity

The difference in width of the transverse sidebands gives a measure of the chromaticity.[15] An example of this was seen in figure 7.

Beam Energy

The revolution frequency for a given orbit in a given machine yields β. One can also determine beam energy by measuring the synchrotron frequency of a bunched beam knowing η and the RF voltage.[16]

Ion Trapping

Anomalous tune shifts can indicate the presence of trapped ions.[17] Figure 20 shows the horizontal tune shift in the presence of ions.

Beam Profile

Unfortunately, Schottky signals (presently) do not tell us much about the beam profile.[18]

RF Bucket Population

The amount of beam seen outside of the synchrotron lines in a bunched beam gives a measure of the beam not captured in the RF bucket. Figures 21, 22 and 23 show the result of various levels of RF voltage.

```
P38  DIAGNOSTIC RECEIVERS           SET      D/A      A/D  Eng-U  ♦COPIES♦
-<FTP>+  *SS♦  X-A/D    X=TIME     Y=A:EMITH ,A IBEAM ,A:EMITV ,A IB
COMMAND  ----  Eng-U    I= 0        I= 0     , 0      , 0      , 0
-< 3>+   One+  AUTO     F= 460      F= 40    , 8      , 40     , 1200


    MULT:2        AMPLITUDE
    A:RCVOU1      RECEIVER OUTPUT              -123.9   dB
    A:RCVOU2      RECEIVER OUTPUT               -82.28  dB


    MULT:2        FREQ ERROR 0V=CNTR
    A:RCVOU3      RECEIVER OUTPUT              -.155    VOLT
    A:RCVOU4      RECEIVER OUTPUT               .11     VOLT


    MULT:2        NEW MONITORS
  → A:EMITH2      New Horizontl Emittance      -67.95   dBm
    A:EMITV2      New Vertical Emittance      *-100.7   dBm


    A:EMITHO      OLD RECEIVER OUTPUT          -.469    EmA
    A:EMITV       RECEIVER OUTPUT               .033    EmA
  → A:EMITH       Horizontal Emittance       *  41.38   EmA
```

Figure 18.
Emittance Monitor Console Parameters

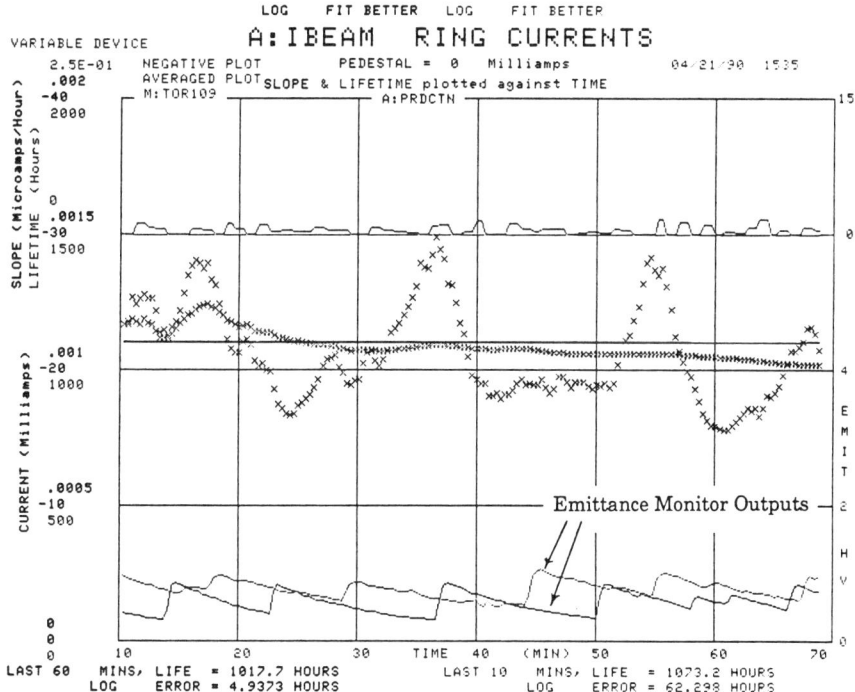

Figure 19.
Beam Lifetime Display

126 Schottky Signal Monitoring at Fermilab

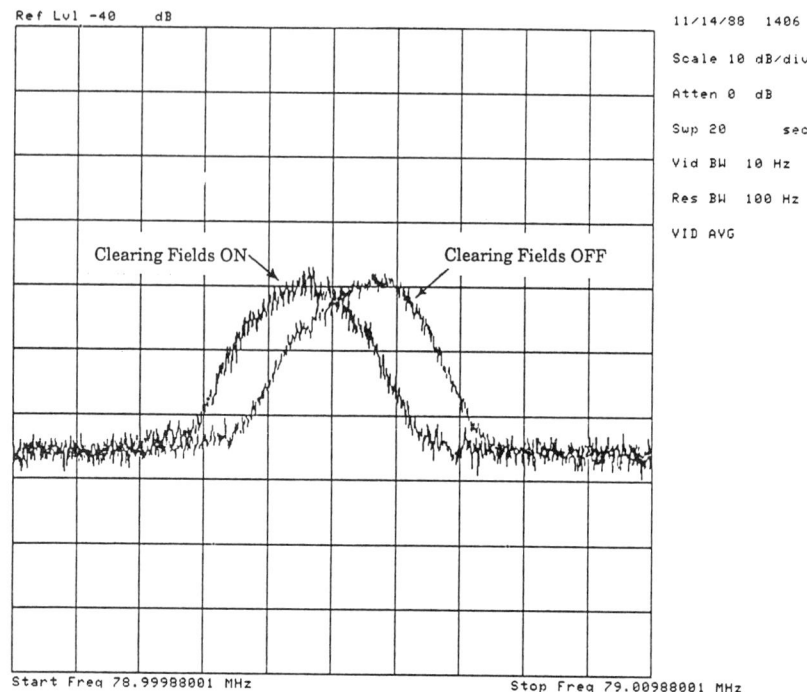

Figure 20.
Tune Shift Caused by Trapped Ions

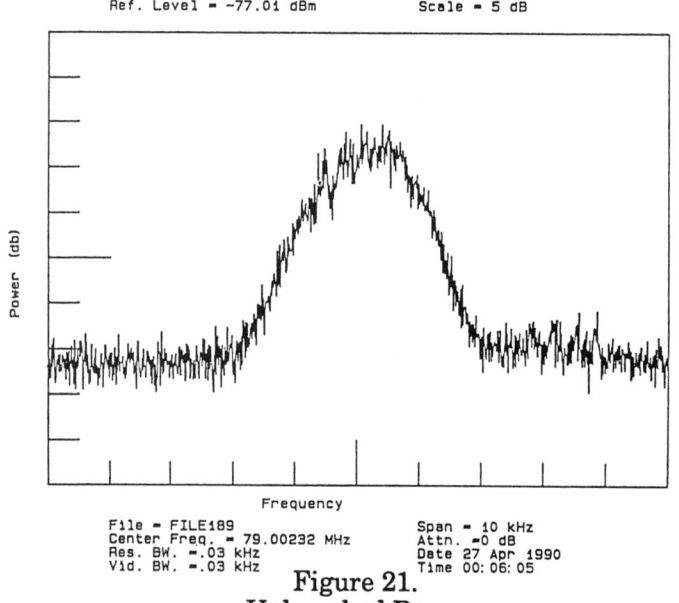

Figure 21.
Unbunched Beam

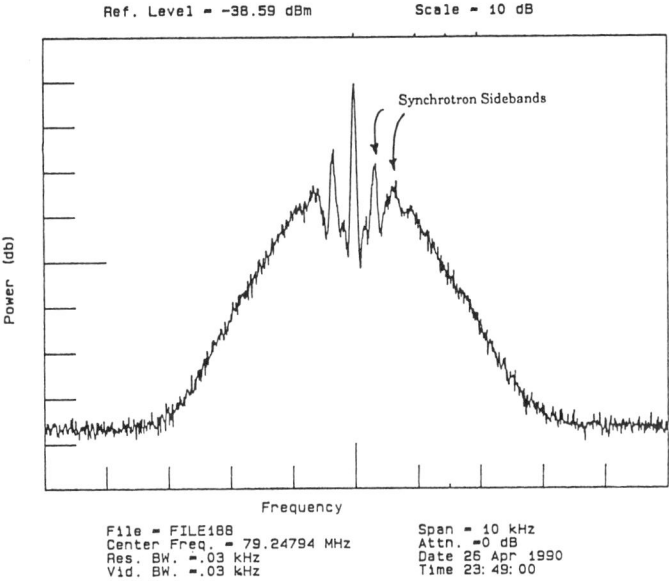

Figure 22.
Some Beam Bunched with 10 kV on ARF1 Cavity

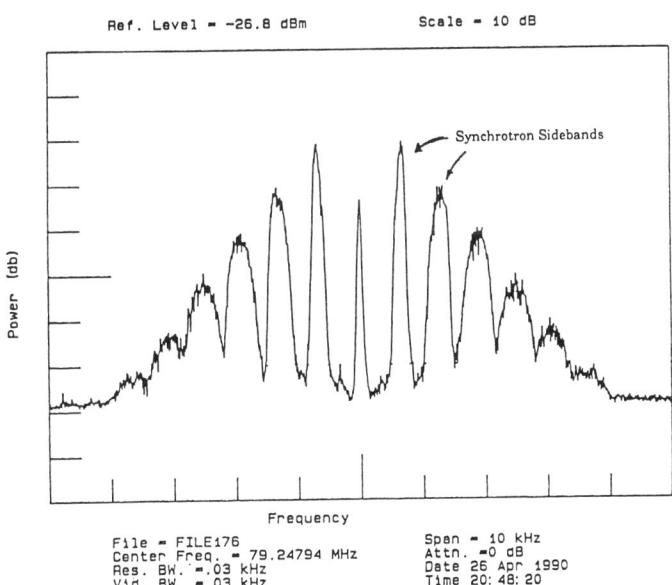

Figure 23.
Beam Bunched with 36 kV on ARF1 Cavity

Other Systems

Stochastic Cooling

Longitudinal and Transverse signals are used in 7 Stochastic Cooling systems in the P-Bar Source covering octave ranges from 1 to 8 GHz. [19,20] These systems apply correcting kicks to the particles to push antiprotons into the stack, reduce the momentum spread or reduce the transverse emittance.

Bunched Beam Cooling

A variation on Stochastic Cooling. A 4 to 8 GHz bunched beam cooling system is planned for the Tevatron. [21]

Conclusions

An overview of Schottky signal monitoring at Fermilab has been presented. Pickups, amplifiers, distribution systems, signal monitors and analysis results have been discussed. The references cited will provide a starting point for further investigation of Schottky signal systems. The U.S. and European particle accelerator schools provide excellent courses in the fundamental properties of particle beams. Particle accelerator conference proceedings include many papers on monitoring systems.

Acknowledgements

Many thanks to John Marriner for his helpful explanations of beam phenomena and his realization that it is perfectly fine to mix Schottky signals down to DC. Thanks to the members of the Fermilab P-Bar group for building an accelerator to hold the Schottky detectors. Thanks to Gerry Jackson and the Accelerator Instrumentation group for their assistance in creating this presentation. And thanks to the many unnamed people who have described and analyzed the phenomena of Schottky signals.

References

1. Allen G. Debus, Editor, World Who's Who in Science, (Marquis Who's Who, Chicago, Il), 1500.

2. D. Martin et al., A Resonant Beam Detector for Tevatron Tune Monitoring, Proceedings of the 1989 IEEE Particle Accelerator Conference, IEEE cat. no. 89CH2669-0, 1486-1488.

3. C.D. Moore et al., Single Bunch Intensity Monitoring System using an Improved Wall Current Monitor, Proceedings of the 1989 IEEE Particle Accelerator Conference, IEEE cat. no. 89CH2669-0, 1513-1515.

4. D. Goldberg and G. Lambertson, Proc. XIV International Conference on High Energy Accelerators, Tsukuba, Japan, 1989

5. Hewlett Packard Company, Palo Alto, California, Test & Measurement Catalog 1990, p. 120.

6. Hewlett Packard Company, Palo Alto, California, Test & Measurement Catalog 1990, p. 133.

7. Yaesu Electronics Corporation, Paramount, California, FRG-9600 data sheet.

8. Electronic Equipment Bank, Vienna, Virginia, ICOM R7000 data sheet and accessory filter information.

9. V. Bharadwaj et al., The Use of the Fermilab Antiproton Accumulator in Medium Energy Physics Experiments, Fermi National Accelerator Laboratory TM-1527, presented at the European Particle Accelerator Conference, Rome, Italy, June 7-11, 1988.

10. Qualcomm Incorporated, San Diego, California, Direct Digital Synthesis Question and Answer sheet and Q2334 Direct Digital Synthesizer data sheet.

11. Stanford Telecommunications Inc., Santa Clara, California, STEL-1375 Direct Digital Synthesizer Hybrid data sheet.

12. Analog Devices Inc., Norwood, Massachusetts, AD536A True rms-to-dc Converter data sheet.

13. D. Martin et al., A Schottky Receiver for Non-perturbative Tune Monitoring in the Tevatron, Proceedings of the 1989 IEEE Particle Accelerator Conference, IEEE cat. no. 89CH2669-0, 1483-1488.

14. J. Fitzgerald, A Tune Measurement System for the Tevatron using a Phase Locked Loop, to be published in the Proc. of the 1990 Accelerator Instrumentation Workshop, Fermilab, Batavia, Illinois.

15. D.A. Edwards and M.J. Syphers, An Introduction to the Physics of Particle Accelerators, AIP Conf. Proc. No. 184 (1987-1988 Particle Accelerator School), Ch.3.

16. R. Webber, Longitudinal Emittance, Proc. of the 1989 Accelerator Instrumentation Workshop, Brookhaven National Laboratory, Upton, New York.

17. J. Krider, Ion Profile Monitor using Microchannel Plate, to be published in the Proc. of the 1990 Accelerator Instrumentation Workshop, Fermilab, Batavia, Illinois.

18. A. Poncet, Neutralization Experiments with Proton and Antiproton Stacks - Ion Shaking, Fermilab P-Bar Note #481, March 1989.

19. J. Peoples, Antiproton Source, AIP Conf. Proc. No. 184 (1987-1988 Particle Accelerator School), 1845-1877.

20. J. Marriner, Review of the Physics, Technology and Practice of Stochastic Beam Cooling, Proceedings of the 1987 IEEE Particle Accelerator Conference, IEEE cat. no. 87CH2387-9, 1383-1387.

21. J. Marriner et al., Bunched Beam Cooling in the FNAL Antiproton Accumulator, to be published in the Proceedings of the 1990 European Particle Accelerator Conference.

Panel Discussion on
Where Does the Instrumentation Engineer's Function End and That of the Computer Engineer Begin?

Scott Miller, Moderator
The Superconducting Super Collider Laboratory, Dallas, TX
Jean Borer
The European Center for Nuclear Research (CERN), Geneva 23, Switzerland
Gerald P. Jackson and Peter W. Lucas
Fermi National Accelerator Laboratory, Batavia, IL 60510
Richard Witkover
Brookhaven National Laboratory, Upton, NY
Gergory Stover
Lawrence Berkeley Laboratory, Berkeley, CA 94720
Marc Ross
Stanford Linear Accelerator Center, Stanford, CA 94309
John Perry
The Continuous Electron Beam Accelerator Facility, Newport News, VA 23606

[Editor's Note: This is the transcript of the discussion session which took place at 3:30 pm Tuesday October 2, 1990 at Fermilab. It has been edited for clarity. Unfortunately, not all of the speakers from the audience could be identified.]

Scott Miller: On the panel today we have Jean Borer, from CERN, Gerry Jackson from FNAL, Peter Lucas from FNAL, Richard Witkover from BNL, Greg Stover from LBL, Marc Ross from SLAC, and John Perry from CEBAF. Please give us a brief introduction of what you are doing so that we can spend most of the time on questions.

Jean Borer: I would like to describe the BOM [Beam Orbit Measurement at LEP] control system a little more. It is a system, directly derived from the overall control system of LEP, which is based on token ring, a common development project of IBM with CERN. This is the first time that token ring has been used for a large control system.

We were developing our own control computers for a project called DLX(VME), but this project collapsed about 1.5 years before the end of the project. It was then decided to use Unix-based PC's, the heart of the LEP control system. The PC's are all connected directly to the token ring through a VME chassis. PCA (Process Control Assemblies) are used mainly to coordinate the transmission in the system. The system uses the MIL 1553 multidrop bus to interface directly to the equipment.

At the equipment level we have VME crates with MC68000 processors. Under the recommendation of another wise CERN committee, we are using the RMS 68K operating system and the Pascal programming language. So we are in quite a mess today because RMS 68K has been abandoned by Motorola, and we are just about the only people using it.

We are now in the process of changing part of the BOM and instrumentation-controllers to OS9. We will try to bypass the MIL 1553 part because it is not well adapted to communication between a clever local processor and the central processors in the control room. We want to have direct connections through a gateway between an ethernet connection, which exists for VME OS9 system, and the token ring. The top-level processing is done with consoles in the control room, which are Appolo computers with Unix.

Gerry Jackson: I am Gerry Jackson from Fermilab. I have been the head of the Instrumentation Department here in the Accelerator Division for two years. In those two years I've worked closely with Peter Lucas in a sometimes love sometimes hate relationship, trying to get instrumentation into the system. Sometimes the instruments are new, sometimes they are instruments I have inherited.

The interface between the instrumentation and the control system at Fermilab runs along both digital and analog routes, as one would expect, either through simple digital bytes or through analog signals fed into an ADC provided by the Controls Group. Some of our systems are in a VME crate, which are awkwardly interfaced to the control system through CAMAC. We also use IEEE 488 (GPIB), again through a very cumbersome interface system. So whenever a new project comes up we always have a battle on how to get the information into the control system. This is especially hard when a project which involves arrays of words, rather than a single 16 bit word. In the case of Main Ring and Tevatron, you have to go through the normal control system because most of your equipment is 3 or 4 miles away. You cannot be sitting at a PC parked outside the control room and play with that.

It has been an interesting two years trying to understand what to do. The basic bottlenecks are usually software, as mentioned before, and also manpower, especially in the era of SSC where we have a constant flow of people moving in and out now.

Peter Lucas: I am Gerry's counterpart here in Fermilab Accelerator Division, Peter Lucas, Head of the Accelerator Controls Department. (I feel compelled to show one transparency and I think it will help clarify what Gerry has said. I have been accused many times of having only one transparency really in the whole world, and this is it actually. If you've got a good one you really don't need any more.) By and large the computers in our control system talk to each other over token ring. In older times Fermilab controls was done differently but now we have a token ring connection amongst all the computers. Those computers are consoles of both the old type PDP-11 and the new type MicroVAX, host VAXen and various front ends driving links out into the field.

Front ends are generally distributed geographically: one for the anti-proton source, one for the linac, etc. The instrumentation interfaced to this system, reviewing slightly what Gerry has said, is through an ADC in CAMAC if it is an analog signal. This signal is digitized and gets to the consoles via one of these CAMAC serial links in a quite straight-forward way. I think we support this very well. If your equipment

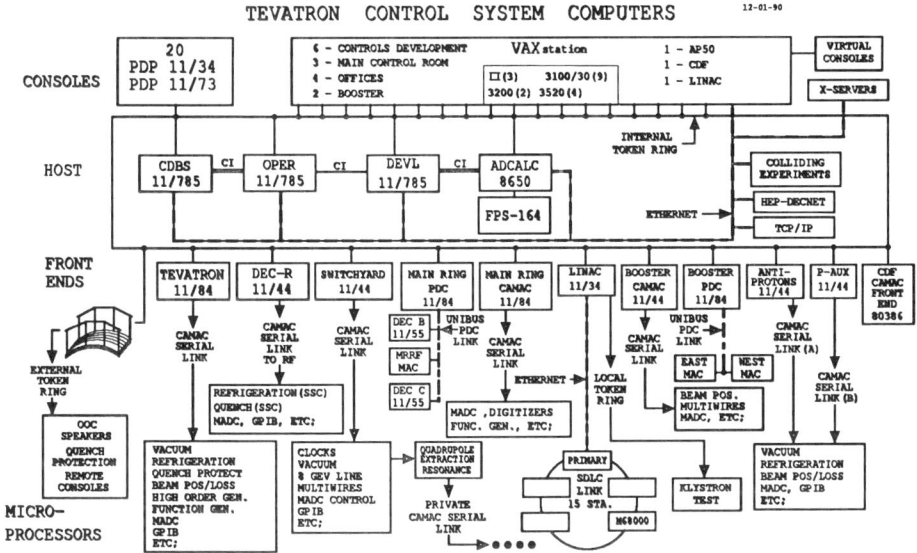

Figure, Lucas's slide of the Fermilab Accelerator Computer/Data Acquisition Network, ACNET

interfaces through GPIB, which we frankly wish it wouldn't, we support that through an intelligent CAMAC card which speaks a protocol called GAS, sitting on top of ordinary CAMAC. If it is an older system, an older BPM for instance that resides in Multibus-I, that interface is done in what some people think is a slightly strange way: we go from a (smart) computer out the (dumb) CAMAC link, back into a (smart) computer. But this system has been around since the early days of the Tevatron and has worked for a variety of systems.

We are migrating these days towards attaching the various nice pieces of instrumentation and smart subsystems directly to our token ring. So the software which does network communication deals with the same protocol whether it is inside the front end computer, host computer, console computer, or inside an instrumentation crate. The two new systems which are are actually doing this are the quench-protection system and the CBA, a time and chromaticity feedback controller for the Tevatron.

Richard Witkover: I am Dick Witkover from Brookhaven National Laboratory, specifically from the AGS Department. I have been at Brookhaven since 1965 and have survived many different controls systems. (I'm not sure about the latest one though.)

I'm presently responsible for the instrumentation on our new booster program. I'll describe how that control system is structured. Our instrumentation is located close to the detector. We go through analog and direct digital signals to an instrument controller. It is a Multibus-I crate with dual processors, one of which services the actual process and the other sends data back and forth as a communications processor.

That goes back and forth on IEEE 488 to the next level up, an Apollo workstation. This computer collects information from many different instrument or power supply controllers and sends the information up to a ring of host computers, which are larger size Apollos. They communicate over Domain, a proprietary protocol from Apollo (I believe it is an ethernet equivalent type system). All of these Apollos are hooked together and can gather information from any of the local computers.

The coding for the most part is done in C++, of which I know nothing. I think the controls group purposely did it this way so that the instrumentation engineers would not be able to code it! This has not always been the case however. In previous implementations, the engineer has usually done everything, including assembly language programming in local processors. We evolved to the point where these engineers are now totally divorced and couldn't even get into it if we wanted to. Whether that is better or not I don't know.

Greg Stover: My name is Greg Stover and I'm with the Bevatron at LBL, Lawrence Berkeley Laboratory. My experience originally was with the HILAC, the heavy ion linear accelerator, which was connected to the Bevatron. Then I had a short stay with the free electron laser at Livermore before returning to the Bevatron. My responsibilities include instrumentation design. I have contact with the controls group regularly, in fact, I have even done some programming with them. But you kind of move in and out of their realm. Our controls group is probably similar to others: a tight group that sometimes sticks together.

The Bevatron in the HILAC control systems were joined together in about 1975 when the transfer line was installed. Their respective control systems evolved similarly. They originally used PDP computers, which first got incorporated in the 70's. There are even some PDP's being used now. We use Modcomp 2's and 4's in the central with Remac data collection at the various instrument positions. Overlayed on all of this are machines with local intelligence that were developed by engineers. This intelligence has been used, typically, to overcome the problems that the early Remax computers had in acquiring data.

Since that time there has been a very aggressive and capable advocate for the Sun-based ethernet systems, named Steve Lewis. He is now our systems man at the Bevatron. These Sun workstations are at the highest level. We also are looking at VME crates, with a lot of local intelligence, which talk to the Suns. They also provide the instrument interface to as many signals as possible from the Bevatron.

The conversion process is gradual; we are basically trying to phase out the Modcomp systems. We presently have both systems up. The Modcomp does a certain amount of controlling certain parameters for the slower functions. The Sun systems are gradually moving in on the Modcomp system.

New Controls for New Machines at the ALS

There is another new machine that is being built, the Advanced Light Source, just up the hill at the old 184-inch. As might be expected, there is another controls philosophy there! There is a control systems group that I worked with when I was at the HILAC. They started with a very talented mechanical engineer named Steve Magnerie. He developed a control system for the upgrade to the new injector for the HILAC which was based on Intel Multibus. Its main strength was that it had local microprocessor power; I would tend to call it a distributed intelligence system. It was more capable than the Remax system; it was used for the local injector. Since then Steve's theories have evolved. He is now the main system designer and project head for the control of the ALS. Their concern is that they have a lot of data that has to be processed quite rapidly. They developed a highly specialized microprocessor on a card for many of the applications. I think it has an 80386 Intel chip on it. It has a number of digital I/O and analog I/O options, and these options can be selected in a number of ways. Instrumentation engineers can build their instruments with this card in one of the slots. Many of the engineers, but not all, program their devices using Basic, but there are several options. The data from these smart instrument controllers are then shipped back to a central memory which is seen by various graphics processors.

So that is kind of the philosophies at LBL. There are kind of two machines and two ways of doing things and they have certain similarities.

SLC Controls Philosophy

Mark Ross: I have worked on the SLC project for seven years now, primarily working on instrumentation, but also working on general accelerator physics problems and simple machine operation.

One thing that I think makes SLC different from other machines is the extent to which active intervention by the control system is required above and beyond what the operator normally does simply to make the machine operate. As such, we place a lot of emphasis on the control system, particularly the software. We have a fairly rigid framework which we can continually build upon. Typically what happens is this. Someone finds a problem, figures out how to solve it and then wishes the control system to solve their problem forever. If the solution is derived in an ad hoc fashion, you would then have to go back and redevelop the whole thing before you could integrate it properly. If you start with a device or with an idea which fits into an existing framework, then it is much easier to apply. SLC presently has several dozen feedback loops that are model driven and operate on the very highest level because of that. There will eventually be many more, amounting to 50 or more in the next year or so.

There is a controls workshop series which is similar in spirit to this workshop, although it has been going on for a few years longer. They say that the costs of these control systems are completely dominated by the software. This is surprising to some people, but it is especially true at the SLC where a large emphasis is placed on the high level control. This means fancy software in which a great deal of effort has been put. We have had to be very careful that this software need does not sink the whole effort. Because of that, as I pointed out this morning, I think it is extremely important to do everything within the existing software channels, few though they may be, to make sure that it fits in, that it has a place and that it can be built upon.

For instance, I can have my wire scanners read out at the same time as some BPM's are read out and maybe some screens are read out and at the same time the controls system looks at all of this and decides what to do. Then you can write a new application which uses all of these systems in some way which wasn't even imagined when any of the systems were built.

John Perry: I am John Perry from CEBAF. I work in the instrumentation and controls group there. The primary strength of CEBAF's controls arrangement is that CEBAF recognized very early that the control system was going to be THE key to making the accelerator run. As a matter of fact we have had a fully functional controls system for quite some time now. Our control system was running before we had any hardware for it to control. I think that illustrates probably the most powerful method of making sure that the controls problems don't arise. I think the reason controls problems so frequently do arise is exemplified by the title of this discussion group.

You can build, in principle, any accelerator no matter how big, no matter how complex, with nothing more than ADCs, computers, DACs and transducers. This is obviously impractical. By the same token, in principle you can build any accelerator no matter how complex with nothing but hard-wired logic and amplifiers and analog computation. That also is obviously impractical, at least it is now. It wasn't in the earliest days of the accelerator technology because there weren't any computers in the earliest days.

Problems arise because we have splintered our specialties to such a degree that communication, even though it is possible, is typically not acceptable. The controls group does not think of themselves as the servants of the overall accelerator designers and neither do the instrumentation people. But if a project is to be successful, that is the way it has to be. Accelerator physicists know how the accelerator operates, what is needed and what kind of knowledge is required to adjust correctly and to run an accelerator. They are the ones who should design not only the accelerator but also the user interface. When the accelerator physicists ignore their responsibility to design this user interface, they pretty much deserve the mess that usually comes out.

This is not to say that there is no place for the software groups. As an example, a software engineer is trained to know about list structures, how these relate to a tree structure or to graph structures, the various types of loop control structures, etc. If you know these things, you have just barely scratched the surface of what is available to the well-trained software engineer.

But even the best trained software engineer doesn't have more than the vaguest idea of what is necessary to run an accelerator. That sort of thing should not be turned over to any software group or person. It is responsibility of the accelerator physicists to define and design all of that. It is also their responsibility to ask the software technologist what is available to make this easier. The controls group should be refining, they should not be designing. The same with the instrumentation engineers. They should be designing and building the connections between the lowest levels and the intermediate levels.

As I said at the beginning of this, the strength of the CEBAF controls system is that our controls group looked upon themselves as building tools so that the accelerator physicist could design the system. The fact that before we even had a functioning injector, we had a functioning controls system that all we had to do was plug the injector into the controls system, demonstrates the strengths of that kind of philosophy.

Specifying Who Makes the Interface

Scott Miller: I'd like to ask a question of Peter Lucas. If one has a software intensive instrument, how is it specified and developed so that the instrument will operate and be interlinked with the software? How is that done in a timely manner rather than just developing the instrument, tossing it over the wall over to the controls group and say "Here it is, it is your job to make it interface"?

Peter Lucas: Obviously the answer to that question differs depending on the kind of instrument we are talking about. Let me make it something sophisticated like a BPM system that would be developed in-house. (I don't want to talk about a GPIB instrument where the answer would be quite different.) The controls group specifies, especially for the new ones which we now say we will attach directly to our token ring, the protocols, the type of information that passes over that token ring, and we supply to the builders of that instrument a collection of useful subroutines. We call this OOC, Object Oriented Communication. OOC is a level of communication which resides inside the microprocessor of the instrument. Software is developed which uses this protocol. It is hard to draw the line, administratively, between the controls group and the instrumentation group here. But someone else other than the people who did the communications will write software that will use that communications package and thus get themselves out to consoles, get control functions in from consoles, etc. So that we draw the line sort of in software at a particular level of the communications hierarchy inside the microprocessor.

User Friendliness

Scott Miller: I guess we will open it up to the audience now. Bob, you had a question I'm sure you like to ask.

Bob Webber, in Audience: Several years ago there was a movement afoot, especially in the SPS control system, to make software which is "user friendly." Are people working on that concept now or is it regressing back to the old days where you put a shield over the software people and say "don't bother them"?

Jean Borer: We are now leaving the situation which is named "user friendly software" where almost everybody could go to a console and type something and it would move the instrument somewhere and the data would come back. Not so long ago at the beginning of LEP project the official attitude was that NODAL was perfect and that everything in the SPS-NODAL system would use CAMAC for connection to instruments would be applied to LEP. That was the early LEP design. Then the controls group in charge of NODAL and SPS control system started to work and they decided that Unix and C were very interesting. Soon everyone noticed that you cannot do file serving or you cannot do all kinds of very complicated database and those kinds of things which are essential in large machine with NODAL to keep track of what is going on. From themselves, the specialists of that group--who were officially recommending NODAL--brought out a very different system, the one I was describing before.

Today we speak C for application programs both in consoles and stations as well as in hardware. The severe difficulty of this project has bee that the concepts of the control system happened much too late. Therefore the tools, the ones you were describing necessary for writing software for communication, were not available, or at least coming in lumps with very poor documentation and changing almost every day. You really had a problem doing things with your own instrumentation.

I believe that this shows that a user friendly system is ineffective for today's machines. But it would be nice to have the "friendliness" of systems like Windows for PC's or like Macintoshes. That kind of easiness should come up for the user. But it means a tremendous effort in software for the controls group and controls groups generally do not have the manpower to do this sort of thing.

Lessons From Industry

You can only rely on industrial products which are well supported and mature. So from all this experience I've seen in the different machines on which I was working, I would say that the dream would be to transfer what one usually calls controls into accelerator processing. Controls would be done by engineers specializing in the accelerator process, similar to what you find in industry to drive plants and things like that. This would be a professional specialty not present in the accelerator world. Then there would be another group called Digital Data Communication and Processing with all the hardware. That is in fact the actual task of the controls group and therefore the name is not well used.

Then finally the third level where we deal with instruments, would be instrumentation process software. To use the example of BOM again, BOM uses the communication and it uses global data processing. It does it also in local crates as a single instrument like a Q meter, and has a single connection to a station. But a BOM system has many connections, so you have to first collect data from all stations and then to do more processing prior to getting the results for an application like orbit correction. The digital data communication and processing needs to be within the instrument. Therefore one can very well think of the instrument engineer as being responsible for the process and communications system dedicated to the functioning of that instrument. That would be my distinction.

Jay Heefner in the Audience: I'd like to comment on that. I think John alluded to it in his introduction about the CEBAF control system. That is exactly what CEBAF has done. In some respects they shrunk the number of people who actually write software, it is a very small group at CEBAF that actually writes C code. But then expanded by orders of magnitude the number of people who can "program a system." The reason is that the software group provides a bag of tools for you to use. This bag of tools includes the ability to program without ever writing any software, to program on the screen as much as you would program a programmable logic controller. It allows a system specialist such as an injector engineer or physicist to design his control system using this bag of tools that is provided by the software engineer. He can design it himself, software, displays, etc., all with very little intervention from the software group other than to provide some specific tools that may not be available at the time such as a frame grabber or something like that. But it shows up as a block on the screen that says "frame grabber." I think it has been very successful so far at CEBAF.

John Perry: I'd like to point out that CEBAF's controls system is unique only in the versatility with which you can build your own displays and your own logic blocks. The people who designed this came from industries in which these concepts were developed. It has been shown many times that this is an extremely successful way to build large scale and small scale systems.

Knowing What Others are Doing

In reference to his comment, I'm glad to hear that somebody once thought of it. In the two years I've been in the accelerator business I seen not the least evidence that any progress has been made in it. But I do feel very strongly that a hardware engineer's education is not complete without a good strong introduction into software technology and that no software engineer's education, if he is working in a controls type atmosphere, is complete unless he has had a good strong introduction to technology. I feel the same way about accelerator physicists, although they do have to turn over the details of building their system to the service type groups. But they should at least know how to talk to these people so that they can understand what is being said when the service type groups ask them for their input. There are some cases where the service type groups think that they are all there are, but that sort of tendency should be absolutely and diligently squashed.

Both the controls and the instrumentation groups should be required to understand they are serving other people's interests and that it is their job to talk to and interface with these other people. An instrumentation engineer who thinks he can build a subsystem without active participation from the people who are actually using it, from the lower end and upper end, is kidding himself. The same with the control system person who thinks he can build a control system without the active participation of the other people involved. The accelerator physicist that thinks that he can design an accelerator and dump the user interface on the controls group is kidding himself.

James Holt in the Audience: What about the SSC, what sort of philosophy do they have?

Scott Miller: The SSC philosophy is computer intensive. One puts as little intelligence on the front end as one can kind of get away with. One puts things like diagnostics, etc. on the front end, but you give the central host computers the ability to manipulate anything that it is possible to manipulate in the machine. Part of this has to do with issues of geographic extent. You can't run down the tunnel and go and unplug this card or change that cable from that plug to another plug. So the requirements of the computer system are that everything is computer driven.

General Controls Ideas at the SSC

Unix seem to be the universal operating system, but as one works down toward lower level machines it gets rather fuzzy. If you are looking at embedded microprocessors, it does not seem appropriate to have Unix or C forced on you. At the embedded microprocessor level, it seems appropriate that perhaps something else, like

Assembly, might be appropriate.

Communications are being handled by a time domain multiplexing system over fiber optics. To handle all the data communications that are required for this kind of top end control, there is something on the order of 15 giga bytes per second worth of information that can flash around the 10 fiber optic cables that supply every place around the tunnel.

VME is a common protocol, but we are intrigued by VXI. In VXI one can get RF shielding built into the crates, but it is rather expensive. But if you state this sort of standard, you probably will not find exactly the module you want already made. But this is a problem whether it is VXI or CAMAC.

The workstations in the system will be heterogeneous. There is not going to be one particular type of controls workstation.

So what's going to be required here is having a large number of different components being used by different groups and all having to work together. That is a very complex task.

Person in the Audience: Is there an ongoing project in which "user friendliness" is getting more emphasis, as Jean Borer mentioned?

Scott Miller: One concern about that is that by the time the software is friendly enough for the instrumentation engineer to use it, he will have finished his job.

Keith Jobe in the audience: This question is controversial, but this forum encourages controversy, so I'll continue. I am terrified of the notion that an injector engineer is going to build his own custom display using a toolbox full of powerful, general-purpose tools. I actually can walk both sides of the street as a software engineer, instrumentationist, and accelerator physicist. As a user, I insist that I be able walk up to an accelerator without any firm knowledge of the heuristic display patterns of a particular area, and to be able to help diagnose problems in instrumentation, power supplies, controls and all sorts of good things without being completely versed in the color pattern in a window, or the shape or the texture of that particular system engineer's favorite family of displays. Is it appropriate to start asking for some sort of unified display and control on a consoles within an accelerator? I realize it is hopeless across the industry. I realize that accelerators are requiring more complicated and more sophisticated control systems.

Making Application Programs Look and Act the Same

Gerry Jackson: I'd like to take a shot at that because I to my name have fifteen controls programs, I am an accelerator physicist and I am in the instrumentation group so I think I can speak to all views.

When I first came to Fermilab four years ago, I was told to write the program which controls all the waveform transmitters in the low level rf for the main ring. So I was given a book. This book was written by the Controls Group and it contained an explanation of how you do file sharing, what colors you use, what colors you don't use, what red means, what green means. Finally, in the back of the book there is a yellow index, which was about one-half the thickness of the book, which lists of all the subroutines and how to use them. Everything was laid out--standards were given. After you write your program, the operators look at your program. They try to use it, they give you suggestions. Then you go back and try to include their suggestions. What you get is a very nice standard program which the operators like. It doesn't take much. Just a few written guidelines.

Rigidity in the Software Specifications

Marc Ross: I would add to what Keith said. The problem with SLC, possibly a problem unique to the SLC, is that it has such a rigid framework. But this is good, not only because I, as a specialist in one part of the machine, want to be able to diagnose something somewhere else, but because eventually there will be an application that is written which rides on top of the old ones. As a fictitious example, imagine developing a feedback system or some type of higher level application which requires data acquisition from rf systems, BPM systems and, perhaps, various other specialized systems. For the person who has this job, it is nice that the code which lies underneath his has been written with that rigid framework in mind. It makes it possible for him to write the code quickly, and it makes it run a lot faster. This is because the person who wrote that part of the code was thinking "Hey, this piece of code that I'm writing is going to be used by somebody else's program someday. By golly, I don't want my stuff to be the slowest bunch in the lot." But, because of this rigidity, it makes it much harder for the novice to contribute to the controls software in a substantial way.

I believe that a book really can be written which defines how to make this sort of application. This book exists for the SLC. But it eventually gets too thick and the guidelines becomes more and more rigid. You find that there are very few accelerator physicists with the stamina to go through it. So most of the software ends up being done by the professional.

Richard Witkover: In the AGS, particularly the Booster, we are trying to do a similar thing by standardizing the application program presentations. One difference is that users, so far, have not had access to the coding; the Controls Group handles all of that. They are sticking to a standard because much of the presentation coding is from commercial products and the rest is locally written in the form of "tools". I am told that these are fully documented and the Controls Group programmers are strongly encouraged to use them. It certainly makes it easier for the user to see the same menu tree format for any application program.

We have been talking about the interface that you see in the Control Room. The other part of the coding is down at the detectors, at the instruments themselves. The same question applies here: Can this coding be standardized? Where do you get at the software? In our case the people who do the coding are in the Controls Group and the ones who build the beam diagnostics are in the Instrumentation Group. and have absolutely no access to the software. This is not the best arrangement. The instrumentation designers are constrained to live at the end of "analog" lines with no intelligence in our own instrumentation. There are many things which are quite sophisticated which we would like to do but have been prohibited under the policy of standardization, which has been interpreted as meaning that there is only one source of code and it is exclusively in the Controls Group. We tried to break into the programming, even by suggesting that software routines be written by the Controls Group which we could access with simple code of our own, but there was never sufficient manpower available in the Controls Group for this to happen. I endorse the need for standardization in coding but in some cases it can impede progress. In this case the standardization can work to reduce the programming load on the Controls Group but the burden of enforcement of those standards still exists. The instrumentation designer benefits in several ways too. He has access to the documented and tested standard code, reducing his work, and also can put improved capabilities in his equipment through better use of the intelligence in the front end processors.

New versus Old Accelerators

Ralph Pasquinelli in the Audience: There are two different worlds I think amongst everyone here. There is the established accelerator and the brand new one that is on the books. The established accelerator, Fermilab, is still trying to figure out how to use 64k of memory cleverly (which you can now get in a pocket calculator with no difficulty!). So since you have to interface to the world with 64k, you usually cannot be very fancy.

Additionally, you are constantly improving your accelerator, so you always have to live with the old way things are going. Furthermore, the change over is always a slow process.

People who have brand new accelerators are going to start with mega bytes of memory where they can be a little less clever, perhaps, and a lot more fancy from day one. Trying to standardize it amongst everybody is impossible because we do not live in the same world. It would be great to take all the PDP-11's, throw them in the dumpster and get some serious computers. But that would require a major amount of money and create a potentially disastrous cross-over period in which you would say "Oh well, we don't want to run the accelerator for a few months while we do it."

Jean Borer: First of all I agree fully with you but I would like to continue the discussion of standardization, which I feel is at the heart of the matter. What was said before was on the mark; it encourages good coordination and ease of communication among the different people working on a project. But strong standardization implies a lot of rigidity and this blocks the introduction of new techniques or new technologies in the computing world. At CERN, we are always having new experimenters come in with new equipment which needs to connect to the control system. So far, we have been able to accept these new technologies and to make an interface for them. So we are pushed in that sense! But one tends to forget how long the accelerator instrumentation systems will be used. You do not consider that this system, with its software, will be around forever. Initially, it was an investment, exactly like the hardware, and when it reaches the end of its useful life, one has to come up with new equipment. Industry is doing this. You see a beautiful factory full of machine tools absolutely able to continue the production, but one is changing it because the overall factory is not adapted to modern production techniques. I think in the accelerator is exactly the same thing.

Ralph Pasquinelli: But the point is that if you want to change something over, you must stop running. It is not like General Motors: If they decide to make a Corvair again, they start a different line. Here we have to stop everything because the control system is down. And now you need to convince the physics community to shut down for a certain period of time to do it.

Controls is Harder than You Think

I think a real problem, in defense of control people, is that the accelerator community does not realize how important those people are and that there are not enough of them. If you really want to get something done right in industry, you hire the people needed to do it. For example, at Johnson Controls, if they are controlling the ventilation for a 100 story building with windows that do not open, that air conditioner damn well better work or they will have a lot of dead people around! That system has a lot of high level controls type software, and Johnson uses a lot of people to make sure it works. The accelerator has got a lot more problems than a 100 story building but it doesn't have any where near the reliability that the building does. The accelerator community just wants to keep the experiment going so they can take data. Two years later, the experiment is done. The accelerator lives for more than two years, as Jean just said. If you want it to be a very high integrity system, then you have to invest in an immense amount of manpower to make it work that way. From my experience, software is typically never reviewed by anyone other than the person who wrote it. It needs more people if you want to make something that has high integrity.

Richard Witkover: This is one of the big mistakes that I was referring to.

David Peterson in the Audience: I would like to make a comment from the hardware engineers' point of view on standardization and adaptability. When we are looking at things you can install in a machine, you do not build the signal generator with the intent that it will plug into the CAMAC slot; you go out and buy something and a lot of times the manufacturer defines a different interface, unfortunately. There are, at the instrument level, a variety of interfaces that can be used and there are a few things supported by the controls system. But it seems like it would be nice to have a particular interfaces supported at that level, sort of at the crate level, with building blocks that other people have implemented, so that you can plug in your instrument and get your acquisition parameters and then present that data in a suitable format to the control system for distribution.

One of the things that hasn't been mentioned yet but seems to be popular around here is a thing called LabView [from National Instruments, Austin, TX], which runs on a Macintoshes and other PC's. It is a set of building blocks. It is really a graphical high-level language. You can select boolean functions; you can select graph type; you can select a fast-fourier transforms, etc. Even hardware-oriented engineers can create a "virtual instrument" this way. Are there plans to implement this sort of system anywhere?

Gerry Jackson: I hope not. I have a Macintosh on my desk and I actually use LabView. And I have hooked up to some HP equipment over GPIB through their GPIB controller. To be perfectly honest, you do not want to run a machine that way because the bandwidth is extremely low. With just the LabView connection, you watch it "clickety ... clickety ... click." (I have seen some controls systems which actually are that slow too!) You have to make sure the thing runs at a speed that is commensurate with the requirements of the accelerator.

David Peterson: That was an unfair statement. A Macintosh computer is much more software intensive than an IBM PC. But it has a much better user interface. So from the start of a project to the end of a project a computer runs slower but the user is much happier through that project. I have had projects where ...

Richard Witkover: But that is not necessary. That's what I've been trying to say but.

Audience comment: That's the whole point of the problem. Some say that you ought to have the controls system set up from the very beginning so that we know what the problem is. They ought to have the BOM system set up so that it is all ready so that the machine works. They have to have the rigid framework so that the instrumentation can be added without any problems. Then you ask the question "Is there an ideal set of systems that can then be added to very flexibly later?" In my experience I would say no, you have the system running and then you start these interactions. Then the question comes up, whose responsibility is it when you want to add stuff? Is it his responsibility not to buy that instrument or is it the main control system's responsibility to be able to accept it? The accelerator wants a new display and they cannot get that kind of display, so is it his responsibility to write the software or is the control systems responsibility to do it? The CEBAF system is a very well developed system. We are using it now. As opposed to the tools allowing us to do things, it confines us to do certain things and there is no way to go into that system to change it.

I want to know some opinions about answering that question. Whose responsibility really is it?

Steve Peggs in the Audience: There is difference between looking at a LabView-like product as being a standard way to program application problems and looking at it the way CEBAF does, the way I think is very successful, in a very constraining way to let you, for example, look at where cards are in a CAMAC crate. Tools both enable you to have more power and constrain you to do things in a given way. When you know what that given way is, when you know there are only so many slots in a CAMAC crate, it is very sensible to use a LabView type of product to do that. But if you want to program, if you want to do applications code development, you should use a standard program not a virtual program language.

Adding Resources to a Computing System

John Perry: I'd like to point out that many of the problems discussed in the last few seconds here derive from the idea that there is a fixed amount of computing resources. The CEBAF system, for instance, has been explicitly designed to allow you to divide your system among a number of different processors. We have 30-40 processors planned already. We are currently using eight or ten. Not being a member of the controls group I can't give exact figures on all of this stuff. But in our situation when we run into a place where a particular computer is locked up we add a computer and divide our functions between the computers. In accelerators that is a very easy thing to do because the requirements for processing are easy to distribute. It requires one thing:

a very rigid, well specified program interface. As I mentioned at the beginning, since our controls system was already running before we had any hardware to run it on, our engineers don't need to write the drivers, all they need to do is set up the displays and set up the logic. Our controls people write the drivers for our shiny new spectrum analyzers or our hot new wave form digitizers. We turn the manual over to the controls group and they write the driver for us and then they tell us the driver is available and such and you can get it by clicking such and such an icon.

Steve Peggs: But writing drivers is not the same as writing application codes. It is very different.

John Perry: The applications code is already fully written. And as I said ...

Ralph Pasquinelli: That can't be true, you don't have a working machine yet.

John Perry: That is simply not true.

Paper Machines or Real Machines?

Gerry Jackson: It is true. I see a big dichotomy here between people who have paper machines, machines in construction, and people who have real machines. I have yet to hear someone with a real machine say that their control system works real well.

John Perry: If you analyze any system it breaks down ultimately into a certain rather small set of functions. Whether they are inner products or outer products, whether they are simple linear translations or whatever they are, any system ultimately breaks down to a certain rather small set of functions. And if you have carefully analyzed your system, then you can break down any system into this small set of functions. If the resulting system is too expensive in computing power to go down to this level then where a particular element of your accelerator bogs the system down or runs too slowly, then you can write a special code that combines these functions. This is all straightforward, there is nothing new about this, there is nothing obscure about this, this has been worked on and successfully implemented for two decades now.

Gerry Jackson: Where, I'd like to know. What accelerator?

John Perry: NASA does it. The space shuttle is running right now on a purely software driven system.

Audience comment: It seems to me, John, that eventually your hardware gets more expensive. If you start adding new software to other systems, workstations in your control room, then you have something that is not an efficient device to run an accelerator.

Peter Lucas: I'd like to comment from somebody who runs a real control group, that runs a real accelerator. I'd like to present you a real problem. Lightning just knocked the power out of the whole site and you've got to get everything back to where it was. All 40,000 settings. That is a real problem. That is the type of thing that we really face.
 And that doesn't break down to a small group of inner products, outer products. It breaks down to literally hundreds, if not thousands of separate things that have to be written down somewhere and kept track of.

Steve Peggs: As a recent CERN experience testifies, an accelerator is not like a factory.
 You cannot think ahead of time what the processes are that have to go on exactly to

make this thing work. You don't know until you turn the accelerator on what the problems are and even ten years later you are still writing codes to make things better.

Marc Ross: SLC is similar to that.

Jean Borer: I don't agree fully with you, I agree partially only. Sometimes we don't think enough before hand. Foresight is missing sometimes in controls.

John Perry: This is what I've been saying from the beginning.

Adding to an Existing Machine

Ralph Pasquinelli: The foresight was missing before because you had 64k of memory to do things with, so you couldn't do everything. Right now I agree, but some thought that brand new and super-powerful computers would let you do wonderful things that didn't exist in the past--many of us are working with machines that grew out of the past. Take for instance the p-bar source, which was once a swamp. There was already a control system and an accelerator here. To make something new you had to develop new things for it, you had to add things to it, you are constantly developing things as you go along. It is much easier for something new than for something that is established.

Jean Borer: I would just like to answer this point. The SPS has to change its system carefully, too. It is the main heart of all the production of particles. It cannot stop, but they are they are undergoing a full change. They go from the present ring system to the token ring and abandon slowly the NODAL system. It is still maintained in parallel and made compatible. Then you can invent changes all the time. But this happened because of the age of this equipment. When the producer of most of the hardware is not maintaining it, you have to change it.

Audience comment: People who are using accelerators quickly find the limits of the accelerators, and they are not the kind of people that stop there. They want to patch it and then push a little bit further. You use a patch in the space shuttle and Voyager or something like only in an emergency. This is quite a different situation, those things are essentially self-running systems. They just run forever. Nobody has a desire to patch those systems to enhance performance. There are fundamentally different ideas about how they work. Eventually you get the accelerator full of software patches and other thing like that to make it run better.

John Perry: The latest NASA launched experiments have been patched in flight.

Audience comment: That's not an operating system.

John Perry: No, both the Hubble telescope and Magellan probe have been ...

Audience comment: The shuttle doesn't take off on a patch.

John Perry: I'm not talking about the shuttle, I'm talking about the ...

Where is the Interface Programmer?

Jim Zagel in the Audience: I want to know more about where instrumentation actually interfaces to controls. Peter pointed out one way, even though it may not be completely bug free yet and it is still evolving. To Ralph I say, at one time 64K was a lot of memory--we're evolving. Instrumentation engineers are designing systems that have intelligence in them, they are building them in VME chassis, Multibus chassis, or buying them from HP. Someone has to write some really low-level code that sits in a very far front end. I'm real curious to know where the ideal place for the programmer, the computer science specialist that writes that code, to work? Does he work within the Instrumentation Group or does he live in the Controls Group? Or is there a pool of these specialists that is circling around? If you have a pool, you never have enough of them. And that is exactly the same thing that he is saying.

Peter Lucas: From an administrative point of view we don't have an answer, as Jim well knows. Every time we come up with a new project we all scratch our heads and say, "Is this controls, is this instrumentation, is this accelerator physics or is this safety? Who should do it?" And so that's the place in which the line is certainly fuzzy for us.

I can at least give a partial answer these days. Whoever does it you don't want to do it in Basic and you don't want to do it in Fortran. I think you want C++ or something related to that. You have to have the type of person who is able to program efficiently and properly in that language. He also needs to know at some point the details of the operation of the instrument. You want a very good engineer-C programmer. Wherever that person resides on the organization chart is for us always difficult.

Al Jones in Audience: From Peter's comment it sounds like you have to be intimate with the electronics. Doesn't that automatically say that the software people in the instrumentation group need to sit there and work at a desk next to the engineer? They can pass the specs back and forth until the piece of equipment gets done. I don't think engineers can work with memos from one group to another group. I don't think there is communications there sometimes.

Peter Lucas: I think that is an ideal answer. I'm not sure that anybody can afford to have all the people to do this job.

Gerry Jackson: I'd like to know where you get those people to begin with. It's not like these people are growing on trees and they are just begging at your door for a job.

Software and Hardware Engineers Together?

Keith Jobe: That sounds like a perfect way to make the software engineer's productivity close to 0, putting the engineer right next to them so that the two can be shouting back and forth about the behavior of the instrumentation. What we haven't established in any place I am familiar with is a set of reasonable specifications the accelerator physicist to make the following gizmo do that, engineer coming back with a set of controls, and the software engineer taking those and also talking to the engineers and the physicist. Everywhere I have ever seen software engineer taken out of his X-windows terminal-equipped, quiet, and dark office and put next to the engineer, you've taken a full-time equivalent and turned him into a one-tenth equivalent software engineer. You can't waste software people by just putting them out in the in the field to gab with engineers that don't know what they want to do.

Greg Stover: I disagree.

Gerry Jackson: I disagree too, strongly.

John Perry: As I pointed out in the beginning, the software needs to be written by someone who has software training. You don't get good software by people who don't have more than the vaguest conception of what software is all about. You can make good use of a software engineer in that context by making sure that the software engineer involved knows some hardware. If it is one of the MIS graduates, he shouldn't be involved. If he is a true software engineering type and his software education has been supplemented with at least a minimal hardware education and he works regularly at this interfacing thing, he will be far more productive than any combination of hardware alone or software engineer alone. But you've got to have somebody that is dedicated to the interface, and that is where too many organizations are totally lacking.

Audience comment: I think both Peter and you have talked about the type of person that we need and where he sits. It could very well be that the controls group itself doesn't want to add new drivers and tell the instrumentation people not to buy that piece of equipment. Then what does this interface person do? He is not needed. On the other hand are you all acknowledging that we do need an interface person and that drivers and new ways of interfacing to the control system are allowed. Then you are acknowledging that this part of the controls group is not needed.

Hardware vs. Software Interfaces

John Perry: There are such things as program interfaces. They are what should be rigidly specified. Not hardware interfaces. Restricting hardware interfaces can be useful to a certain extent but it is also extremely restrictive. But rigidly and well-specified software interfaces can overcome that to a great extent. Because you still have the option of trading off computer power for the number of computers. If your system is designed for that.

Marc Ross: We have a similar experience to Dick Witkover's. The hardware interface is not quite as restrictive, it is simply a set of analog signals but it is fairly restricted. When it comes time to develop a new instrument, we find that not a lot of processing power can be put into it and that it needs to conform with the existing software because the software is so expensive. Software resources are far and away much more expensive than any of the other engineering resources. For example, with the wire scanner I talked about this morning, the software constraints talk to virtually every aspect of the design including where the screws are. And that's certainly a problem. However, I feel with a little bit of work you can come up with a decent device that does the job.

For us it is a benefit not to have a lot of on local processing similar to the SSC. We are going to bring the raw data up to the main host computer and use it with the on line model and the on line model can change for any one of a variety of reasons. I don't want to have to redistribute the on line model down to some local processor every time it gets changed. I want to bring the raw data up and have high band with synchronized data acquisition and fill it with these things at the highest level. In the final analysis its a restrictive hardware interface but we get around it.

Software Costs

David Peterson: Can you explain why the software effort is so intensive?

Marc Ross: Because, as I pointed out in response to Gerry's comment, the book is very thick and the people who write the code are experts. They are professionals and we expect that their code should be highly exception proof. In other words virtually anything can go wrong with the device and the code will say something intelligent to the naive user so he knows what to do to fix the device, the whole business is logged and that can be buried in the bowels of some unrelated piece of code, and another programmer can write on top of this piece of code, and so forth. It is expensive to take software people and train them at that level. It is expensive for them to write code which performs these functions as rapidly as we need to have them performed. For example, I want my wires to scan continuously, on a sub-minute basis, doing a lot of high level processing all over the place and I want to be able to add another one tomorrow with no sweat.

To bring that kind of software functionality on line and get it going and then to have it function nicely, build upon it with feedback loops and automated tuning procedures requires a professional job, and that's what makes it expensive.

David Peterson: Does that mean instrumentation engineers are cheap?

Marc Ross: It means that in comparison, his time is less of a problem. The controls hardware engineer certainly has plenty of time on his hands compared to the software. The software dominates the job at that level.

David Peterson: One of the things that happens is this. Let me use GPIB as an example. In my presentation, there were some spectrum analyzer plots which displayed the tune number. That single number requires GPIB to transfer something like 2000 points across the whole network to a front-end computer. It calculates that number. It would seem more reasonable to use a 68000 sitting out in the service building to calculate that number and transfer just that one number over the network.

Marc Ross: For an instrument like that, you're absolutely right. However I will point out that in the HP 71000 spectrum analyzer, you can write a code which runs in the analyzer. All that the controls people have to do is read that number.

David Peterson: Right. I wish I could do that right now!

Gerry Jackson: Here's where I'm going to jab Peter a little bit. We recently made a request to do that, and they said "Go away... Don't bother us... We are busy."

Peter Lucas: Let me answer that. We do precisely that in our secondary emission monitor beam profile monitors. We compute a gaussian, we see if there are any bad points, we see if there are any missing wires, we correct for those, and we recompute the gaussian until the whole process converges. We send you up a mean and standard deviation which is better calculated, better conditioned than you will find anywhere else. One number shows up to any console only once.

David Peterson: That's a custom device and ...

Peter Lucas: It's not a custom device, I programmed it myself.

David Peterson: What I'm saying is I'd like to see some sort of application protocol carried out. We should be able to use the commercial standards that are available, like VXI, so that the instruments provide standardized information. You talk about resetting 40,000 devices after a power outage. You get distributed instruments out there that are smart enough to remember their settings though the outage. If your instrument goes down and you power it back up again, it just rests itself to where it is supposed to be.

Scott Miller: That is all the time we have for now. This discussion will continue at the banquet. I would like to take the time to thank the panelists.

ADVANCED, TIME-RESOLVED IMAGING TECHNIQUES
FOR ELECTRON-BEAM CHARACTERIZATIONS*

Alex H. Lumpkin
Physics Division, Los Alamos National Laboratory
Los Alamos, New Mexico 87545

ABSTRACT

Several unique time-resolved imaging techniques have been developed to address radio frequency (RF)-linac generated electron beams and the free-electron lasers (FEL) driven by such systems. The time structures of these beams involve a series of micropulses with 10 to 15-ps duration, separated by tens of nanoseconds. Mechanisms to convert the e-beam information to optical radiation include optical transition radiation (OTR), Cherenkov radiation, spontaneous emission radiation (SER), and the FEL mechanism itself. The use of gated, intensified television cameras and synchroscan and dual-sweep streak cameras to time-resolve these signals has greatly enhanced the power of these techniques. A brief review of the less familiar conversion mechanisms and electro-optic techniques is followed by a series of specific experimental examples from the RF linac FEL facilities at Los Alamos and Boeing (Seattle, WA).

I. INTRODUCTION

Using advanced, electro-optic imaging principles, several time-resolved diagnostic techniques have been adapted to the characterization of electron beams accelerated by radio frequency (RF) linear accelerator (Linac) structures to drive free-electron lasers (FEL).[1-3] Both submicropulse (10 ps) and submacropulse (100 μs) effects have been addressed. The e-beam information can be converted to optical radiation in the visible regime by the optical transition radiation (OTR), Cherenkov (CH), synchrotron radiation (SR), spontaneous emission radiation (SER), or even the FEL mechanisms. Besides gated, intensified cameras we have used streak cameras operating in the fast, slow, synchroscan, and dual-sweep modes.[4-6] The latter two modes have been shown to be particularly useful for understanding RF-linac related systems due to the relatively low jitter of the phase-locked synchroscan unit when compared with a 10-ps pulse or to the jitter of a single, fast streak sweep.

It should be pointed out that these techniques have been developed within the Los Alamos FEL program and were optimized for such systems. One recalls that the FEL simply consists of the RF-

―――――――――――
*Work supported and funded by the U.S. Department of Defense, Army Strategic Defense Command, under the auspices of the U.S. Department of Energy.

linac, the wiggler, and the resonator cavity as shown in Fig. 1.[7] Also, to obtain FEL signal buildup, a series of electron pulses is brought to the wiggler at a spacing equal to the round trip transit time in the cavity. This results in somewhat specialized pulse trains (macropulse) composed of many micropulses as schematically shown in Fig. 2. This time structure then dictates the diagnostic requirements.

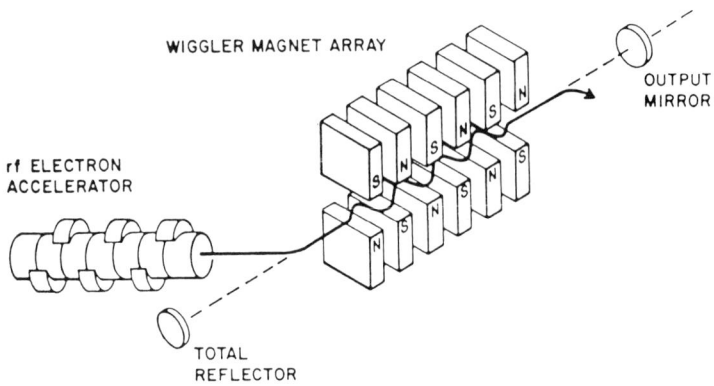

Fig. 1. Schematic of a RF-linac driven FEL.

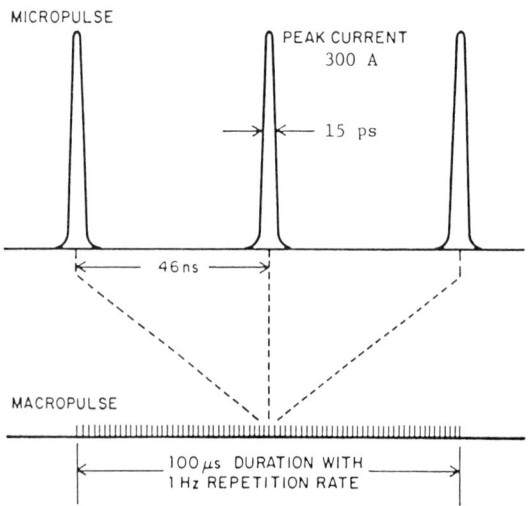

Fig. 2. Schematic of the e-beam pulse structure for an FEL with a 7-m long cavity.

II. ELECTRON-BEAM PARAMETERS AND DIAGNOSTIC TECHNIQUES

In this section, an outline of the various e-beam parameters that can be addressed and the particular electro-optic techniques that have been demonstrated within our RF-linac driven FEL programs will be described.

Beam Parameters

Most of this work has been done via my involvement in programs on the Los Alamos FEL facility (~20 MeV linac) or the Boeing visible FEL facility (~110 MeV linac). As an example, Fig. 3 shows a schematic of the High Brightness Accelerator FEL (HIBAF) Facility diagnostics at Los Alamos.[6] After the linac (only two accelerators at this point) a series of beam position and profile screen (aluminized fused silica) stations (0-3) are included along the beam line before and after the quadrupole. Each was viewed by an intensified television camera [either silicon intensified target (SIT) or gated, intensified charge-injection-device (ICID)]. The sensitivity of these cameras was such as to image via OTR a single e-beam micropulse of about 3 to 5 nC charge and 15-17 MeV in energy. All screens used were designed for OTR experiments except the fourth station where a 45° port also provided Cherenkov radiation imaging. The beamline was terminated with a 90° spectrometer and beam dump whose focal plane had a fused-silica screen to generate Cherenkov light. This scene was imaged by two other intensified cameras. The parameters investigated are shown in Table I and include beam position, beam profile, divergence, emittance, pointing, charge, energy (spread, jitter, slew), micropulse duration, drive laser micropulse duration (jitter and slew), and beam spill. The table provides the nominal value, its span, the technique, estimated resolution,

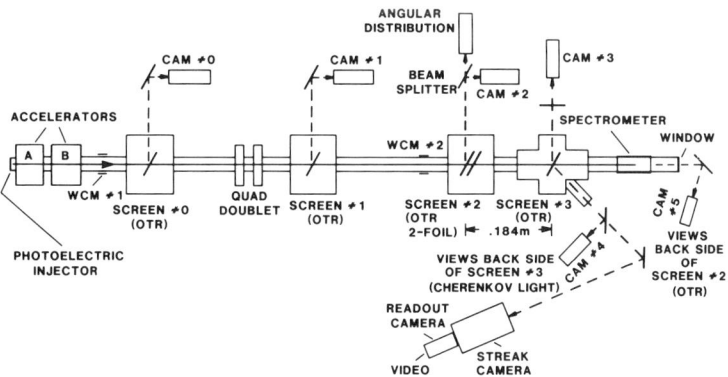

Fig. 3. Schematic of HIBAF beamline and diagnostic stations (OTR; CH).

TABLE I. ELECTRON BEAM PARAMETERS AND THEIR DIAGNOSTICS (HIBAF)

Parameter	Nominal Value	Span	Technique	Resolution	Error	Timescale
Position	Reference to laser line	±10 mm	OTR screens	~50 μm	<100 μm	Single-micro, or streak
Transverse Profile	1 mm, FWHM	0.5-15 mm	OTR screens	~50-100 μm	5-10%	Single-micro or streak
Divergence	1-2 mrad	0.5-3.0 mrad	Two-station OTRI	Few tenths mrad 0.4 mrad limiting	15% 15%	Single-micro macro
Emittance	50π mm-mrad	10-150π	Two-station OTRI	Depends on beam transport	(30%) (30%)	Single-micro macro
Pointing	on laser line	0-5 mrad	Two-station OTRI	~0.2 mrad ~0.2 mrad	20-30%	Submacro
Charge	5 nC	0-10 nC	Wall current monitor	0.1 nC	20%	Submacro
Energy Spread Jitter Slew	15 MeV 0.5% ±0.2% 0.24%	5-17 MeV	Spectrometer 90° bend, Cherenkov screen	~0.2%	(10%)	Submacro
Micropulse Duration	10-15 ps	5-35 ps	Streak camera Cherenkov rad	2 ps (fast) 6 ps (synchro)	10-15%	Submicro
Drive Laser Jitter Slew	10-15 ps	5-25 ps	Streak camera Synchroscan Synchroscan	2 ps (fast) 4 ps over seconds <4 ps over seconds	10-15%	Submicro
Beam Spill	Relative to no beam	--	X-ray detector	Field emission		Submacro

error, and timescale. All but the charge and beam spill were directly assessed via imaging techniques. Recalling our pulse structure of a series of micropulses of 10 to 15 ps duration separated by 46 ns over about 100 µs, the last column addresses whether the time-resolved information separates the individual micropulses (submacropulse) or resolves information within the micropulse (i.e., submicropulse).

Diagnostic Techniques

There are two aspects to these imaging techniques that bear attention. The first involves the necessity of converting the electron-beam parameter information into optical (generally, visible) radiation that can be detected by these electro-optic devices. Mechanisms that have been employed include OTR, CH, SER, SR, and the FEL itself. Additionally, the photoelectric injector (PEI) at HIBAF is driven by a ND:YLF oscillator, which has been frequently doubled into the green at 527 nm. This wavelength is directly detectable with our techniques. The second aspect involves the rather novel application of streak/spectrometers, synchroscan, and dual-sweep streak techniques that have been applied to the imaging problems and are somewhat unique to our programs.

Conversion Mechanisms

Our motivations for exploring OTR techniques include the following:

1. Most relevant e-beam parameters measurable at a single axial location,
2. Need for better spatial resolution on beam profile (100 µm or better),
3. Need for on-line emittance measurement,
4. Improved emittance expected with photoelectric injector,
5. Competition of sources with fused silica plus metal mirror geometry,
6. Technique for future beam profile, intensity, pointing, emittance, energy,
7. Reduced beam scattering and x-rays, and
8. Wiggler diagnostics position and angle.

OTR was first predicted by Ginzburg and Frank in 1946,[8] but its practical usefulness as an e-beam diagnostic was not explored or demonstrated until 1975 by Wartski[9] with further development in the 1980s by Fiorito and Rule.[10,11] The first demonstrations of applications to the actual linacs driving FELs has been reported by Lumpkin, et al.[12,13] and they are part of our discussion here. OTR is generated when a charged particle beam crosses the interface, or transition, between two media of different dielectric constants. This means that the effective spatial resolution limit in diagnostics becomes the optical sensor, and one is no longer limited by the thickness of the fused silica or phosphor. In a poster paper at this Workshop, Rule and Fiorito discuss the micron-resolution imaging potential.[14] The OTR angular distribution pattern can be exploited to measure e-beam divergence and energy, although these two parameters are somewhat convoluted. One of its more useful features is that the backward lobes of radiation are emitted around the angle of specular reflection so that a foil at 45° to the beam emits at 90° to the beam, a standard port geometry for accelerator beam lines (see Fig. 4). An approximate expression for the angular distribution from a single foil given by Wartski[9] for high γ and $|\epsilon| > 1$ is:

$$\frac{d^2W}{d\omega d\Omega} \approx \frac{e^2}{\pi^2 c} \cdot \frac{\theta^2}{(\theta^2+\gamma^{-2})^2} \times F(\psi,\theta,\omega) \quad , \tag{1}$$

θ is angle with respect to particle velocity $\bar{v}$,
F is the Fresnel reflection coefficient for backward lobe,
$d\omega$ is unit frequency,
$d\Omega$ is unit solid angle,

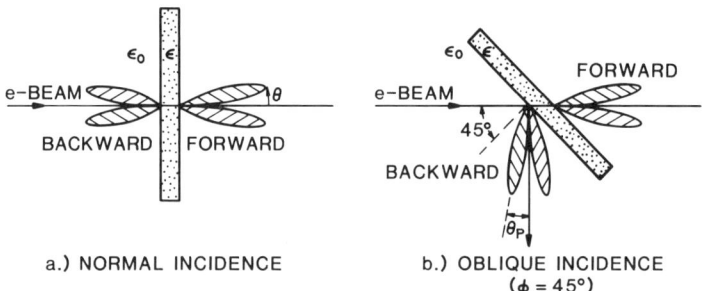

a.) NORMAL INCIDENCE

b.) OBLIQUE INCIDENCE (ϕ = 45°)

CHERENKOV RADIATION PATTERN ($\theta \sim 46°$)

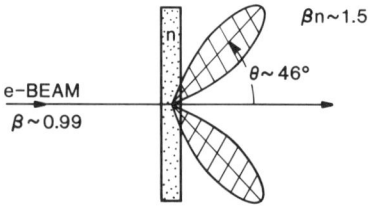

Fig. 4. Schematic of OTR and CH patterns: (a) normal incidence, (b) oblique incidence (ϕ = 45°); Cherenkov radiation, $\theta \sim 46°$.

and the number of photons in the frequency interval $(\omega_1-\omega_2)$ is

$$N = \frac{2\alpha}{\pi} \left|\ln(2\gamma)-1/2\right| \ln \omega_2/\omega_1 \qquad (2)$$

where $\alpha = 1/137$, the fine structure constant (cgs).

The latter expression indicates that the mechanism is not bright in an integral sense (~1 photon per 100 e⁻) but the radiation is relatively concentrated in a cone of opening angle $\pm 1/\gamma$. A schematic of this is shown in Fig. 5, which also shows that the lobe structure valley depends on the e-beam divergence, the lobe peak intensity goes as γ^2, and the spectral intensity goes as $1/\lambda^2$. The radiation is radially polarized in the observation plane.

The assessment of modeling the divergence effects has been described by Rule,[11] where the measured intensity of the parallel ($I_\parallel$) and perpendicular ($I_\perp$) components of the angular distribution pattern is

$$\bar{I}_{\perp,\parallel} = \int d\alpha P(\alpha) I_{\perp,\parallel} (\theta-\alpha) \qquad (3)$$

with an assumed Gaussian distribution for the angles such that

$$P(\alpha) = \frac{1}{2\pi\sigma^2} e^{\frac{-(\alpha-\alpha_0)^2}{2\sigma^2}}, \quad \sigma \text{ is the rms divergence.} \qquad (4)$$

Also, the polarization (χ) is given by

$$\chi = \frac{I_\parallel - I_\perp}{I_\parallel + I_\perp} \qquad (5)$$

Sensitivity to divergence effects in certain regimes can be enhanced by using a two-foil OTR interferometer initially suggested by Wartski.[9] This is schematically indicated in Fig. 6 where the forward lobe from the first foil can interfere with the backward lobes of the second foil for relativistic charged particles passing through the assembly. The previous single lobe angular distribution pattern then becomes modulated in such a way that the fringe location or relative amplitude depends on beam energy (or γ, the Lorentz factor), and the visibility of the fringe depends again on divergence and energy. Proper choice of the foil spacing depends on γ, λ, and the divergence regime of interest. At large divergences the interference fringes are obscured and the pattern reverts to the single foil angular distribution envelope. At lower γ's we have found that we must also consider the effects of the clear foil amplitudes from the first interface of the two-foil system. This effect was described by Rule, et al.,[15] and illustrated in Fig. 7, where the fringe

158 Advanced, Time-Resolved Imaging Techniques

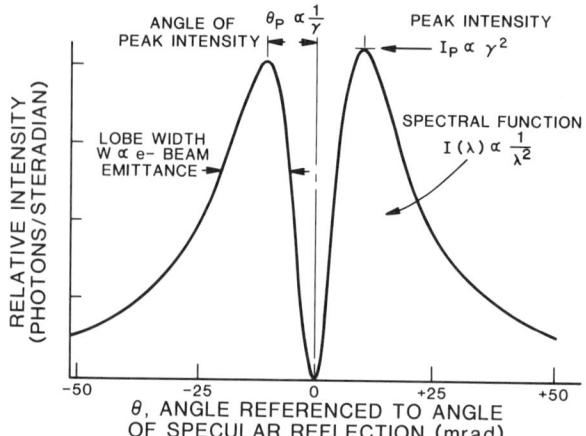

Fig. 5. A schematic representation of the single-foil OTR angular-distribution pattern dependence on e-beam parameters, where γ is the Lorentz factor.

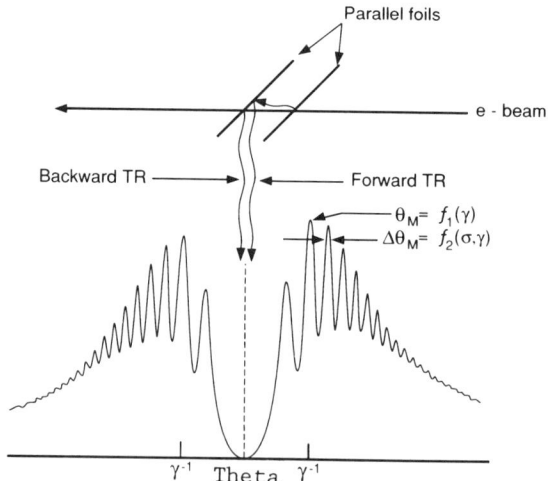

Fig. 6. Schematic of two-foil OTR interferometer.

visibility is seen to be dramatically changed for certain regimes of parameters. Finally, we comment in passing that the mechanism is generated quite promptly compared to 10 ps so it has a good chance of being used also for temporal measurements.

Cherenkov radiation[16] has been used more routinely for imaging relativistic charged-particles, which transit a medium of index of refraction n with velocity $\beta = v/c \rightarrow 1$ and will not be reported in detail here. Its opening angle $\theta = \cos^{-1} 1/\beta n$ (where

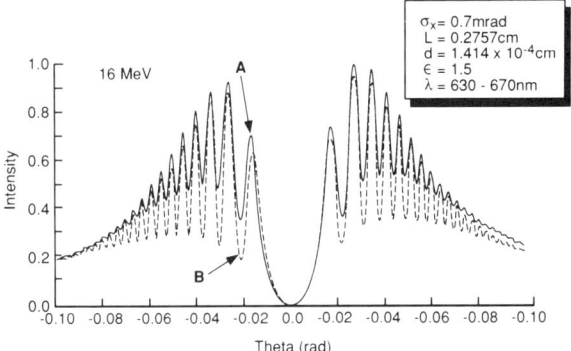

Fig. 7. A comparison of calculated interferograms with and without the clear foil's first interface coherent amplitudes.

βn > 1) is very large compared to OTR's 1/γ for highly relativistic beams. Other aspects are that its angular distribution and intensity saturate rapidly as β -> 1, its spectral intensity goes as $1/\lambda^3$, and it is a prompt mechanism compared to 10 ps. Its emitted photon number in the visible is more like 10 per electron per mm of material. Therefore, it should be noted that to avoid thickness effects in the converter when imaging submillimeter beams, one would be led to use thinner screens whose photon yield starts approaching that of OTR in a specific solid angle.

Synchrotron radiation is emitted by charged particles under the effects of accelerations/forces in bends (usually ascribed to circular accelerators) and can be used to determine properties of those charged-particle beams as described elsewhere in this Workshop.[17]

Spontaneous emission radiation (as it is called in the FEL community) is actually enhanced synchrotron radiation via an undulator or wiggler. A static, alternating, and periodic magnetic field structure produces "instantaneous" circular orbits that can provide reinforcements in the angular distribution patterns at specific angles and wavelengths. The standard resonance relationship in such a situation is given by the following relationship,

$$\lambda_{obs} = \frac{\lambda_w}{2\gamma^2} [1 + A_w^2 + (\gamma\theta)^2] \quad (6)$$

where

λ_w is the period of the magnetic structure,
γ is the Lorentz factor,
A_w is the wiggler field parameter, and
θ is the angle with respect to the particle velocity.

160   Advanced, Time-Resolved Imaging Techniques

Many synchrotron rings use these undulators to provide enhanced sources of x-rays, but we have found at $\gamma \sim 220$ the visible radiation from the Tapered Hybrid Undulator (THUNDER) at the Seattle, WA, FEL facility[18] can be used to great advantage to characterize the electron beam. Because the SER spectral breadth is $\sim 1/N$ for an ideal electron beam and wiggler, and $N = 220$ in that system, we have actually deduced e-beam energy centroid, energy slew, and energy spread from SER spectra. This is in addition to monitoring the beam position and profile (in a convolved sense), the brightness, and the micropulse duration.[3] Examples will be given in Section III.

Advanced Electro-Optic Techniques and Adaptations

Although some of these techniques may exist in other laboratory research experiments, the adaptation of the following techniques to RF-linac-driven FELS, and in particular the characterization of e-beam properties is in the development/demonstration stage.

The Los Alamos streak/spectrometer system was installed on the Boeing visible FEL experiment in January 1988. Although initially only designed to measure the e-beam micropulse duration via SER, it was soon extended to cover other time-resolved aspects of the SER properties. A schematic view of the oscillator cavity is given in Fig. 8 showing the resonator leg, the end mirrors, the wiggler, the e-beam spectrograph, and the pop-in mirror (which allows the interception of the SER from the single passage of the electrons through the wiggler). This radiation is relayed optically to a diagnostics table in the control room over 30 m away. Figure 9 shows a layout of the streak/spectrometer, which is based on a Hamamatsu C1587 synchroscan streak camera mainframe, a 1/4-m Jarrell-Ash monochromator, a SIT readout camera, and a microcomputer for data acquisition, analysis and display. The mirrors M1S and M2S direct the radiation through the monochromator (which is used as a spectrometer), and the wavelength dispersed information is directed onto the entrance slit of the streak camera. This information can be time-resolved on the submicropulse or submacropulse domains via the fast and slow sweep plugins, respectively. With these mirrors removed, direct time-resolved spatial information can be obtained.

Additionally, the synchroscan plugin unit, which can be phase-locked to a reference 108.33 MHz frequency, provides the capability of synchronously summing the micropulses with about 4-ps (FWHM) jitter and 5-ps (FWHM) resolution. The jitter specification is three to four times better than that of the single fast sweep unit and allows critical phase information to be monitored. This is further extended by a dual sweep attachment, which adds a slow, horizontal deflection ramp to the streak tube during synchroscan operation as shown in Fig. 10.[19] Then one can measure micropulse length and phase <u>during</u> the macropulse. The data to be presented are the first (to our knowledge) on a FEL

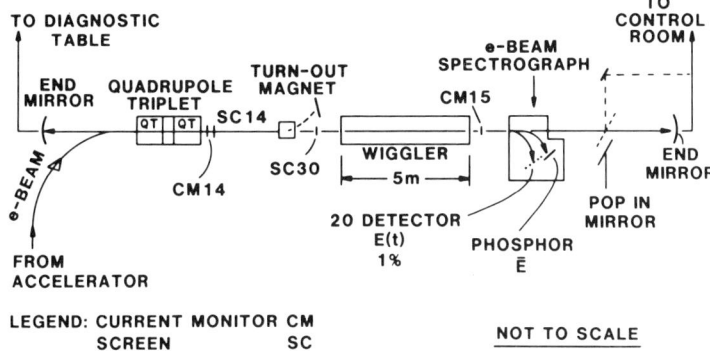

Fig. 8. A schematic view of the Boeing burst-mode oscillator cavity (not to scale). CM: current monitor, SC: screen.

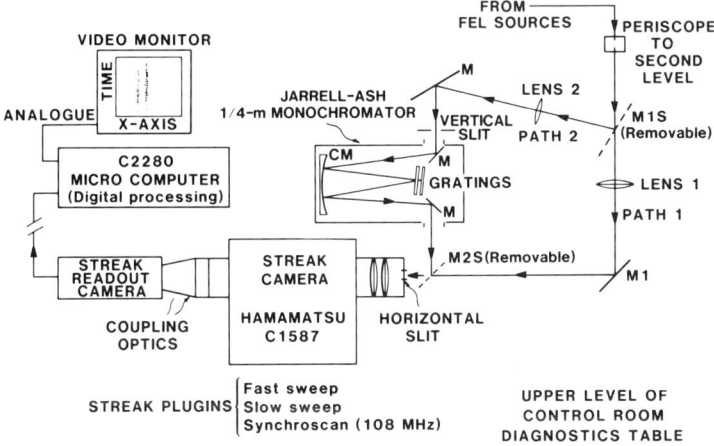

Fig. 9. Schematic of the Los Alamos streak/spectrometer system located on the burst-mode control room diagnostics table.

anywhere, and the first on an accelerator in the USA. The issue of phase stability (jitter and/or slew) is particularly critical for RF-linac e-beam parameter stability and even more so if used to drive a FEL.

III. EXPERIMENTAL RESULTS AND DISCUSSION

Having provided some background of experimental procedures in the previous section, it is hoped the series of experimental

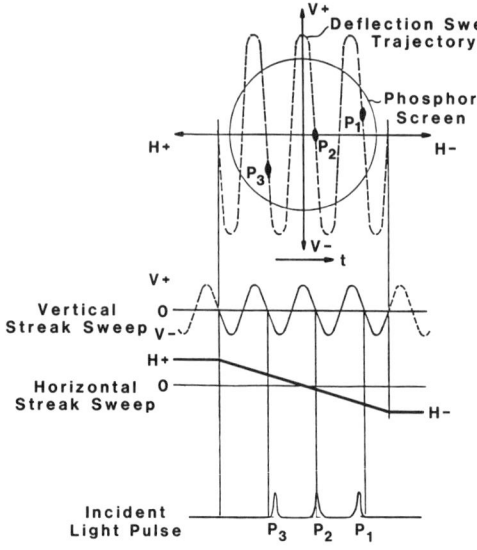

Fig. 10. Schematic of the synchroscan and dual sweep features of the streak camera.

results that will now be presented will graphically illustrate our techniques.

OTR Demonstrations

For electron beams of $\gamma \simeq 30\text{-}35$, we routinely imaged the beam spots and determined position and transverse profiles. Examples of this are shown in Fig. 11, where a single micropulse from different shots on two different screens are shown. The x- and y-profiles are given below.

Comparison of such profile data to a Gaussian distribution in Fig. 12 led us to use such a description as a working assumption. We then tracked emittance versus transported charge from our HIBAF photoinjector as shown in Fig. 13. We see strong evidence for HIBAF's meeting design goals of $\epsilon_N < 50 \pi$ mm mrad at 5 nC. These data were subsequently supplemented by macropulse measurements under a variety of conditions.[6] Figure 14 illustrates the need for coherent amplitude terms in the divergence measurements of values less than one mrad. Table II supports these with a summary of divergence and emittance measurements for three experimental cases of solenoidal field strengths at the PEI. The trend in emittance and divergence as compared to the integrated numerical experiment (INEX) calculations is very encouraging.

At higher beam energies ($\gamma \sim 215$) in experiments at Boeing, we had several successes. For the particular fused silica-polished metal mirror geometry, the predicted competition of sources that could blur the total beam spot was observed. This

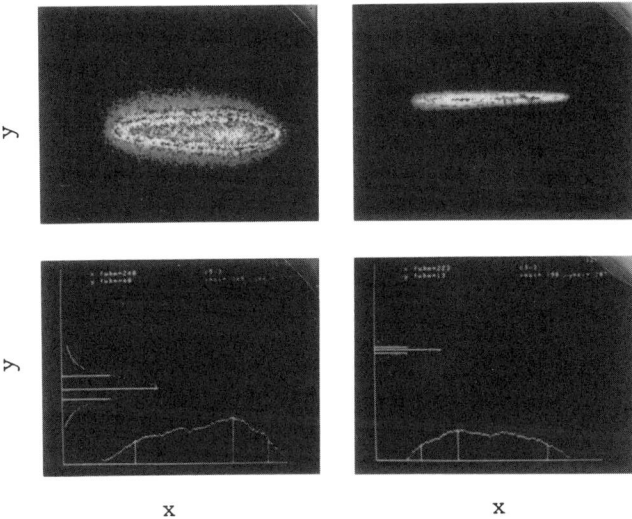

Fig. 11. A composite set of beamspot images and profiles for stations 2 and 3 at HIBAF.

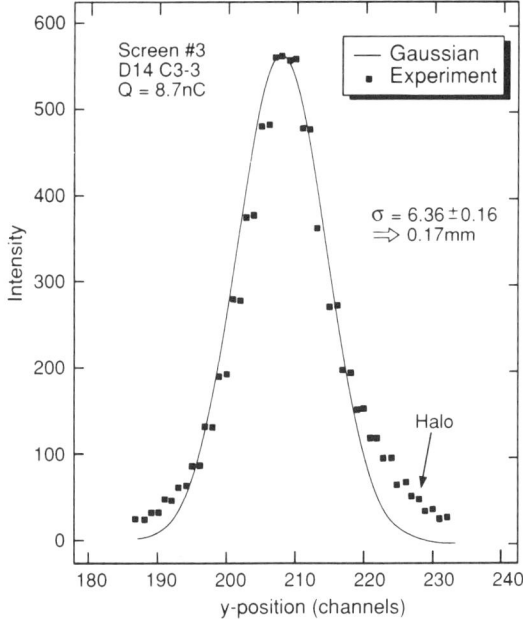

Fig. 12. Gaussian-shaped profile compared to experimental spatial profile (box average of single micropulse image).

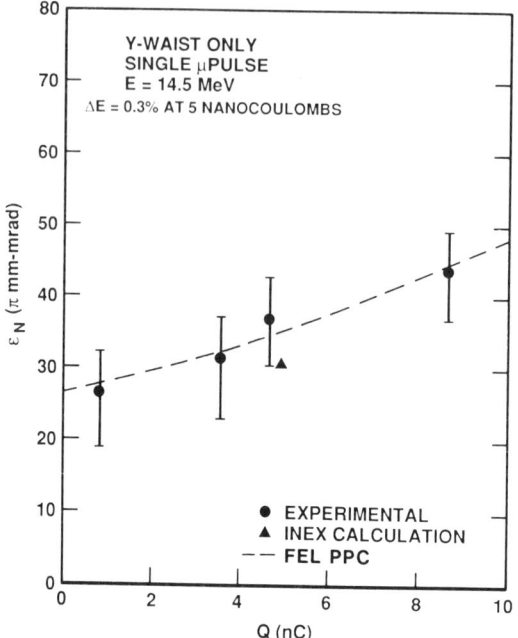

Fig. 13. Emittance versus charge for a single micropulse with comparison to INEX and FELPPC codes.

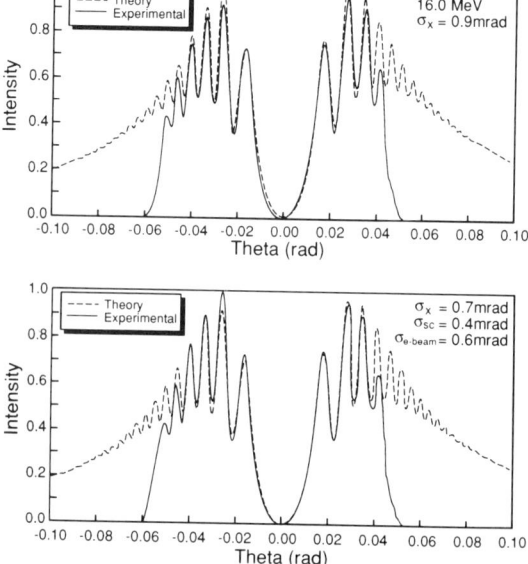

Fig. 14. Comparison of theory and experiment with calculated patterns using rms divergence of 0.9 and 0.7 mrad.

TABLE II. SOLENOID VARIATION EFFECTS, SIMULATION AND EXPERIMENT (OTRI)*
(May 17, 1990)

Case No.	Laser Phase	Solenoids(A) 1	Solenoids(A) 2	Charge (nC)	X-FWHM (mm)	Divergence (mrad) OTRI*	Divergence (mrad) INEX	Emittance (πmm-mrad) HIBAF**	Emittance (πmm-mrad) INEX
1	$20°$	25	500	3.0	0.8	1.5	2.0	65	62
2	$20°$	50	500	3.4	0.7	1.5	(1.5)	58	50
3	$20°$	75	500	3.1	0.7	0.9	1.3	36	43
4	$20°$	55	550	3.1	--	--	1.0	--	39

*Optical Transition Radiation Interferometry (OTRI).
**Assumed Gaussian Distributions for spot size and particle directions.

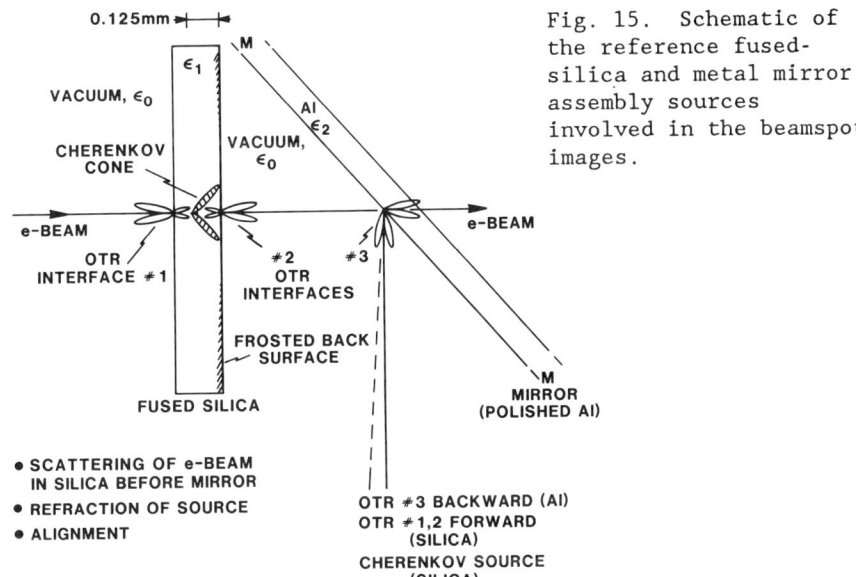

Fig. 15. Schematic of the reference fused-silica and metal mirror assembly sources involved in the beamspot images.

effect is shown in Fig. 15 schematically and Fig. 16 experimentally. A very successful interferometer experiment was also performed. As shown in Fig. 17, the interferogram for a calculated divergence of $\sigma_{rms} \sim 0.3$ mrad and E = 107 MeV is a good match to the data. Additionally, an energy sensitivity is shown in the amplitudes of the two inner peaks as illustrated by comparing calculated curves at 107 and 109 MeV to the same data in Fig. 18. The energy and divergence effects are somewhat separated in this manner, and thus gives hope to having the 1% energy accuracy for the technique. It is also noted that the centering of the OTRI pattern relative to the reference alignment laser in angle and transverse position provides a pointing capability from a single axial location. This feature might be used to help match the e-beam into a FEL wiggler. Another graphic example on spatial resolution is shown in Fig. 19. In this case, nine of the eleven

166 Advanced, Time-Resolved Imaging Techniques

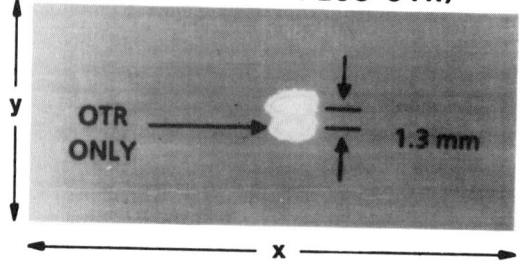

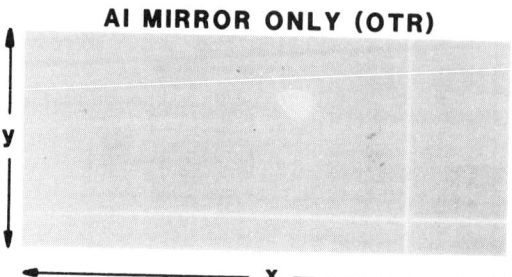

Fig. 16. Experimental example of the multiple source problem on a beamspot.

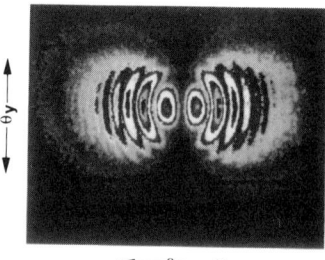

Fig. 17. Initial interferometer image (top) and match to its horizontal profile (bottom) with an electron beam energy of 107 MeV and a divergence of ~0.3 mrad.

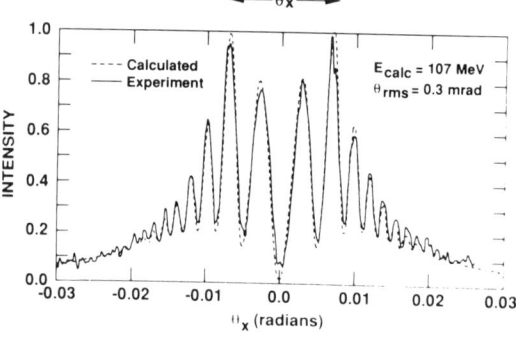

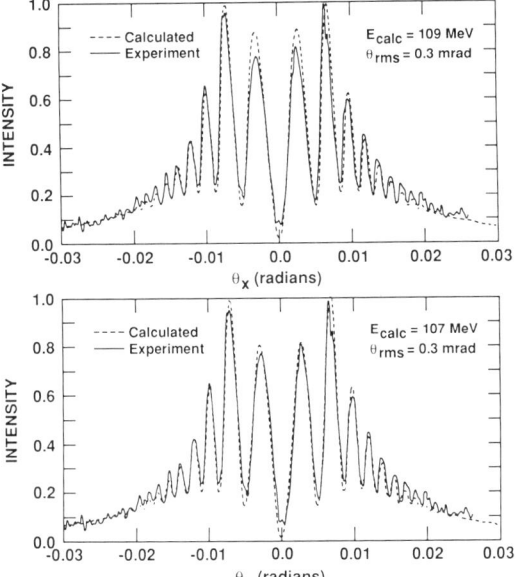

Fig. 18. Calculated interference patterns for 107 and 109 MeV compared to experiment show e-beam energy effect.

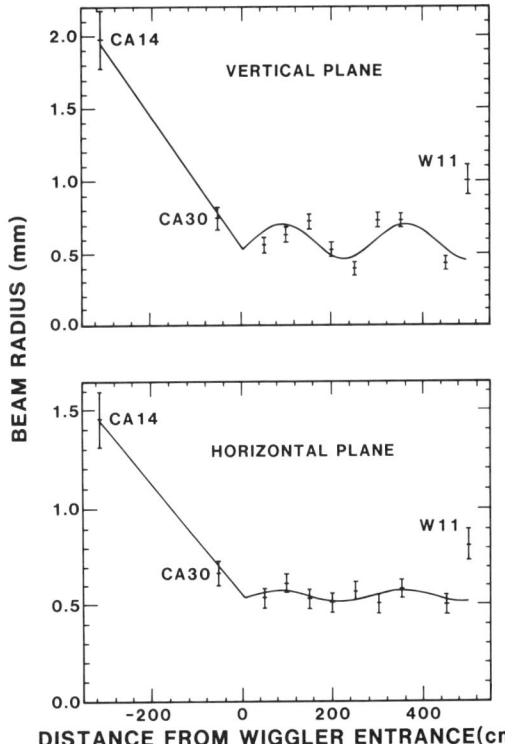

Fig. 19. Plot of spot sizes obtained before and within the 5-m wiggler: (a) vertical plane, and (b) horizontal plane. The resolution blur caused by the alumina screen W11 is noticeable compared to the OTR-screen spots.

alumina screens in the Boeing/STI 5-m wiggler were replaced with diamond-polished aluminum screens with a thin metal mask in front of them. In this situation, the visible SER had to be blocked from the path to the cameras to avoid overwhelming the OTR visible radiation. The limiting resolution (~0.5 mm) is indicated by the observed radius from the alumina screen (W11) being much larger than the OTR-measured spots. Improved pointing and matching in the wiggler led to an improved small signal gain for the FEL.

Before leaving this category, I note that at a SLAC workshop[20] I reported the possibility of 50 μrad type divergence measurements on those electron-beams at very high γ. Also, some profiling measurements of 450 GeV protons at CERN have been done with OTR techniques and the reader is referred to those articles.[21,22] In that work, thinner OTR foils (~3.5 μm) were possible then those used for secondary emission foil techniques (~100 μm thick) so beam scattering was reduced.

The OTR technique compares favorably with standard e-beam emittance diagnostics that rely on measurement sequences (quadrupole field variation) at one screen position or two. In closing this subsection, a list of some of the advantages and disadvantages is presented.

Several Advantages:

1. Measurements are possible on a single macropulse,
2. Data structure and theory allow on-line evaluation of emittance,
3. A single position in the beamline can be used for e-beam profile, divergence, and angle (pointing) measurements,
4. Thinner screens (foils) reduce beam scattering and x-ray production,
5. OTR provides a simultaneous e-beam energy diagnostic (~1% accuracy), and
6. Thinner screens and single shot mode should reduce vacuum degradation near the photocathode of injectors.

Disadvantages:

1. Source brightness,
2. Required careful optical alignments and parts, and
3. Convolution of divergence and energy effects.

Spontaneous Emission Radiation Results

These experimental results include the nonintercepting measurements of e-beam position, profile, charge, pulse length, and energy. They are intertwined with the streak/spectrometer, synchroscan, and dual sweep measurements.[3] As an example, Fig. 20 shows the integrated effects of beam centering change *during* the macropulse. In Fig. 21 beam bunching is measured/diagnosed before

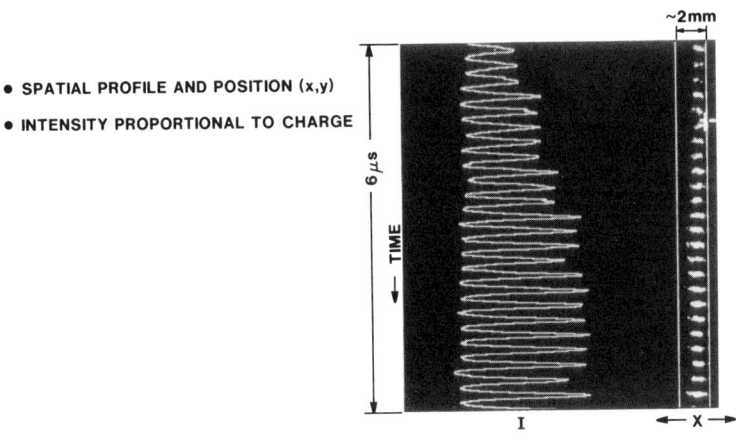

Fig. 20. A 6-μs sampling of the e-beam spatial position versus time using SER.

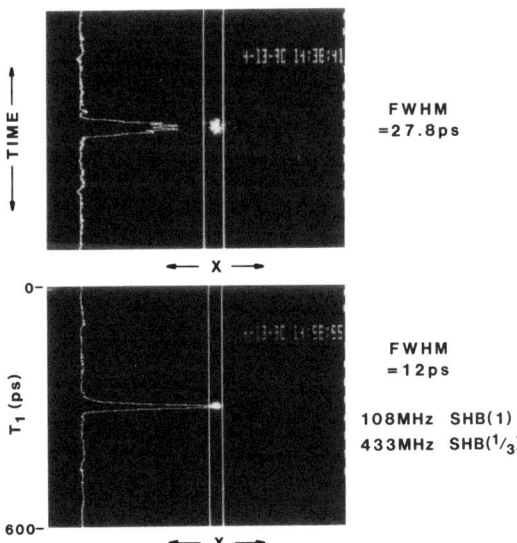

Fig. 21. Electron beam micropulse bunching measurement using visible SER in Boeing Facility before (upper) and after (lower) tuning the injector.

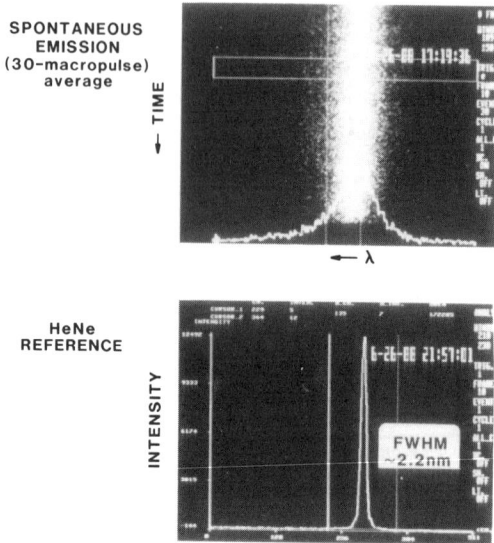

Fig. 22. A 30-macropulse average of the time-resolved SER spectrum (upper) and HeNe reference (lower). The e-beam energy and spread can be determined.

(upper) and after (lower) adjustment of the subharmonic buncher phases in the injector. The decrease in observed pulse length from 28 ps to 12 ps (FWHM) is important to FEL operations. The e-beam energy centroid, width, and slew can be deduced from the SER time-resolved spectra given in Figs. 22 and 23. The HeNe laser reference illustrates the instrumental limiting resolution of ~2.2 nm.

Streak Spectrometer Applications

A few additional demonstrations of streak/spectrometer results are shown by Fig. 24 and Fig. 25. The first one shows the first direct measurement of a wavelength shift within a FEL micropulse (~3 nm in 8 ps).[2] The second shows an example of the suppression of a long-wavelength sideband by a simple 8-10 μm cavity length detuning as monitored <u>during</u> the macropulse (~100 μs).[23] Such time-resolved information was critical to understanding the FEL evolution.

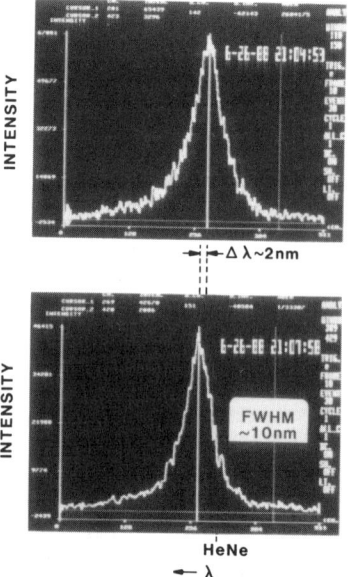

Fig. 23. Selected spectra from macropulse time intervals 30-μs apart in Fig. 22. A small centroid shift is attributed to e-beam energy slew.

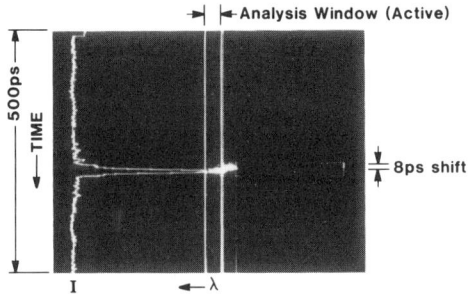

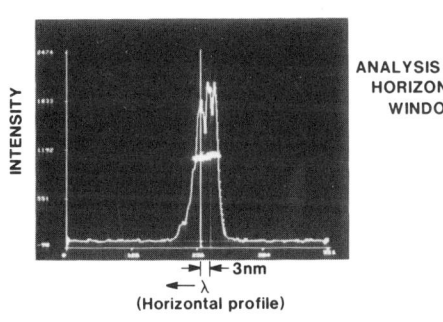

Fig. 24. Streak/spectrometer data for a single lasing micropulse (<10 ps) showing a wavelength difference in 8 ps.

Fig. 25. Time-resolved lasing spectra for a 63-μs span for cavity detuning of 0 and 8 μm.

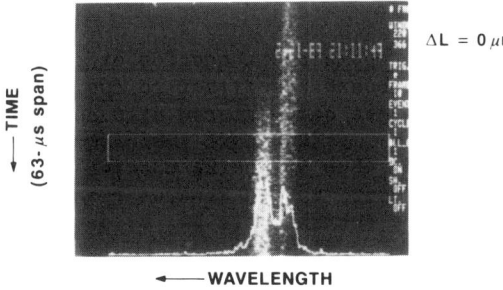

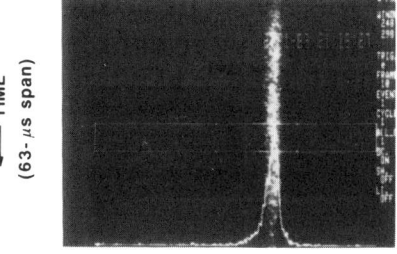

Synchroscan and Dual-Sweep Streak Techniques

It has been determined that low-jitter synchroscan streak measurements can be used to diagnose phase and pulse length <u>during</u> the macropulse (with dual-sweep). A number of examples are briefly cited.

Our ability to evaluate some key dynamic issues in the HIBAF photoelectric injector was greatly enhanced by using the synchroscan feature.[4] Figure 26 shows a schematic of the PEI accelerator with photocathode inserted into the wall of the first cell. Upon irradiation by the pulsed drive laser operating at 527 nm and 10 to 15 ps pulse length, the electrons are released and immediately accelerated by a 26 MV/m gradient to relativistic velocities. At a downstream screen, the electron beam micropulse information is converted via Cherenkov radiation and detected in the streak camera. Using path-length matching, a fraction of the drive laser light is also relayed to the same streak camera and streak sweep. Figure 27 then shows the simultaneous measurement of these two pulses. By adjusting the phase of the drive laser relative to the RF cycle, we mapped out the effects on e-beam elongation and transit time as shown in Figs. 28 and 29. These were the first successful measurements of their kind on a PEI and contributed to the qualification of the space-charge calculations in our simulations program (basically PARMELA).

Another aspect involved the ability to track shot-to-shot phase stability relative to the 108.3 MHz.[5] Figure 30 shows a series of 16 macropulse averaged temporal profiles indicating quite noticeable jitter is occurring (~1.6 ps/ch). In response to suggestions that this was streak camera jitter, we acquired a coordinated set of temporal positions versus e-beam energy centroid as detected in the electron spectrometer. As shown in Fig. 31, a strong correlation was observed with a sensitivity of

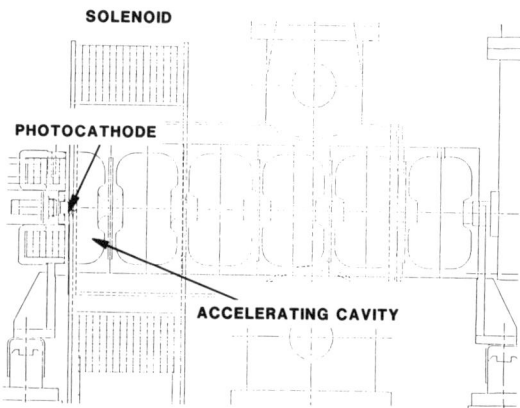

Fig. 26. Schematic of photoelectric injector accelerator.

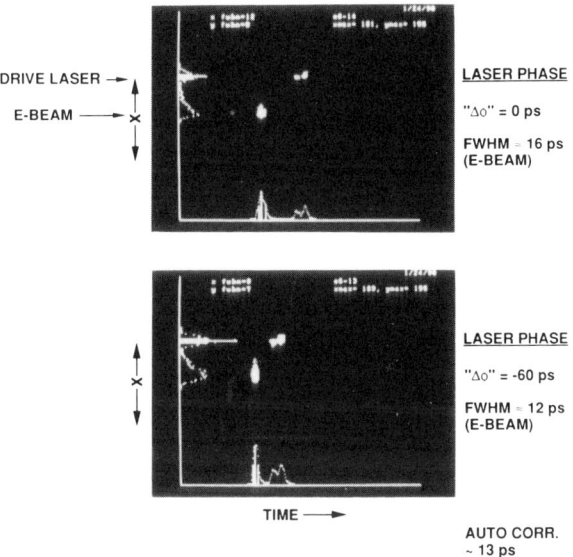

Fig. 27. "Simultaneous" synchroscan streak images of 17-MeV e-beam (Cherenkov converter) and drive laser micropulses.

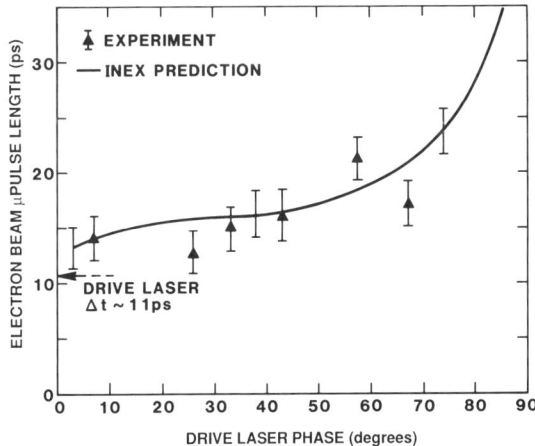

Fig. 28. Electron-beam micropulse length variation with drive laser phase changes.

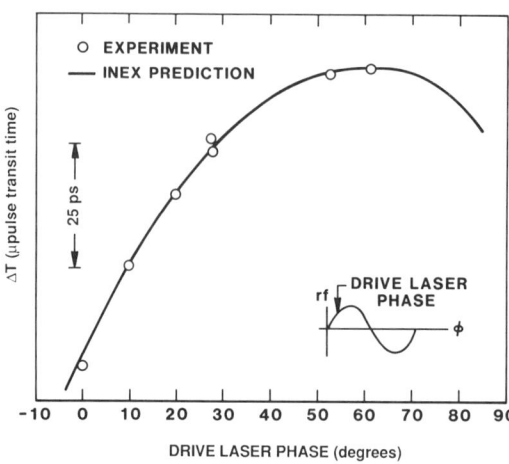

Fig. 29. Electron beam transit time effects versus drive laser phase in the photoinjector.

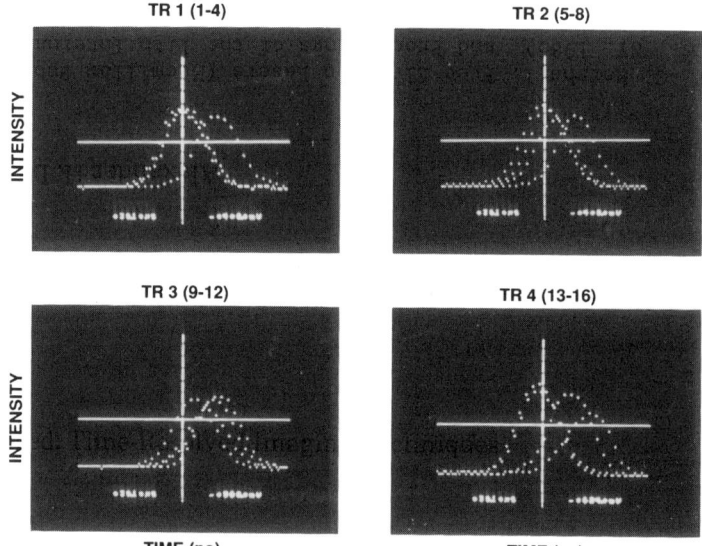

Fig. 30. Synchroscan streak profiles tracking drive laser phase jitter.

about 0.1% energy shift per picosecond of phase. The extreme cases of phase shifts greater than 5 ps were difficult to track due to the span of energy covered and the loss of image brightness. Significant improvement in the drive laser phase stability was demonstrated with the addition of a commercially available phase stabilizer to the ND:YLF oscillator system. Figure 32 shows the reduction of the rms jitter from 10.9 ps to

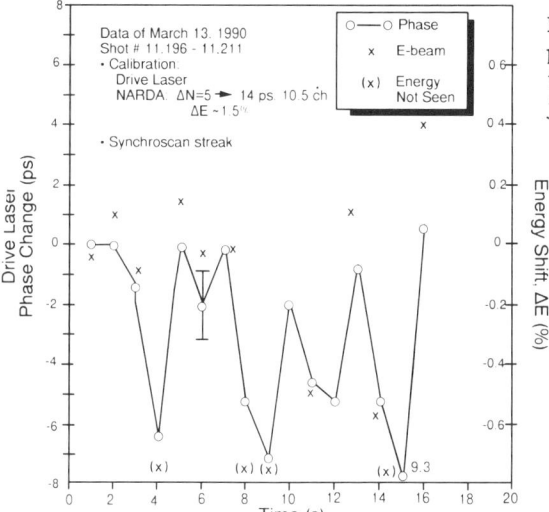

Fig. 31. Correlation of phase jitter of Fig. 30 to electron beam energy jitter at spectrometer.

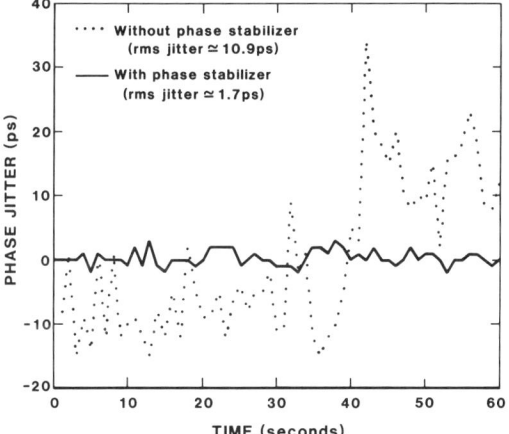

Fig. 32. Synchroscan streak measurement of drive laser intermacropulse jitter with and without phase stabilization.

1.7 ps (about the camera limit). These data further exonerated the streak camera jitter specification of 4 ps (FWHM).

Another unexpected use of the synchronous summing of micropulses involved our ability to provide temporal information in the 10-ps domain about field-emission electrons. The picocoulomb magnitude charge generated at the peaks of the RF cycle were synchronously summed into usable images as shown in Fig. 33. Transport conditions could result in only some of the electrons being sampled, and these had a surprising 20-ps FWHM structure.

Finally, referring back to the SER discussion, the final synchronous sum of 12 ps (FWHM) still includes intramacropulse jitter and/or slew. The nature of these were assessed by turning on the dual sweep streak feature.[5] As shown in Fig. 34, most of the macropulse shows the effect of the buncher phase adjustment. These results are tabulated in Table 3 and indicate that in the tuned case, the synchronous sum of 10 ps may include jitter and/or slew that expands the individual 8-ps width out to 10 ps. Additionally, in Fig. 35, the visible lasing macropulse has a

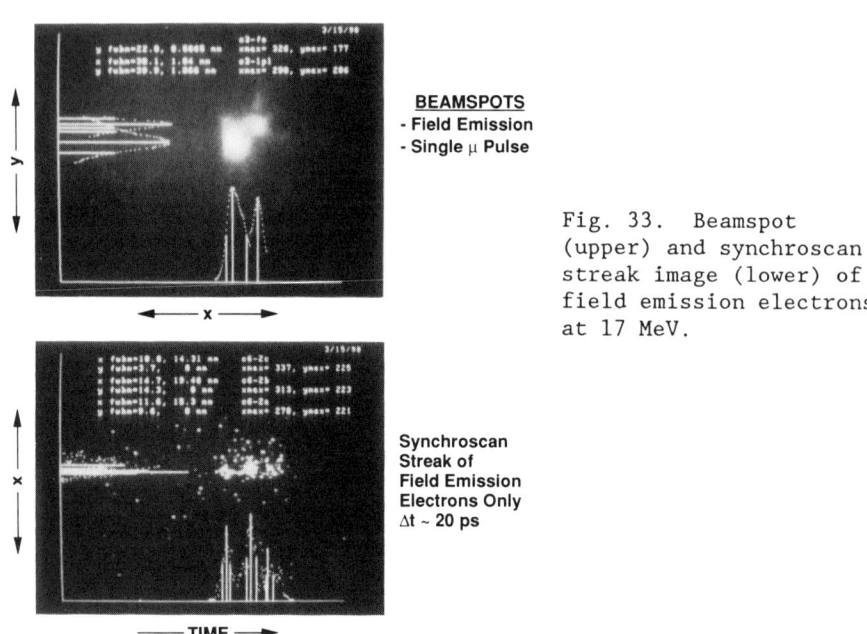

Fig. 33. Beamspot (upper) and synchroscan streak image (lower) of field emission electrons at 17 MeV.

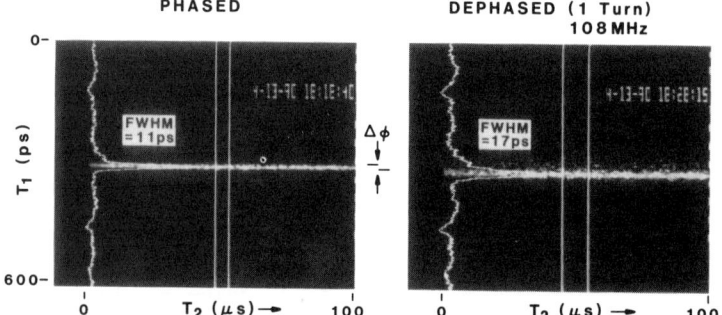

Fig. 34. Initial dual-sweep streak images of 109-MeV e-beam using SER with injector phased and dephased.

TABLE III. SUMMARY OF SYNCHROSCAN AND DUAL SWEEP
STREAK MEASUREMENTS OF e-BEAM AT BOEING
(Data of 4-13-90)

Measurement Technique	Accelerator Condition	Pulse Length* Observed (ps)	Pulse Length* e-Beam (FWHM, ps)
Synchroscan	100 µs, as found	28	27
Synchroscan	100 µs, 108 MHz tune	17.5	16
Synchroscan	100 µs, 433 MHz tune	12.3	10
Dual Sweep	Phased, 100-µs span, ~10 µs sample	10.6	8
Dual Sweep	Dephased, 108 MHz ~10 µs sample	17.2	16

*$\Delta t_{obs} = \sqrt{(\Delta t_{eB})^2 + (\Delta t_{Res})^2 + (\Delta t_{jitter})^2}$

where:

Δt_{eB} includes intrinsic pulse length for a single micropulse plus jitter and slew for a macropulse
Δt_{Res} = 6 ps of synchroscan sweep
Δt_{jitter} = 4 ps (FWHM) over seconds specification

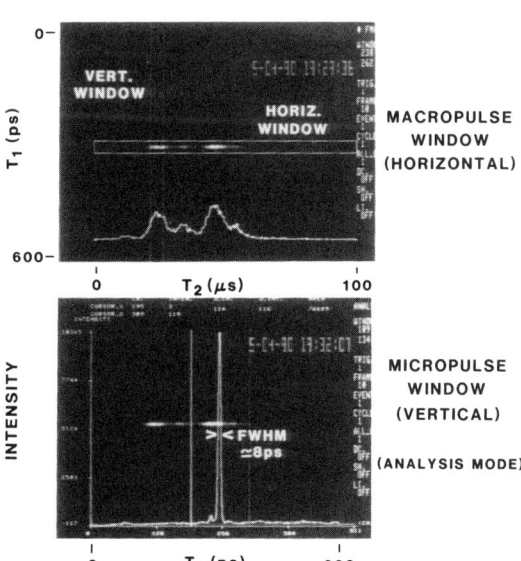

Fig. 35. Initial dual-sweep streak image of the visible FEL output from the ring resonator showing macropulse temporal structure (upper) and micropulse bunch length measurement (lower).

modulation of ~13-μs period during the macropulse, while simultaneously we measured the 8-ps lasing pulse length.

IV. SUMMARY

In summary, a number of graphic demonstrations of these advanced imaging applications have been cited, albeit somewhat briefly. Of particular note in our investigations have been the use of OTR and SER conversion procedures to reveal e-beam (or charged particle beams) parameter information. The radiations from these conversion processes then have been imaged in time-resolved modes on a wide range of timescales that are generic to RF-linac driven electron beams (and FELs). We have particularly found useful the low time jitter of synchroscan streak cameras and the extension to multiple time domains with dual-sweep techniques. These techniques have been presented to the Workshop participants in hopes of stimulating their adaptation to other accelerator-based systems for charged-particle beams.

ACKNOWLEDGMENTS

The author acknowledges the assistance of N. S. P. King and M. D. Wilke of Los Alamos and S. P. Wei of Boeing for integrating the streak/spectrometer at Boeing, the collaboration with D. W. Rule and R. B. Fiorito on OTR experiments and analyses, and discussions with J. C. Goldstein of Los Alamos on spontaneous emission radiation phenomena.

REFERENCES

1. A. H. Lumpkin and D. W. Feldman, Nucl. Inst. and Meth. A259, 49 (1987).
2. A. H. Lumpkin, N. S. P. King, M. D. Wilke, S. P. Wei, and K. J. Davis, Nucl. Inst. and Meth. A285, 17 (1989).
3. Alex H. Lumpkin, Nucl. Inst. and Meth. A296, 134 (1990).
4. Alex H. Lumpkin, Bruce E. Carlsten, and Renee B. Feldman, "First Measurements of Electron-Beam Transit Times and Micropulse Elongation in a Photoelectric Injector," presented at the 12th International FEL Conference, Paris, France, Sept. 17-21, 1990.
5. Alex H. Lumpkin, "The Next Generation of RF-FEL Diagnostics: Synchroscan and Dual-Sweep Streak Technique," presented at the 12th International FEL Conference, Paris, France, Sept. 17-21, 1990.
6. A. H. Lumpkin, et al., "Initial Observations of High-Charge, Low-Emittance Electron Beams at HIBAF," presented at the 12th International FEL Conference, Paris, France, Sept. 17-21, 1990.

7. T. C. Marshall, Free-Electron Lasers (Macmillan Publishing Co., NY, 1985), and Proceedings of the 11th International Free-Electron Laser Conference, Nucl. Inst. and Meth., A296, (1990), and references therein.
8. V. L. Ginzburg and I. M. Frank, Sov. Phys. JETP 16, 15 (1946).
9. L. Wartski, et al., J. Appl. Phys. 46 3644 (1975).
10. R. B. Fiorito, et al., Proc. 6th Int. Conf. on High-Power Beams, Kobe, Japan (1986).
11. D. W. Rule, Nucl. Inst. and Meth., B24/25 901 (1987).
12. A. H. Lumpkin, et al., Nucl. Inst. and Meth. A285, 343 (1989).
13. A. H. Lumpkin, et al., Nucl. Inst. and Meth. A296, 150 (1990).
14. D. W. Rule and R. B. Fiorito, "Imaging Micron-Sized Beams with Optical Transition Radiation," Poster Paper this Workshop.
15. D. W. Rule, R. B. Fiorito, A. H. Lumpkin, R. B. Feldman, and B. E. Carlsten, Nucl. Inst. and Meth. in Physics Research, A296, 739 (1990).
16. J. V. Jelley, Cherenkov Radiation and Its Applications, Pergamon Press Ltd., London (1958).
17. R. J. Nawrocky and J. Rogers, "Beam Profile Measurements Using Synchrotron Light," These Proceedings.
18. K. E. Robinson, et al., Nucl. Inst. and Meth. A259, 62 (1987), and IEEE J. Quantum Electronics, QE23, 1497 (1987),.
19. Y. Tsuchiya, et al., "Two-Dimensional Sweeps Expanding Capability and Application of Streak Cameras," SPIE, Vol. 693, 125 (1986).
20. Alex H. Lumpkin, "Imaging Techniques for Charged-Particle Beam Diagnostics," International Workshop on Next Generation of Linear Colliders, Stanford Linear Accelerator Center, Palo Alto, CA, Dec. 6, 1988.
21. J. Bosser, J. Dieperink, G. Ferioli, J. Mann, L. Wartski, IEEE Trans. on Nucl. Science, Vol. NS-32, No. 5, 1905, October 1985.
22. J. Bosser, J. Mann, G. Ferioli, and L. Wartski, CERN/SPS/84-17 (AB).
23. A. H. Lumpkin, et al., Nucl. Inst. and Meth. in Physics Research A296, 169, (1990).

Fiber Optic Links for Instrumentation

Ralph J. Pasquinelli
Fermi National Accelerator Laboratory*, Batavia, Illinois 60510

INTRODUCTION

There has been a great deal of progress in the use of fiber optic systems over the past ten years. Granted they are not always justified, but in many cases they are the only method of providing low power information transmission in complex systems. First I will cover the questions one might ask in selecting a transmission system followed by the available hardware on the market. A short discussion of optical techniques and actual hardware systems will finish the report.

SELECTION OF A TRANSMISSION SYSTEM

In many projects the selection of a transmission system is not given a great deal of attention until the end. This can of course be a fatal mistake if the transmission media is incapable of handling the data load. For the purposes of instrumentation, we will limit the choices to transmission lines and fiber optical systems. Certainly free space transmission is another choice for the communications industry but rarely presents an alternative to the instrumentation community. The following is a list of questions that the engineer might ask at the beginning of a project followed by the pros and cons of either transmission lines or fiber optics as a solution.

1: *What is the bandwidth requirement of the system?*

Transmission lines: Traditionally transmission lines such as coax provide very high bandwidth for short haul transmission lengths. Typical bandwidth for a low loss foam dielectric 1/2" coaxial cable is DC to 8 GHz. As the length of the transmission line gets long, say 100 feet or so, cable dispersion becomes a problem and limits the useful bandwidth. (see Figure 1) At higher frequencies it is possible to launch wave guide modes in a coax which travel at a velocity that is different than the TEM mode. For instance a 1- 5/8" foam dielectric coax has a maximum usable bandwidth of approximately DC to 2 GHz due to the larger diameter.

---

* Operated by Universities Research Association Inc. under contract with the United States Department of Energy

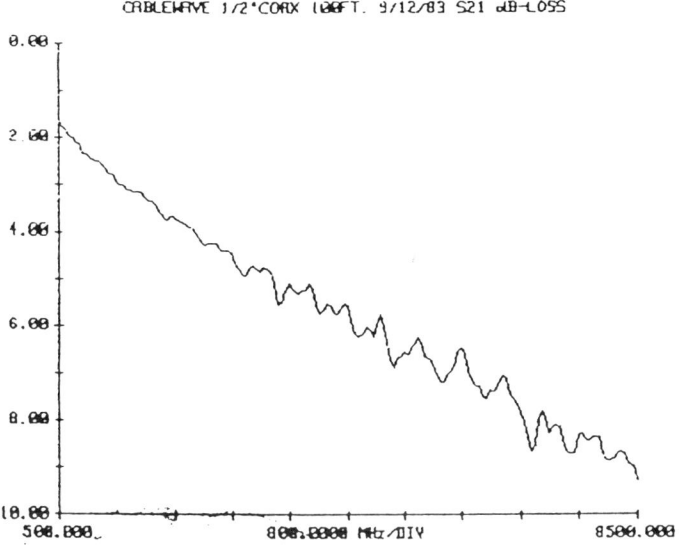

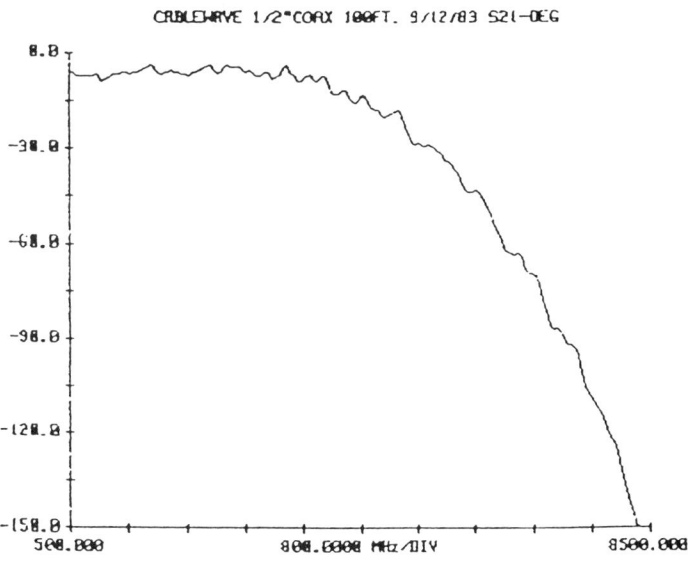

Figure 1. Amplitude and Phase response for 100 feet of 1/2 inch foam coax.

Fiber Optics: With a fiber optic system, the information is translated in the frequency domain from baseband to light! Bandwidths of 12 GHz are presently possible and limited by the transmitter not the fiber or the receiver. This wide band signal is now centered at light frequencies and is considered to be "narrow band". It is important to note that in most optical links DC coupling is eliminated.

2: *What losses are tolerable in the system?*

Transmission lines: Losses can be quite low for very short runs of a few feet on the order of a couple dB into the GHz range. But as length of the line increases so does the loss. Referring to Figure 1 shows 2.5 dB of attenuation per 100 feet at 1 GHz for low loss 1/2" foam coax cable. As frequency increases so do the losses. A 500 foot length of this coax at 8 GHz would provide approximately 40 dB of insertion loss. It is important to note that the insertion loss slope increases with length for wider bandwidths. This might require some form of amplitude equalization. Lower loss is achievable by going to larger diameter cables but at the expense of bandwidth. Insertion loss for a 1-5/8" foam coax is 0.8 dB per 100 feet at 1 GHz.

Fiber Optics: Insertion loss in fiber optic systems is dominated by the conversion losses of the transmitter and receiver. The fiber itself provides only a small loss. Present state of the art fibers typically have an optical insertion loss as low as 0.19 dB per kilometer! Typical transmitter/receiver insertion losses for wide band (12 GHz) optical systems is approximately 40 dB. Substituting a fiber optic link for the coax in the example above will provide very similar performance at 8 GHz with two very important differences. The fiber can be many kilometers long and will add only a few dB insertion loss to the link. In addition the fiber system provides a flat frequency response. (See figure 2) A one kilometer length of 1/2" coax will have an insertion loss of 274 dB at 8 GHz compared to 40 dB for the optic link!

3: *What degree of linearity is required of the transmission system?*

Transmission Lines: Short runs of transmission lines provide good linear performance. As Bandwidth increases, linearity suffers on long runs due to dispersion and skin effect gain slope. Examination of figure 1 shows a 6 dB gain slope and 150 degree phase change for an 8 GHz wide signal on the 100 foot length of 1/2" coax.

Fiber Optics: Linearity of the system is very good over the entire bandwidth of the system. Optical transmitters such as laser diodes have high Third Order Intercepts on the order of +20 to +30 dBm. Amplitude flatness can be maintained to better than +- 2 dB and phase linearity of +-15 degrees.

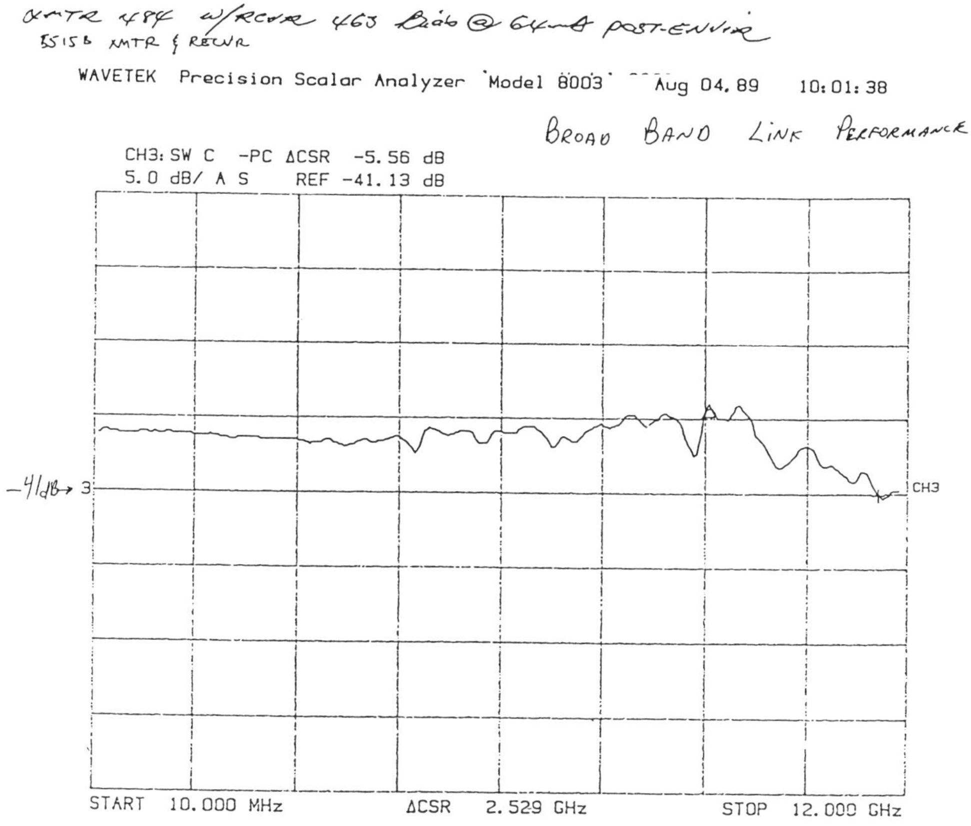

XMTR 484 w/RCVR 463 Bias @ 64mA post-envir
5515B MTR & RCVR

WAVETEK Precision Scalar Analyzer Model 8003   Aug 04, 89   10:01:38

BROAD BAND LINK PERFORMANCE

CH3: SW C   -PC ΔCSR   -5.56 dB
5.0 dB/ A S   REF -41.13 dB

-41dB→ 3   CH3

START 10.000 MHz   ΔCSR 2.529 GHz   STOP 12.000 GHz

Figure 2. Frequency response of Ortel 5515B Optic Link

4: *What are the dynamic range requirements?*

Transmission Lines: Transmission lines provide excellent dynamic range over short hauls. As length and insertion loss increase, a reduction of Signal to Noise ratio results. This corresponds directly to dynamic range. Obviously in the example above of 1 Km at 8 GHz the system has no dynamic range in that the signal at the output end of the cable is buried well beneath the thermal noise floor.

Fiber Optics: Dynamic range in an optical system is governed by the noise characteristics of the transducers. Typical laser system dynamic range is on the order of 130-140 dB per Hertz of Bandwidth so that a 1 GHz wide system may have a dynamic range of 40-50 dB.

5: *Is High Voltage isolation or EMI a concern?*

Transmission Lines: Radiation from coax cable ranges from -60 dB for braided shields to -90 dB for solid copper shields. Coaxes can be used for the transmission of High Voltage but provide no High Voltage isolation.

Fiber Optics: Fiber optic links are considered to be impervious to ElectroMagnetic radiation Interference. Because glass is such a good insulator, optic systems can be used for data transmission between points with a very high potential difference such as Pre-accelerators.

6: *Is propagation velocity important in the system? Is variation in propagation delay vs. temperature important?*

Transmission Lines: There is a wide range of velocity of propagation in coaxial cable. The velocity is proportional to the reciprocal of the root of the dielectric material used to support the center conductor. This ranges from 67% c (the velocity of light) for solid teflon and polyethylene to 99% c for air dielectric. Higher dielectric constants can make "slower" cables. Typical temperature coefficients range from +10 ppm to -10 ppm per degree F depending on the ambient temperature of operation.

Fiber Optics: The refractive index of an optical fiber is defined as the ratio of the speed of light in vacuum to the speed of light in the fiber. The velocity of propagation in fiber is given by the reciprocal of the index of refraction which is typically 67% c for silica glass single mode fiber. Tests at Fermilab have shown the temperature coefficient to be +3 ppm per degree F. To obtain all the benefits of optics, one is stuck with this slower velocity which may preclude its use in some systems.

7. *What is the cost / performance ratio?*

Transmission Lines: Over short hauls transmission lines still offer the best cost / performance ratio. Typical coax price ranges from pennies per foot for such standards as RG-58 to $10 plus per foot for large diameter coax (1-5/8" and above). Transmission lines are passive devices that require little or no maintenance and operating costs.

Fiber Optics: Optical systems become practical when bandwidth is wide, long transmission lengths are required (greater than several hundred feet), flat insertion loss and linear phase, and good dynamic range are requirements. Costs for low frequency optical links using LED's start in the hundreds of dollars. State of the art wide band LASER links cost approximately twenty thousand dollars. Optical links are active therefore increasing operational costs.

OPTICAL LINK HARDWARE CONSIDERATIONS

At this point in the design all of the above questions have been carefully answered and it has been decided to look into an optical link to suit your needs. For the most part, almost all of the available optical hardware uses amplitude modulation (AM). It is possible to make use of phase modulation but this makes the receiver more complicated in that it must have a coherent source to detect the information. Only AM links will be considered.

For a relatively low bandwidth system (less than 100 MHz) it may be possible to use a LED (Light Emitting Diode) as the transmitter. The optical spectral line width of a good LED transmitter is 40-80 nanometers. This "lack" of spectral purity is what band limits the LED. (Figure 3) The fact that LED light is not monochromatic leads to chromatic dispersion on the transmission fiber. Chromatic dispersion is the change in group travel time per unit length of fiber per change of wavelength and is given in units of picoseconds per kilometer nanometer (ps / km*nm). The end effect is a pulse widening of the transmitted signal. This type of dispersion limits bandwidth to length performance of the transmission link. A LED is more cost effective than a LASER diode and should be considered for narrower bandwidths and shorter transmission distances. A wide variety of hardware is available for both Analog and Digital systems[1].

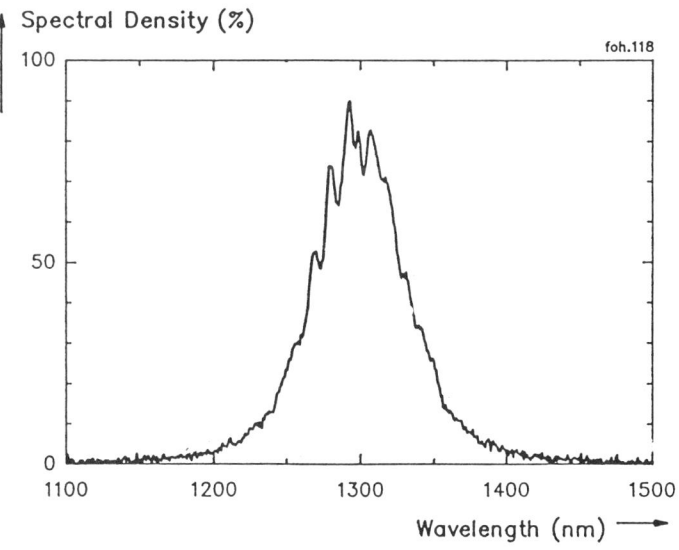

MEASURED SPECTRUM OF AN EDGE-EMITTING InGaAsP LED
(measured at the end of a graded-index fiber)

Figure 3.

For wide bandwidth systems a LASER diode transmitter is required. The optical spectral line width of a LASER diode is 5-10 nanometers. (figure 4) Modulation rates of 12 GHz [2] are available with state of the art systems approaching 20 GHz. The light output of LASER's is very temperature dependent. For this reason many of the higher quality LASER transmitters are equipped with a thermoelectric (Peltier) coolers to stabilize the temperature. Having a constant light output is important in maintaining the dynamic range of the transmitter.

Choice of fiber is also very important. Two main types of materials are used in fibers, silica glass and plastic. Plastic fibers have very high losses and are specified in dB per meter. Glass fibers have loss specified in dB per kilometer. Plastic fibers are used mainly for runs of only a few feet such as in the internal control of a High Voltage power supply. The high losses associated with plastic fiber do not make it a viable choice in most instrumentation systems. For glass fiber there are several choices available that depend on core size and construction. The core is the actual "light pipe" of the fiber. Insertion loss of glass fibers is a function of the optical wavelength with the optimum being in the infrared red range at 1550 nanometers . Figure 5 shows the cross section and refractive index for several available types of glass fibers.

Less demanding applications can take advantage of multimode fiber. These fibers can be broken up into two groups, Step index and Graded index. Multimode fibers are easier to work with due to the larger core diameters but are limited in bandwidth and have higher insertion losses. Core diameter for Step index fibers are approximately 100 microns. Attenuation is in the range of 5-12 dB per kilometer with a modulation bandwidth of 20 MHz per kilometer. Graded index fibers have core diameters on the order of 50 to 60 microns. Attenuation is 1 to 3 dB per kilometer with modulation bandwidths of 400 MHz to 1 GHz per kilometer Both types of multimode fiber are usually used for shorter transmission lengths.

When high bandwidth is a requirement then the choice is single mode fiber. These fibers typically have core diameters of 9-10 microns. This small size makes connecting and fusing of the fiber a precision job. Single mode fiber has the lowest attenuation of the glass fibers and is dependent on optical wavelength. Modulation capacity of single mode fiber is approximately 100 GHz per kilometer. Figure 6 is a graph of attenuation vs. optical wavelength. It is also important to note that the small core size makes coupling light into the fiber more difficult.

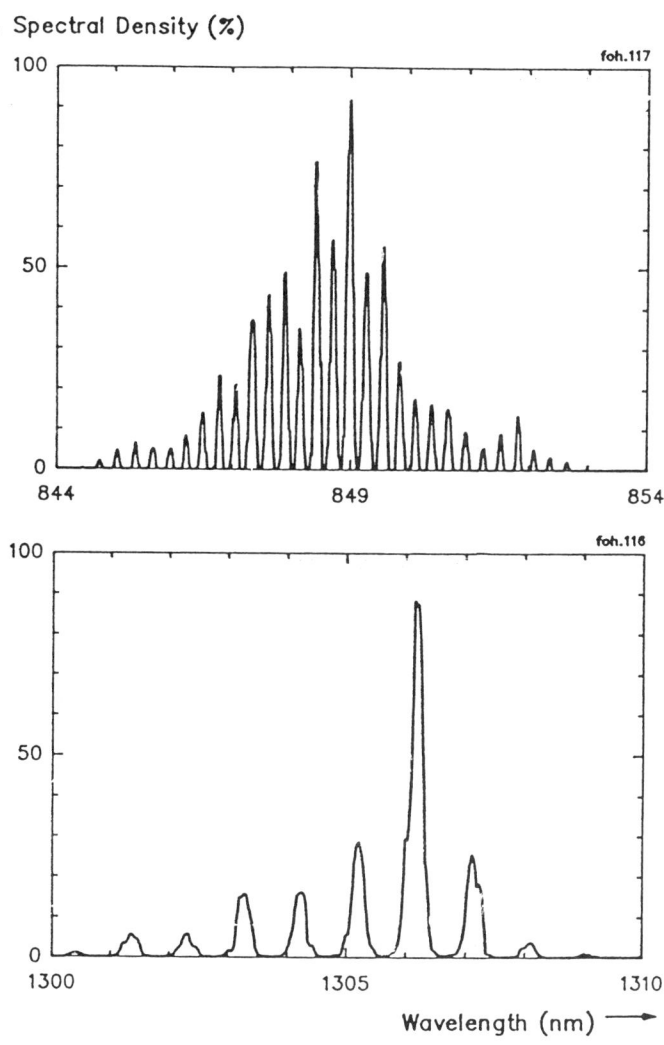

MEASURED LASER SPECTRA
850 nm GaAlAs measured at the end of a graded-index fiber, and
1300 nm InGaAsP laser measured at the end of a single-mode fiber

Figure 4.

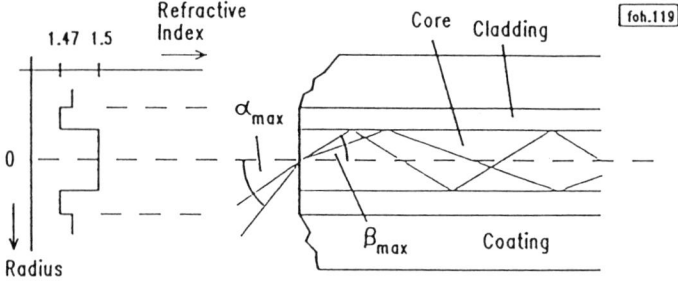

STEP-INDEX MULTIMODE FIBER

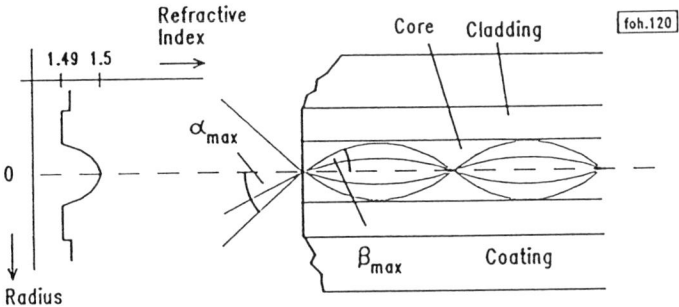

GRADED-INDEX FIBER

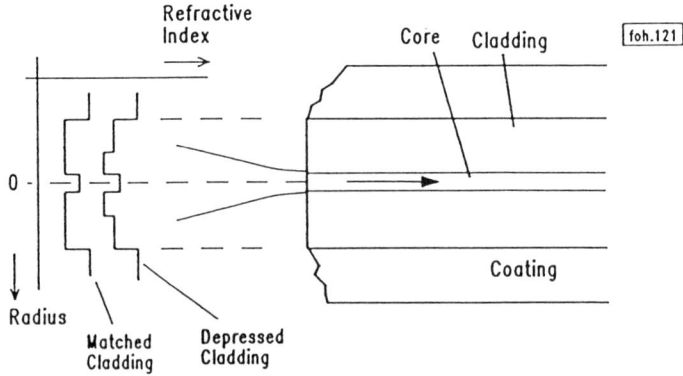

SINGLE-MODE FIBER

Figure 5. Cross section and index of refraction for different types of optical fiber.

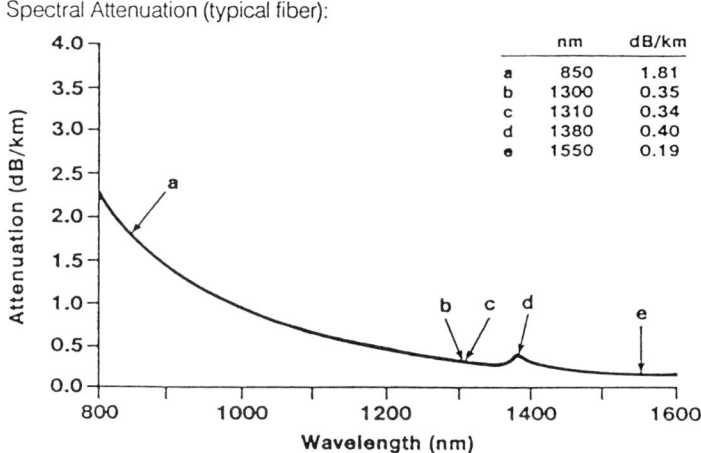

Figure 6. Attenuation vs. Wavelength for single mode fiber.

The receiver function is implemented by a photo diode. PIN photodiodes are widely used due to their exceptional stability and bandwidth. Photo diodes exist with bandwidths of greater than 12 GHz[3]. The material with which a photodiode is made will govern its operating optical wavelength. Figure 7 shows the responsivity of typical photodiodes.

There is a wide range of connectors for use in optical links. No standardization exists as there are many types of fibers and applications. Certainly the multimode fibers lend themselves to mechanical connectors more easily than single mode fiber. The use of a connector can be very beneficial in the field. It is more difficult to obtain repeatable connections (with a connector) on single mode fibers due to the higher mechanical tolerances. For the best system performance it is recommended to use a fusion splicer for making the connection. This of course requires an expensive apparatus to do the fusion as well as a skilled operator.

Fiber Optic Links for Instrumentation

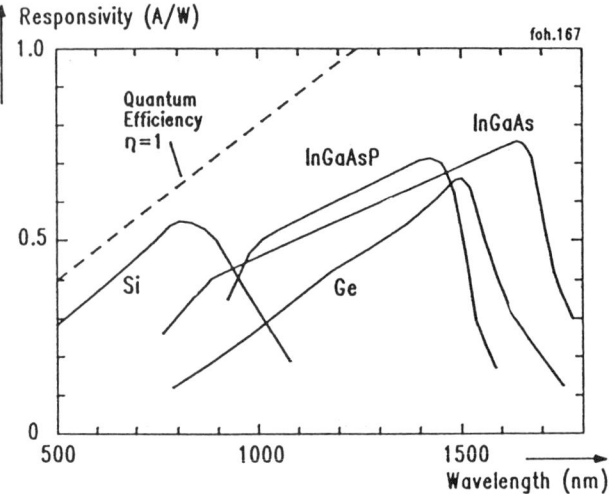

RESPONSIVITY OF TYPICAL PHOTODIODES

Material	Bandgap	Cutoff-Wavelength	
Si	1.11 eV	1.12 μm	Silicon
InGaAsP	0.89 eV	1.4 μm	Indium-Gallium-Arsenide-Phosphide
InGaAs	0.77 eV	1.6 μm	Indium-Gallium-Arsenide
Ge	0.67 eV	1.4 μm	Germanium

TYPICAL PHOTODIODE MATERIALS

Figure 7.

SPECIFIC EXAMPLE OF AN OPTICAL LINK

From this point on I would like to concentrate on the long haul wide band system requirement as it provides the greatest engineering challenge. Such a requirement exists in the implementation of a Bunched Beam Stochastic Cooling system for the Fermilab Tevatron.[4] Stochastic cooling [5,6] is a feedback process which operates on the schott noise of the circulating particles. It is important to apply the feed back with the correct phase and the exact delay of the time of flight of the particles around the ring. The end result is a smaller emittance and momentum spread in the beam. Figure 8 is a block diagram of the system. For further information on Stochastic Cooling see the references. Some of the main systems parameters are:

Bandwidth.. 4-8 GHz

System insertion loss....................... 50 dB max.

Linearity- amplitude........................ +- 1 dB

Linearity- phase.............................. +- 10 degrees

Dynamic range................................ 30 dB min.

EMI. rejection.................................. as high as possible

Propagation delay.......................... 21 microseconds

Propagation delay tolerance........ +- 5 picoseconds

The above conditions can be met with some effort using an optical link but would be very difficult, impractical and more expensive using standard transmission lines. The transmitter is a LASER diode, a single mode fiber for the transmission and a PIN photodiode receiver. This system has in excess of 10 GHz of bandwidth as shown in figure 9 and hence provides the desired performance. A 21 microsecond delay is the equivalent of 4.5 kilometers of glass fiber. The insertion loss of this fiber is approximately 1.8 dB @ 1310 nm and is independent of the microwave frequency information carried on it. The insertion loss of an equivalent length of 1/2" foam coax (5.7 kilometers, largest usable diameter with the required bandwidth) would be 965 dB at 4 GHz and 1535 dB at 8 GHz. Obviously these are ridiculous insertion losses for a continuous link. Repeater stations could be placed periodically to compensate for losses but there would still be the problem of high gain slope and the added system noise from multiple amplifiers.

Some of the system difficulties are stabilization of the group delay to less than a part per million as specified. The +- 5 picosecond requirement is equivalent to +- 15 electrical degrees at 8 GHz. The feedback performance will suffer less than 5 % with this tolerance. In addition this, group delay must track the delay of the particles' time of flight around the accelerator. With a temperature coefficient of the glass at 3 ppm per degree F, every one degree temperature change on this length of fiber would produce a group delay change of 63 picoseconds or approximately 180 electrical degrees at 8 GHz. If the spool of fiber is placed in a oven, the temperature can be maintained to a fraction of a degree. To provide this delay tolerance by means of temperature control alone would require temperature regulation better than 0.05 degree F. This is not an easy task. The distance between the pickup and kicker electrodes in the system is about 300 feet and it is desirable to use the fiber as the connection hence it cannot all fit into the oven.

192 Fiber Optic Links for Instrumentation

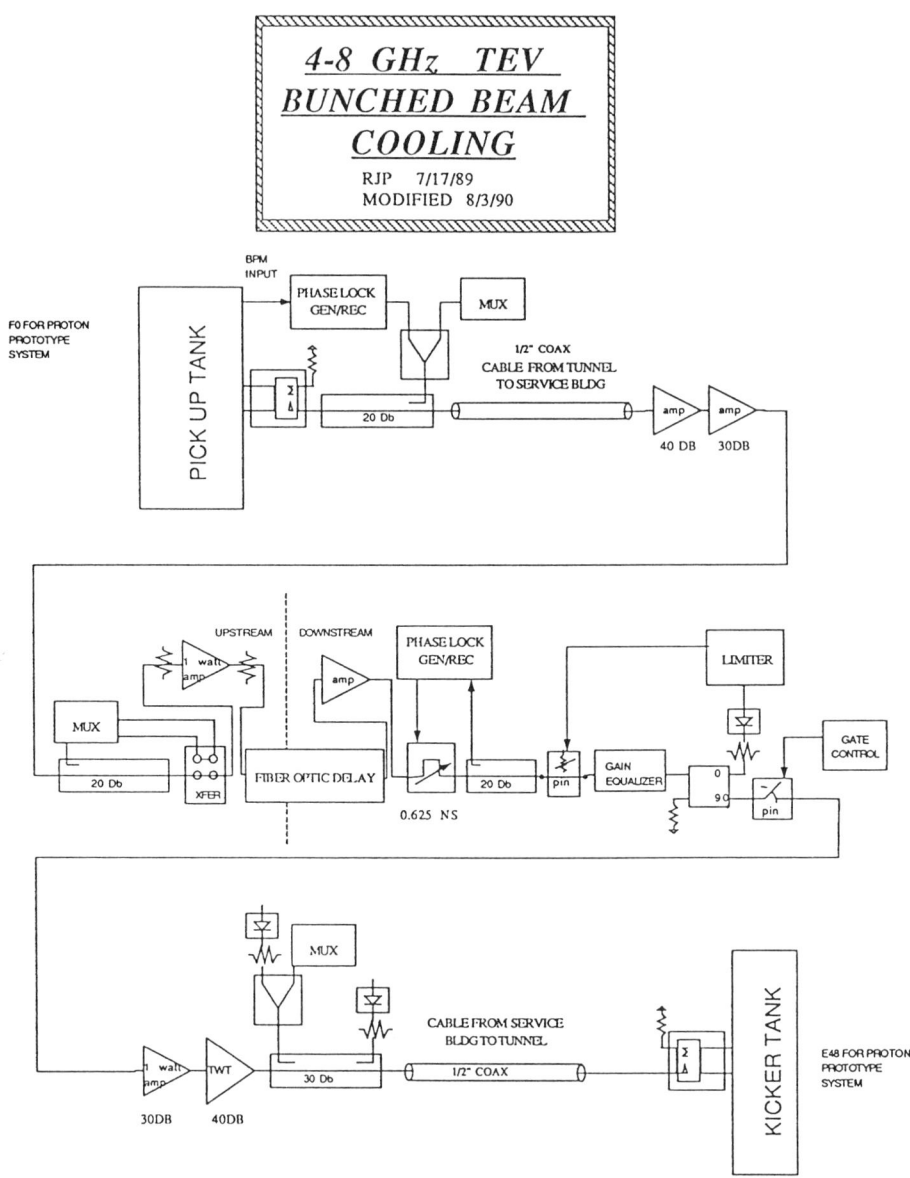

Figure 8. Block Diagram of Bunched Beam Stochastic Cooling system.

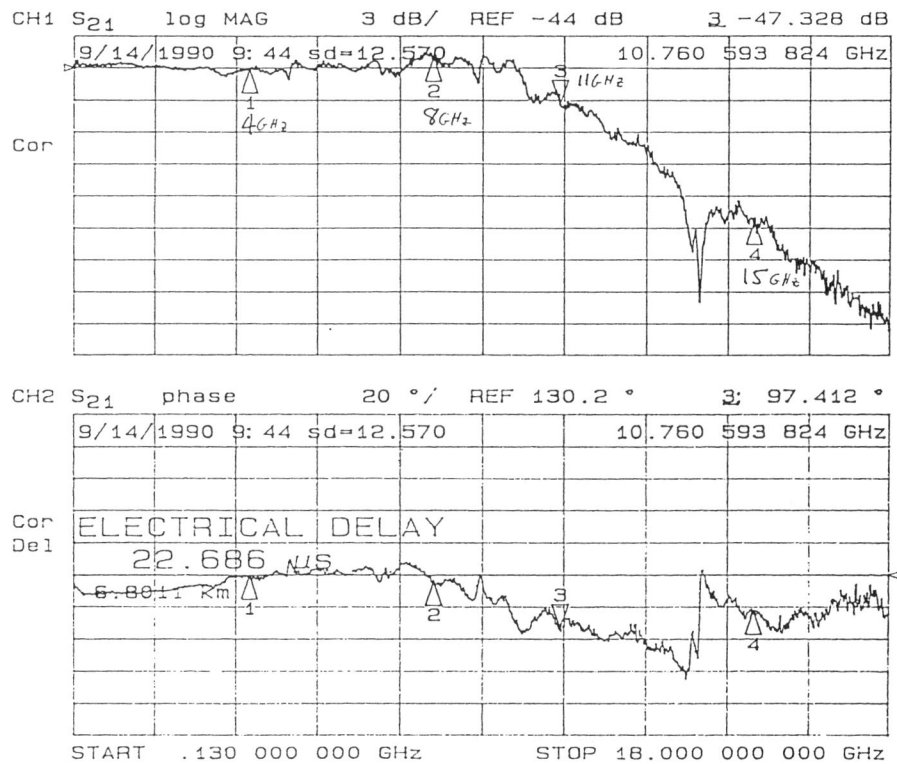

Figure 9. Amplitude and Phase response of 22 microsecond optical delay line.

The solution is to use a combination of temperature control and a servo feedback system to control group delay. A precise microwave signal at 3 GHz is generated from the coherent signal obtained from a beam position pickup. This pilot signal is then transmitted along with the 4-8 GHz information over the optical link to a coherent detector sensitive to the 3 GHz at the receiver end. The error signal developed is then used to move a mechanical line stretcher in series with the fiber optic. If the beam energy changes slightly so will the revolution frequency of the particles. The beauty of this system is that because the pilot tone is developed from the beam itself, the system tracks the change in revolution delay automatically.

The bunched beam cooling system described is presently being installed in the Tevatron. Similar delay systems have been successfully implemented in the Fermilab Antiproton source and have been in operation for several years.[7]

CONCLUSIONS

It has been shown that Fiber Optic systems can be a viable choice for information transfer in instrumentation systems. In some cases, such as the example presented, it is not only viable but also gives the best performance and minimizes cost. In communications, optics are rapidly replacing long haul transmission lines. Their performance and cost savings is superior to free space or transmission line links. Optical techniques are now becoming more attractive solutions for problems in particle accelerator environments.

REFERENCES

1. Hewlett Packard Fiber Optics and Optocouplers for Data Transmission. and PCO Inc. Short form Catalog.

2. Ortel Corporation, 5515 A/B Fiber Optic Link data sheet.

3. Ortel Corporation, 4515B Fiber Optic receiver data sheet.

4. Marriner, Jackson, McGinnis, Pasquinelli, Peterson, Petter. "Bunched Beam Cooling in the FNAL Antiproton Accumulator." European Particle Accelerator Conference, (June 1990).

5. van der Meer, et al, " Physics and Technique of Stochastic Cooling". CERN/PS/AA 79-23, (July 1979).

6. Sacherer, "Stochastic Cooling Theory" , CERN-ISR-TH/78-11, (May 1978).

7. Pasquinelli, Kells, Peterson, Marriner, "Optical Correlator Notch Filters for Fermilab Debuncher Betatron Stochastic Cooling", Proceeding of the IEEE Particle Accelerator Conference, Vol. I, (March 1989), pages 696-696.

8. Hewlett Packard, Fiber Optics Handbook, second edition, (January 1988).

9. Corning Glass Works, Corguide Single Mode Fiber data sheet, (April 1988).

Instrumentation Issues at SSC

Donald J. Martin*
SSC Laboratory, Dallas, Texas 75237

Abstract

An overview of the instrumentation task at the Supercollider complex is presented. The baseline instrumentation for the site can be grouped into a dozen or more major systems, totaling approximately fifteen thousand possible data channels. The budgeted total cost of this instrumentation is $40M[1]. The beam position systems contain the most important instruments for beam control. They are also the most costly. A discussion of Collider ring position monitors illustrates the range of issues faced by instrumentation designers.

Introduction

Beam instrumentation at the SSC complex will eventually include every device type used previously in all existing accelerators. The instruments may be simple, such as pint size paint cans filled with liquid scintillator, while some instruments will be esoteric. This discussion will focus on instruments for which basic engineering issues, rather than new measurement science, are concerned. Therefore, more recent techniques such as OTR, Feshbach resonance, or laser neutralization are bypassed.

More so than any existing machine, the SSC is really an engineer's accelerator. The statement can be justified in the sense that, because of the immense scale, basic engineering issues such as operational availability, reliability, manufacturing control, quality assurance, and cost become overriding concerns. These concerns are the purview of the engineer. The technical components must of course satisfy the physics requirements - they must do so day after day. Only thorough engineering design will result in systems requiring a minimum of tinkering.

Instrumentation Systems

Table 1 lists the planned beam instruments for the SSC complex, as specified in the Site-Specific Conceptual Design[2]. Position monitors, loss monitors, cross-calibration intensity monitors, and TV viewed flags must be available and reliable for early commissioning. Position detectors and loss monitors are numerous devices. Each position detector might measure in two planes, and some 3000 detectors are required in the complex. Some 6500 loss monitors are required by SSC, nearly 6000 of those in

---

*Operated by Universities Research Association under contract with
U.S. Department of Energy, Contract No. DE-AC02-89ER40486.

the Collider Top and Collider Bottom rings. Cross-calibration intensity monitors serve the function of measuring transfer efficiency between machines. Schottky monitors in two varieties provide low perturbation continuous tune measurements. The resonated strip electrodes of the low frequency detectors may be retracted, or inserted to improve sensitivity[3]. This device is a sensitive betatron oscillation and Schottky noise pickup, providing 25 dB gain over a non-tuned detector of similar geometry.

Table 1   SSC Accelerator Beam Instrumentation Systems

Instrument	LINAC	LEB	MEB	HEB	BOT	TOP
Beam Position Monitor (electrodes)	180	300	352	1160	3872	3872
Loss Monitor	30	150	176	290	2904	2904
Cross-Calibration Intensity Monitor		2	2	4	2	2
Schottky Monitor (Low frequency)			2	3	3	3
Schottky Monitor (Cavity)			2	2	2	2
Flying Wires (Accelerator)	10		2	3	6	6
Flying Wires (IR's)					12	12
Multiwire Profile	3	6	6	6	12	12
TV Viewed Flags	6	4	4	8	8	8
Precision Transductor		1	1	1	2	2
Fast Current Transformer	16	3	3	3	3	3
Wall Current Monitor		1	1	1	1	1
Synchrotron Radiation Profile Monitor					6	6
Pinger Magnets		4	4	4	4	4

The microwave cavity Schottky operates at a frequency above the coherent bunch spectrum, so narrow filters in the processing electronics are not required[4]. Flying wires and multiwire scanners produce beam profiles, which help to understand luminosity lifetime, scraping, and losses. Flying wires become especially useful when adjusting the beam crossing near the interaction regions[5]. Should early operations prove difficult, phosphorescent screens viewed by TV cameras serve as a reliable diagnostic.

Three varieties of beam current transformers have proved useful in earlier accelerators. The transductor, an active, non-linear type is effectively D.C. coupled to the beam and operates up to about 100 kHz[6]. The transductor must be stable enough to measure the beam lifetime time constant, which may be several hours, in about ten seconds. The "fast" beam current transformer bandwidth ranges between 100 Hz and 50 MHz. The low frequency corner can be extended to about 1 Hz using feedback. The fast transformer monitors beam envelopes, for example, the batch structure within an accelerator. Wall current "transformers" detect missing bunches, sick bunches, and intrabunch structure. They have been constructed with bandwidths up to 10 GHz[7].

Instrumentation electronics in the Collider tunnel will be located in 161 equipment alcoves or niches spaced every six half-cells, or about 540 meters, as shown in Figure 1. The niche entrance is covered by concrete partitions to provide radiation shielding. Access to the collider tunnel will not be possible during beam on. However, access will be possible at the bottom of the vertical access shafts. The vertical shafts are at 4,320 meter intervals, each shaft serving eight alcoves.

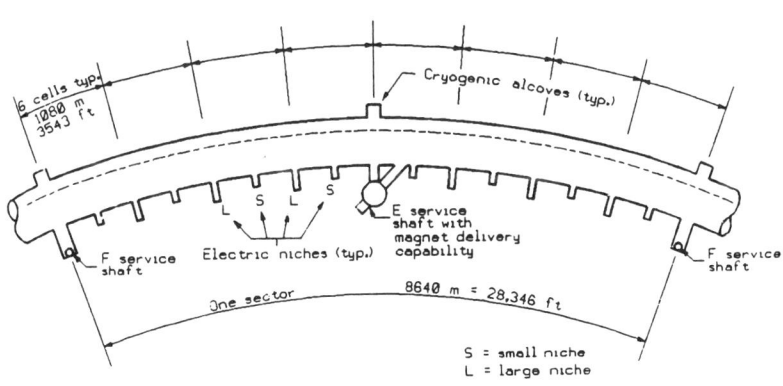

Figure 1. Equipment Niches In SSC Collider Tunnel

A number of important issues concerning the partitioning of electronic systems have not been resolved. What intelligence level should tunnel electronics have? How much electronics should reside in the alcoves? in the vertical shafts? Conflicting issues of data transmission rates, remote intelligence, and access for repair are yet to be resolved. For example, should all digitized position data be shipped back over the fiber optic communication network, or just the averaged closed orbit value? The latter requires a microprocessor in the alcove. A crate and backplane environment that supports analog, RF and digital electronics must be selected. Systems like STD, VME, and VXI are being considered. The attenuation of analog signals over long cable runs, as great as 270 m, is unavoidable. Noise pickup must also be reckoned with. An optimal distribution of cable and electronics must be determined. At this stage of SSC design, decisions have not been made to specify these hardware implementations meeting the data collection requirements. In fact, the requirements are largely unspecified.

Collider Beam Position Monitors

The Beam Position Monitors (BPM) must provide position and intensity signals under the various operating conditions of the Collider: during machine commissioning, at Collider design intensity, in fault diagnosis, and during specialized accelerator studies. They may also help protect superconducting magnets by sending abort signals if the closed orbit at any detector exceeds a preset limit. The position detectors will be located at the sextupole end of each magnet correction package, every 90 meter half-cell length, for four monitors per betatron period. Each sensing device contains four short-circuited, 15 cm long, 50 Ω strip transmission lines, placed above, below and to both sides of the beam center as shown in Figure 2.

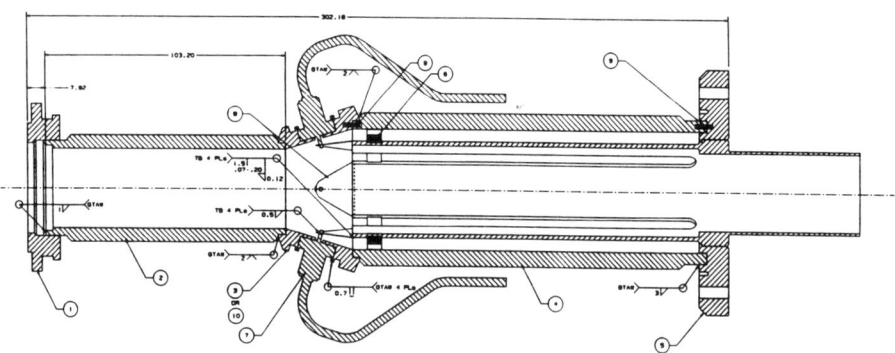

Figure 2. Collider Beam Position Detector Prototype

Strips short-circuited at one end may be used since directionality is not required in the collider, and this measure also saves construction cost, reduces heat leak, and improves reliability. (In the present plan, detectors are installed such that beam in the Top Collider traverses the detector opposite from the direction of beam in the Bottom Collider. This is imagined to be a third or fourth order effect, but needs to be investigated). Individual beam pulses, nominally 14 cm FWHM and separated by 5 m intervals, will produce doublet pulses of 1 ns duration at 16 ns intervals. The peak-to-peak amplitude of the pulses varies from 0.5 V_{pp} at commissioning to 25 V_{pp} at an operating intensity of 7.5×10^9 protons per bunch (ppb). (In at least one scenario for a luminosity upgrade, that level could reach 234 V_{pp}). To maximize the signal, each electrode subtends most of one quadrant. In a circular geometry, a 70 degree subtended angle produces near minimum sensitivity to orthogonal beam displacement[8]. So that the electrodes will not be aperture defining elements, the beam tube bulges out around them. The electrodes are also recessed 2 mm outside the aperture so that synchrotron radiation cannot strike them. Synchrotron radiation impacts SSC Collider design by loading the LHe system with a 9 kW heat burden, photodesorption of gases from the vacuum chamber walls, and the ability to produce noise in BPM electronics. The photon flux is 6×10^{20} sec^{-1} with a critical photon energy of 284 eV[9]. If the photons, assumed uniformly distributed around the accelerator circumference and azimuthally uniform inside the beam pipe, strike the electrodes, and if each photon liberates a photoelectron, the induced signal power could reach -40 dBm at the output port. Since the temporal distribution of the photons would follow the beam distribution, this noise would peak at harmonics of the RF frequency, and be detectable by BPM electronics.

Ionizing radiation will be present in the Collider and other circular accelerators, with the position detectors duly exposed. Estimates of energy deposition in devices close to the beam pipe have been made by several authors[10,11,12]. Radiation exposure results from beam scattering on residual gas, beam scraping on the beam tube, and catastrophic beam loss. The lifetime radiation dose of most beam detectors in the Collider arcs will probably fall in the range 30 to 300 Mrads. A small number of detectors near the Interaction Regions may see doses up to 30 Grads. The radiation problem further restricts beam detector design. Various organic dielectrics used in high frequency work must be excluded from consideration as mechanical support materials. Materials that possess radiation hardness, low outgassing rate, maintain dimensions during machining, and have good RF properties, such as low loss and reasonable permittivity, are few. Ceramics are cautiously being considered in the 4K detector design, as brittleness and dielectric constant are of concern. A ceramic ring on the stripline open end is part of the Collider position detector design, which will make the strips electrically some 9% longer. This causes no problem in a single gap detector. Detector mechanical components are stainless steel, possibly with internal copper plating. The austenitic 300 series stainless steels are frequently used in cryogenic engineering for their strength and corrosion resistance. However, they tend to be unstable at low temperatures, undergoing thermal or stress-induced martensitic transformation, leading to reduced fatigue strength. The martensite form is harder and more brittle than the austenite. This effect is of greatest concern in the smaller metal components, particularly in the vacuum feedthrough, where ceramics are involved. There is some evidence that this transformation caused insidious failures in the cryogenic feedthroughs used in the Fermilab Tevatron[13]. Among the stainless steels, Type 310 is a good choice for cryogenic applications, remaining stable when cold.

The good beam aperture of the collider rings, only 2.5 mm radius[14], suggests a position accuracy specification of 0.1 mm for the whole measurement system. In a purely dead reckoned system, in which the errors in the position detectors, cables, and electronics would not be calibrated out, all mechanical and electrical calibration errors summed in quadrature must be less than 0.1 mm. If half the error budget is allocated to the detector, then the detector electrical to mechanical center error should be less than 0.07 mm. Maintaining these tight tolerances will be difficult, so it will probably be necessary to perform some in situ calibration. By injecting a test signal into the detector, all downstream cables and electronics can be calibrated. Therefore most of the error budget can be absorbed by the detectors. Why can't the tolerances in the detector be relaxed and also be calibrated in situ? That may only happen if the test signal is known to be injected on mechanical center. However, no antenna can be placed there. The most convenient place for an antenna is one or more of the pickup electrodes. Since that electrode must be placed precisely with respect to mechanical center, and orthogonally to adjacent electrodes, the design of a mechanically accurate detector is forced. In other words, error in the placement of the exciter antenna can not be deconvolved from error in the electrical center location, relative to mechanical center.

Machine tracking studies suggest that a vertical position measurement, in addition to horizontal, in the F quads, and a horizontal measurement, in addition to vertical, in the D quads, improves the closed orbit by only 10%[15]. Three possible scenarios for electrodes come to mind. First, position could be measured in one plane only, requiring two electrodes, and a third exciter electrode, with about -25 dB coupling between exciter and measurement electrodes. This design eliminates one vacuum feedthrough, but the detector would have a deflecting mode wakefield. The second scenario would be similar to the first, with the addition of a dummy electrode, properly terminated, to restore the impedance balance. A third scenario using four electrodes measures position in both planes, and electronics would be designed for sensing both planes and orthogonal testing. The conclusion is that it may be impossible to achieve the required position accuracy without an exciter electrode, undesirable from an impedance point of view to have three electrodes, and reasonable to have four. In the construction plan, position electronics for only one plane have been budgeted.

Figure 3 is a diagram used to estimate the positional tolerance required on electrode placement to achieve a 0.005 inch e.c. to m.c. specification. The distance between electrode and grounded body, h, appears in the expression for electrode characteristic impedance. The distance also appears in the expression relating the fraction of beam image current intercepted by a surface at zero potential for an off-axis beam current. The image current density for an off-axis beam is given[16] in Equation 1

$$I_{image}(r, \theta, \chi, b) = \frac{I_b}{2\pi b} \frac{1 - (r/b)^2}{1 + (r/b)^2 - 2(r/b)\cos(\theta - \chi)} \qquad (1)$$

for a beam located at polar coordinates (r,χ), at observation point θ, in a cylinder of radius b. The worst case displacement of centers occurs when the angle χ is set to zero or 180^0.

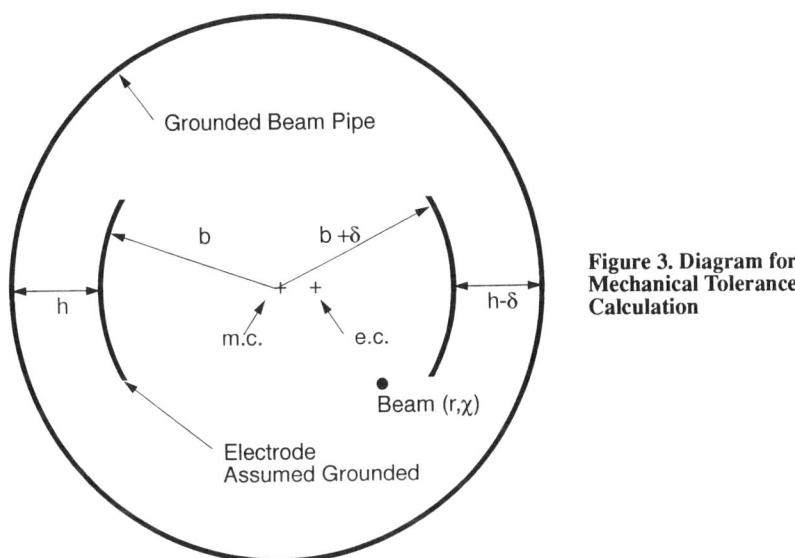

Figure 3. Diagram for Mechanical Tolerance Calculation

The current developed on the right electrode, a sector of the surface subtending an angle ϕ, is obtained by integrating Equation 1.

$$I_r(r, b+\delta, \phi) = \int_{-\phi/2}^{\phi/2} (b+\delta) I_{wall} d\theta = \frac{2I_b}{\pi} \mathrm{atan}\left[\frac{1 + (r/(b+\delta))}{1 - (r/(b+\delta))} \tan(\phi/4)\right]$$

The current developed on the left electrode is:

$$I_l(r, b, \phi) = \int_{(\pi+\phi/2)}^{(\pi-\phi/2)} b I_{wall} d\theta = \frac{2I_b}{\pi} \mathrm{atan}\left[\frac{1 + (r/b)}{1 - (r/b)} \cot(\phi/4)\right] \quad (3)$$

The characteristic impedance of the electrode can be estimated using the empirical formula for flat geometry stripline.[17] The parameter w_e is an effective electrode width that compensates for the non-zero electrode thickness. The constants are $k_1 = 1.393$, $k_2 = 0.667$, and $k_3 = 1.444$. For a given allowed displacement between electrical and mechanical centers, r, a tolerance δ on spacing, h, can be estimated.

$$Z_0(h) \cong \frac{120\pi}{\sqrt{\varepsilon_r}} \frac{1}{\frac{w_e}{h} + k_1 + k_2 ln(\frac{w_e}{h} + k_3)} \qquad (4)$$

$$w_e = w + \frac{t}{\pi}\left[1 + ln(\frac{2h}{t})\right] \qquad (5)$$

When the beam is on electrical center, the voltages on opposing electrodes are by definition equal. The equality is a transcendental relation, Equation 6.

$$I_l(r, b, \phi) Z_0(h) = I_r(r, b+\delta, \phi) Z_0(h-\delta) \qquad (6)$$

Using $Z_0(0.213) = 50 \, \Omega$, $\phi = 70°$, b = 0.728 in., and r = +0.005 in., Eqn. 6 shows the height tolerance to be +0.0068 inches. The calculation shows that the plate location tolerance can be somewhat worse than the 0.005 inch centering specification.

To facilitate accurate and repeatable beam steering, the detector is rigidly attached to the spool piece. The spool pieces are catch all devices that bring magnet bus power leads outside the cryostat, allow gases to vent, hold correction magnets, and bring various instrumentation leads into the LHe cold mass. The scheme for detector attachment to the spool is not entirely resolved. Presently each detector is located by two 0.25 in. diameter pins on the cold mass end plate, the end of the spool piece. One pin is diamond shaped so that a bit of clocking on the round pin is possible. Otherwise the center-to-center tolerance on the mating holes in the BPM end flange would need to be specified very tightly. Three machine bolts tighten the detector flange to the cold mass end plate. The flanges will be tack welded together to fix the transverse alignment indefinitely. Advantages to this method are that the distortion of a heavy vacuum sealing weld is avoided, disassembly for repair is easy, and the bolt arrangement simplifies installation of the detector on a test stand. The detector can be removed from the spool by cutting the small weld joints. The detector is designed with a tailpiece section of beam pipe, which is robot welded to the beam tube during spool manufacturing. This weld joint is readily inspectable before the detector assembly is butted against the cold mass end plate.

Each position detector requires four vacuum feedthroughs, so 7,744 are required by the two Collider rings. Because of the great quantity of feedthroughs and detectors, and the inherent difficulty of replacement, these components must be very reliable. A detector MTBF of about 10^9 hours keeps the downtime of the collider to one week out of five years operation from BPM detector failures alone. Detector failure is defined as that failure that prevents machine operation. It could result from a helium to beam vacuum leak, a stripline electrode collapsing into the beam aperture, or other unforeseen mode. In the present design, the feedthrough and signal cable to the outside of the cryostat are considered as an integral assembly. The feedthrough is not the usual demountable connector, such as SMA. This is done because the outer body of the BPM is exposed to LHe, so a leaking feedthrough would allow LHe into the beam tube. If a connector like SMA were used, it would need to be sealed by a metal sheath, making the connector pair all but inaccessible. Therefore, the connector is eliminated in favor of an all-welded stainless steel feedthrough and cable assembly. A 0.142 in. diameter stainless steel jacketed cable is commercially available. The jackets of the four cables form a part of the sealed coldmass, operating near 4K. At the 4K end of the cable, a seal separates beam vacuum from the cable dielectric. This seal will be of brazed ceramic-to-metal type, and is presently the subject of intense investigation. If the ceramic develops small cracks, beam vacuum would be exposed to the cable dielectric, but not helium. For a while, the failure would likely be transparent to operation. The 25 cm of cable closest to the beam detector is at 4K. The dielectric would outgas into beam vacuum, but slowly, since most gases trapped there would be solidified. The room temperature end of the cable is hermetically sealed using glass or ceramic. This is done for redundancy, and to keep moisture from entering the cable dielectric. Silica in a finely powdered state is proposed for the cable dielectric, which is highly radiation resistant and largely inert. The silica is vacuum baked at 800^0C during cable fabrication, so it is unlikely that significant outgassing through a cracked ceramic would occur. If either cable seal fails, a potential risk to the cable is absorption of water, as powdered silica is slightly hygroscopic. Unlike powered magnesia, also used in MI (mineral insulated) cable, silica absorbs water but does not react chemically with water. Magnesium oxide changes in an exothermic reaction from the anhydride to hydroxide, forming an electrolytic solution. With silica, the absorption is a surface effect, so the conductivity should not greatly increase, but a change in dielectric constant would be observed. Absorption of water into the cables seems a remote possibility, but needs to be explored as a worst case scenario. MI cable is routinely used for instrument purposes in the pressure vessels of nuclear reactors, and in oil well operations. The cable itself, not considering connectors, is extremely rugged. Various manufacturers rate the cable over the temperature range -300^0F to $+1400^0$F, and to pressures of 20,000 psi[18]. The cable is normally sold as a connectored assembly, and in production quantities is fairly expensive at about $300 each.

Isolation between cryostat guard vacuum and air is maintained through an SMA jack-to-SMA jack coaxial feedthrough. This is done to avoid having the signal cable terminate into the tunnel environment, so that were the end damaged, the entire detector assembly would not have to be replaced. Because the detector is at the end of the cryostat, it is susceptible to handling damage, particularly the signal cables. An option being considered is a sheath covering the four cable bundle, and forming an extra shield against LHe exposure. In this scheme, the signal cables do not seal LHe, so a wider range of coaxial cable types may be considered.

Reliability

How often will vacuum feedthroughs fail catastrophically, causing seven to ten day shutdowns of the Collider? We might expect the failure of feedthroughs to show infancy behavior, and hope for a decreasing failure rate as the equipment matures. A failure rate or "hazard function"[19] for vacuum seals which shows infancy failure and then a pedestal constant failure rate is expressed as

$$h(t) = \alpha e^{-\beta t} + \delta \tag{7}$$

where α and δ are in units of failures/ hr. Using the defining relation for the hazard function, which relates f(t), the probability density function of failures, to the hazard function

$$h(t) \equiv \frac{f(t)}{1 - \int_0^t f(u)\, du} \tag{8}$$

a differential equation may be written. The function f(t) represents the likelihood of a failure over some finite interval, and the cumulative function may be calculated by integrating f(t).

$$F(t) = \int_0^t f(t)\, dt = 1 - \exp\left[\frac{\alpha}{\beta}(e^{-\beta t} - 1) - \delta t\right] \tag{9}$$

F(t) is an estimate of the percent of population failed since time zero. No statistics are available for the numbers α, β, or δ, but we can speculate about what they need to be. Taking α = 1/ 500 hr, 1/ β = 3000 hr, δ = 1/ 350 Mhr, the failure rate and the number of accumulated failures are calculated and shown in Figure 4. With the values shown, feedthroughs eventually fail at the rate 1 failure/ 5 years, with 13 total failures after 30 years.

A feedthrough may be more likely to fail because an adjacent one failed and the two had to be warmed to room temperature to make the repair. So failure rate may depend upon the number of temperature cycles each feedthrough experiences. Assume that feedthroughs fail randomly at a low rate, say 1 failure/ 5 years. Cryogenic isolation occurs at the section level of the Collider. Each section contains 12 position detectors and therefore 48 feedthroughs. A single failure causes 48 feedthroughs to be temperature cycled when repairs are made. If the failure rate is low, the analysis is simple, since multiple failures in a given section aren't very likely. Now assume a

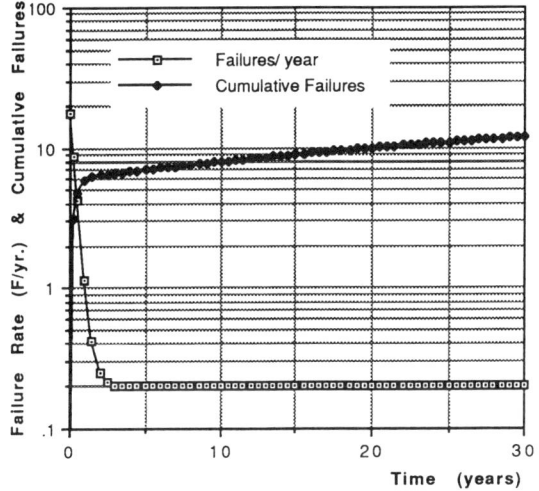

Figure 4. Vacuum Feedthrough Failure Rate and Cumulative Failures versus time.

quench rate of 1 section/ week, and a complete machine warm-up every eighteen months. The Average Number of Lifetime Thermal Cycles for vacuum feedthroughs is calculated considering these three contributions.

$$ANLTS \cong 30yr\,(0.0012 + 0.312 + 0.66\,) \;=\; 29.2 \qquad (10)$$

An increase in the failure rate up to 1 failure/ week, which would leave the Collider in a constant state of repair, results in

$$ANLTS \cong 30yr\,(0.312 + 0.312 + 0.66\,) \;=\; 38.5 \qquad (11)$$

lifetime thermal cycles. So assuming random failures, no one feedthrough in the Collider will experience more than about 40 lifetime thermal cycles (delta T> $100^{0}C$) over the 30 year operating span. This result suggests a methodology for accelerated life testing of the components. The beam detector and vacuum feedthrough assembly must be robust in at least one other respect. The supercritical helium static pressure is 150 psi and may reach 300 psi during quench transients. Using a 50% safety factor, the device has been designed to hold 450 psi.

Electronic Position Signal Processing

The BPM electronics must provide position and intensity signals under varied conditions. Optimal performance in a variety of applications requires front-end processing tailored to the various modes and ring locations. Machine commissioning will be done with as few as 10 bunches of 2×10^8 protons per bunch (ppb). A proposed luminosity upgrade to 1.0×10^{34} cm^{-2} sec^{-1} would require intensities of 7.0×10^{10} ppb. The amplitude of the 60 MHz fundamental frequency component measured at the position detector ranges from about 4 mV$_{pp}$ to 1.4 V$_{pp}$ given the intensity upgrade. So an intensity variations dynamic range of 50 dB is required in the position front-end. A typical beam current pulse is shown in Figure 5, modeled after a raised-cosine or Hanning waveform. The cosine works nicely in beam calculations since it vaguely resembles the shape of real pulses, has a zero derivative at the endpoints, and is time limited. The frequency components decay at -18 dB/ octave, about the spectral content observed in the Tevatron.

All detector signals will be brought to processor racks in the tunnel niches from the adjacent three consecutive upstream and downstream half cells. The longest cable runs are 270 meters. For maximum sensitivity and maximum dynamic range in sensing trains of bunches separated by 5 meter intervals, down converted amplitude-to-phase conversion is being considered[20]. Recent work has made log-ratio processing a possibility as well[21]. An AM/ PM processor with variable or selectable bandwidth IF filter could be implemented. By changing the filter bandwidth, response time is traded against ultimate resolution. For example, with a 15 kHz bandwidth filter and 15 dB noise figure hard limiters, a resolution of 100 nm should be achieved. The good sensitivity of these channels at low beam current will be valuable in steering the beam through first turn commissioning. For this purpose, it is not necessary to instrument every half-cell, and for that reason, AM/ PM processing will be used at the position detectors adjacent to the niches. These locations use 30 m of low loss, 7/8 in. diameter solid copper jacketed cable, making the niche BPM's the most sensitive locations in the Collider. Position monitoring near the interaction regions will also use this system.

At the position detectors most distant from the niches, skin effect losses even in high quality low loss cable remain excessive for the pulse risetime to be preserved. Therefore, located close to the detectors, peak detection circuits which effectively down-convert the high frequency components to baseband are proposed. Since peak detection involves overcoming the cut-in voltage of diodes, it is non-linear and limited in dynamic range. Therefore, locating the peak detectors in the niches would yield a lesser dynamic range than if they are located at the detectors. It would, however, solve two problems. First, the irradiation concern largely disappears. Second, the potential for EMI is reduced since the raw detector signals are skin effect shielded within the long coaxial cable running to the niche. A circuit performing a full wave rectification of the doublet signal, if it could be built, would appear to be the optimal detection network for conserving signal energy.

The voltage appearing across a 50 Ω load from a single position electrode is shown in Figure 6. Figure 7 shows the doublet pulse after ideal rectification and an assumed 8 dB non-frequency sensitive conversion loss. The effect of attenuation and dispersion of a long cable was calculated and is shown in Figure 8. Because of the high diode cut-in threshold, producing position data from the peak detecting BPM's during commissioning may not be possible.

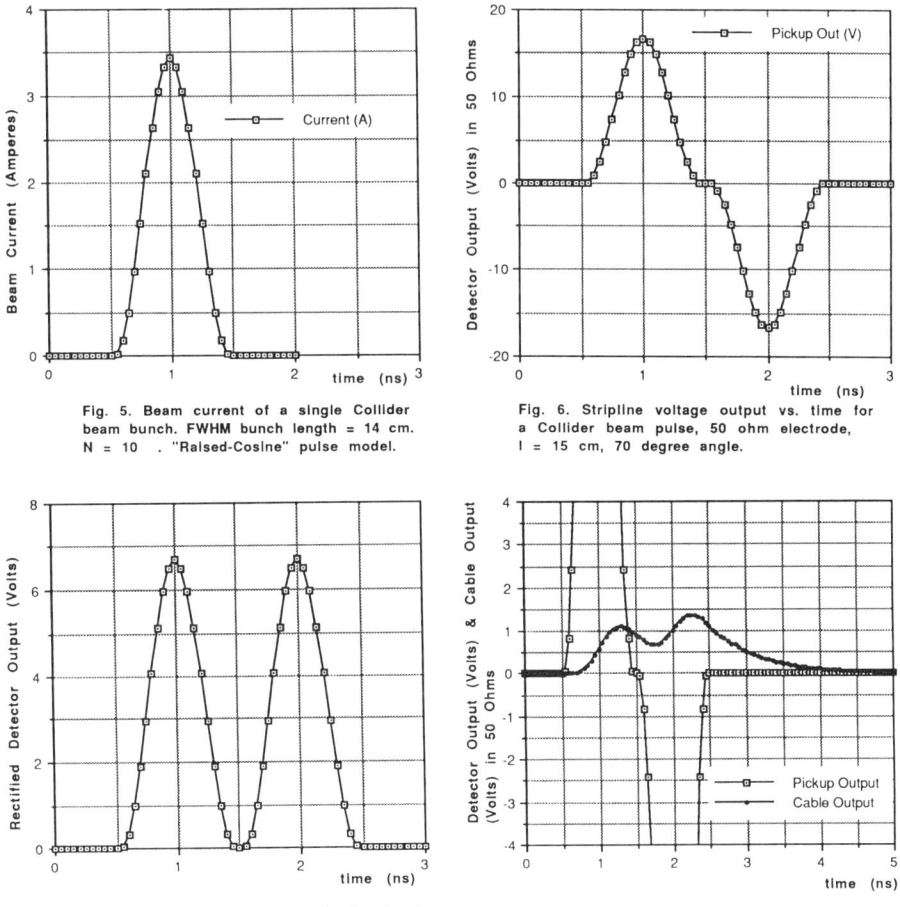

Fig. 5. Beam current of a single Collider beam bunch. FWHM bunch length = 14 cm. N = 10. "Raised-Cosine" pulse model.

Fig. 6. Stripline voltage output vs. time for a Collider beam pulse, 50 ohm electrode, l = 15 cm, 70 degree angle.

Fig. 7. Rectified detector output after ideal rectification and 8 dB rectifier loss.

Fig. 8. Detector voltage output after ideal rectification, 8 dB rectifier loss, atten. and dispersion of 270 m of RG-213 coax. cable.

Position measurements are made over various time intervals. To determine closed orbit, measurements taken at every monitor (every half-cell) are averaged to remove the betatron oscillation envelope. If ten Collider turns are recorded and averaged, 3 milliseconds are required to obtain closed orbit. BPM's should also be able to measure the position of an isolated bunch, any sequence of bunches, and the same bunch on every beam orbit. At a few locations in each machine, these turn-by-turn and bunch-by-bunch measurements will be made. These BPM's could employ wideband front-end circuits and 8-bit 100 MHz flash A/D converters with FIFO memories to record position histories. The flash converters, clocked at 60 MHz, take one sample of horizontal position, vertical position, and intensity of each bunch. The difference to sum ratios are obtained with the analog signals. A 100 Kbyte memory could store the position data of 6 full turns, or the position of a single bunch on 10^5 turns. The three different front-end processors, AM/PM, peak detector, and flash A/D converter, ought to be designed with identical position sensitivity, though resolution, bandwidth, and dynamic range are not the same. Also to be considered are the means of linearizing the BPM digitized data, which could be implemented in hardware. A fast lookup table could contain both gain and offset corrections for each detector, cable set, and analog front-end. All systems must be designed for in situ testing by injection of signals as far forward as the beam detectors, and at one intermediate point for an electronics only test. Historically, in situ testing has been shown to save valuable beam-on time which would otherwise be wasted diagnosing equipment failures. The expense of the specialized automatic testing equipment is always more than justified. In the case of cable continuity checks to the detectors, results will not be entirely definitive because of the shorted striplines.

Conclusion

Design and construction of the SSC beam instrumentation is a task of broad scope, interdisciplinary in technology, and a sizeable manufacturing and installation effort. A Collider position detector design was needed early in the project, since the detector resides within the cryogenic helium and vacuum environment of the spool piece. The detector design therefore preceded the existence of a comprehensive BPM plan.

Schemes for locating the vacuum feedthroughs outside liquid helium are being considered, and various dielectric materials, including Kapton, are being investigated as a signal cable dielectric. The prototype detector is being fabricated by a small machine shop having vacuum and precision machining experience. One of the prototypes will be installed in the small E-sector string test (ERST) scheduled for June at Fermilab. Extensive testing of an additional three prototypes, and evaluation of vacuum feedthrough samples, is planned throughout 1991.

References

1. SSC Baseline Cost Estimate, (8/31/90).
2. Site-Specific Conceptual Design, SSCL-SR-1056, Section 4.2.7, (7/90).
3. D. Martin et al, Proc. 1989 IEEE PAC, 89CH2669-0, (3/20/89),
   "A Resonant Beam Detector for Tevatron Tune Monitoring", pg. 1483.
4. D.A. Goldberg and G.R. Lambertson, Proc. 1987 IEEE PAC, (3/87),
   "A High-Frequency Schottky Detector For Use In The Tevatron", LBL-22273.
5. J. Gannon et al, Proc. 1989 IEEE PAC, 89CH2669-0, (3/20/89),
   "Flying Wires At Fermilab", pg. 68.
6. K. Unser, CERN-ISR-OP/81-14,
   "A Toroidal D.C. Beam Current Transformer with High Resolution".
7. C.D. Moore et al, Proc. 1989 IEEE PAC, 89CH2669-0, (3/20/89),
   "Single Bunch Intensity Monitoring System Using An Improved Wall Current Monitor", pg. 1513.
8. R. Shafer, Fermilab UPC No. 101, (5/18/79),
   "Sensitivity of Beam Position Detectors for the Tevatron", pg. 7.
9. Site-Specific Conceptual Design, SSCL-SR-1056, (7/90), pg. 239.
10. I.S. Baishev et al, Institute For High Energy Physics, USSR, (7/28/90),
    "Beam Loss And Radiation Effects In The SSC Lattice Elements".
11. T.A. Gabriel, SSC-110, (8/20/87), "Preliminary Simulation of the Neutron Flux Levels in the Fermilab Tunnel and Proposed SSC Tunnel".
12. Gilchriese (editor), SSC-SR-1035, (6/88), "Radiation Effects at the SSC".
13. Private Communication, R.E. Shafer.
14. Site-Specific Conceptual Design, SSCL-SR-1056, Section 4.1.3.1, (7/90).
15. Private Communication, G. Bourianoff.
16. R.T. Avery et al, UCRL-20166 (1971),
    "Non-intercepting Monitor of Beam Current and Position".
17. S. Y. Liao, Prentice-Hall (1987),
    "Microwave Circuit Analysis and Amplifier Design", pg. 200.
18. Private communication, Whittaker, ERI Inc.
    See, for example, MIL-T-81490, or "Report of Qualification Tests on Coaxial Cable Assemblies Space Shuttle Program, Test Report No. 1028, ERI (8/4/75).
19. W. Nelson, John-Wiley & Sons, ISBN 0-471-52277-5,
    "Accelerated Testing - Statistical Models, Test Plans, and Data Analysis".
20. R.C. Webber et al, Proc. 1987 IEEE PAC, 87CH2387-9,
    "A Beam Position Monitoring System For The Fermilab Booster", pg. 541-543.
21. F.D. Wells and R.E. Shafer, LA-UR-90-3340,
    "Log - Ratio Beam Position Detector".

Poster Presentations

HIGH RESOLUTION, POSITION SENSITIVE DETECTOR FOR ENERGETIC PARTICLE BEAMS

E.P. Marsh, M.D. Strathman, D.A. Reed, and R.W. Odom

Charles Evans and Associates, Redwood City, CA 94301

ABSTRACT

An imaging position sensitive, particle beam detector is described which is minimally invasive, operates over a wide dynamic range ($>10^7$), and exhibits high spatial resolution. The detector images secondary electrons or ions produced when an energetic particle beam passes through a thin foil. These secondary electrons or ions are transported onto a two dimensional imaging detector using stigmatic ion optics. The detector has been employed as a tuning aid for the Ion Microtomography (IMT) system at Sandia National Laboratories and its performance in this application will be discussed.

INTRODUCTION

We have developed a novel technique for characterizing the spatial and intensity distribution of energetic particle beams. The passage of energetic particles through a thin foil produces secondary electrons and ions.[1] These secondary charged particles can be extracted and imaged onto a two dimensional detector using stigmatic ion optics. These ion optics make it possible to magnify the image of the secondary electron or ion emission so that the image resolution is not limited to the spatial resolution of the detector. In addition, these ion optics increase the overall flexibility of the system since a wide range of detector configurations are possible. This flexibility permits changes in field of view, image magnification, image resolution and detector size.

A detailed description of the detector design and performance has been given in a previous paper.[2] The efficiency for the production of secondary electrons was found to be 5 to 10 electrons per incident ion for a 2.27 MeV He^{2+} beam and a thin carbon foil (10 $\mu g/cm^2$). Both positive and negative secondary ions were also produced with a yield of between 0.005 and 0.01. The negative ions were detected by steering the electrons out of the optical path by placing a 0.1 T magnetic field near the foil, the positive ions were detected by inverting the lens bias voltages. The detector is minimally invasive and typically has an attenuation of less than 5%. The detector has a spatial resolution of about 5 μm for a 1 mm diameter beam and the dynamic range is better than 10^7. The useful operating range varies from a few hundred counts to several nA. The maximum count rate can be increased by changing the channel plate detector configuration. Two imaging beam detectors in series should permit the measurement of beam emittance and time of flight techniques for beam energy measurements up to moderately energetic beams (<100 MeV). The high dynamic range of this device can measure much higher count rates than a surface barrier detector, and, therefore, should reduce data acquisition times dramatically in time of flight energy measurements.

RESULTS

The detector has recently been employed as a tuning aid for the Ion Microtomography (IMT) system on the FN tandem accelerator at Lawrence Livermore National Laboratories. Tests were carried out in collaboration with Dr. Arthur Pontau and Mr. Dan Morse of Sandia National Laboratories at Livermore, CA. The image detector was used to focus the beam, aid in adjusting object slits in the beam, and provide measurements of the particle beam size and intensity profile.

The measurements were performed using an auxiliary transmission electron microscopy (TEM) grid as a shadow mask in front of the foil. Placing this TEM grid (50 μm grid opening) at the sample position casts a shadow onto the beam transducer foil. Figure 1 shows the image on the phosphor screen at the image plane of the detector when the grid is placed in the beam line. The large dark grid pattern shown in figure 1 is the shadow from the TEM grid. The smaller, brighter grid pattern is produced by the TEM grid used as a support for the carbon foil.

The beam used in this experiment was a 21 MeV Li^{3+} at 10,000 to 20,000 ions/sec. The effects of sweeping the magnetic field of the quadrupole lens on the shadow grid pattern can be observed as the lens is adjusted. The aspect ratio of the image can be used to produce uniform focusing along the X and Y axis (Z being the beam axis). The magnification of the shadow pattern is representative of the focus point relative to the sample position. Since the image is being observed in real time on a video monitor, it is a simple matter to place the focus point at the sample position. As the focus approaches the grid, the image magnification increases, Fig. 2, until the image inverts as the focus point passes through the grid. The lens is then adjusted to locate the focus at the point where this inversion of image occurs. Once the focus has been set

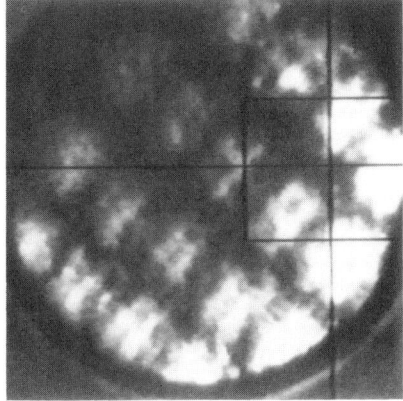 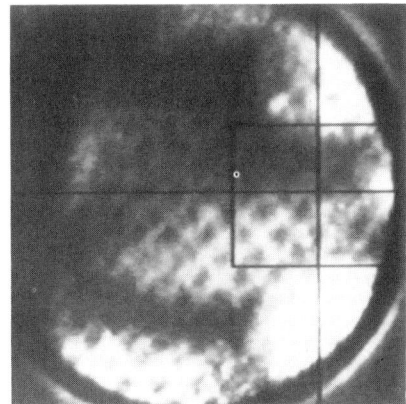

Fig. 1. Secondary electron image of shadow grid and of carbon foil support grid.

Fig. 2. Focusing the beam at the shadow grid.

the beam diameter is reduced by closing the object slits. Figures 3-5 show the progressive reduction in slit width. This process is observed in real time which also simplifies the procedure. The magnification of the image is independent of the object slit setting, so by calibrating the screen with the known spacing of the shadow grid, the beam diameter can be read directly from the screen, Figure 6. From this figure the estimated diameter of the beam is under 2 μm.

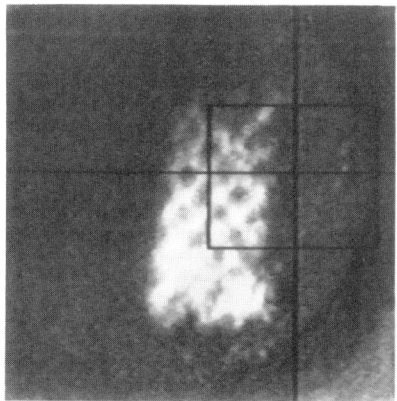 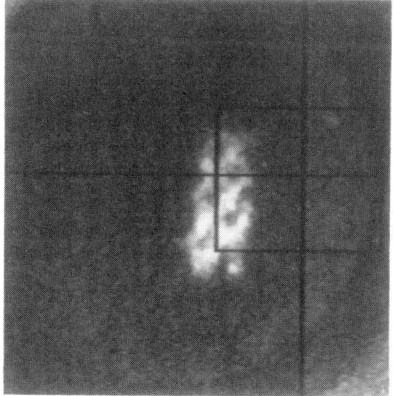

Fig. 3. Image of beam after reduction in slit width.

Fig. 4. Further reducing object slit widths.

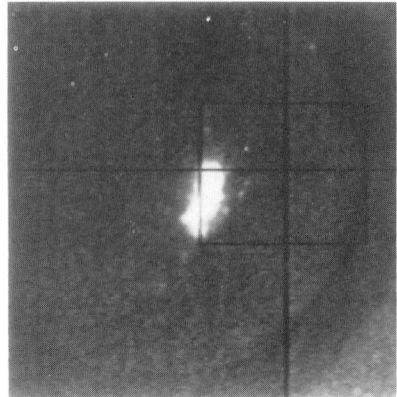 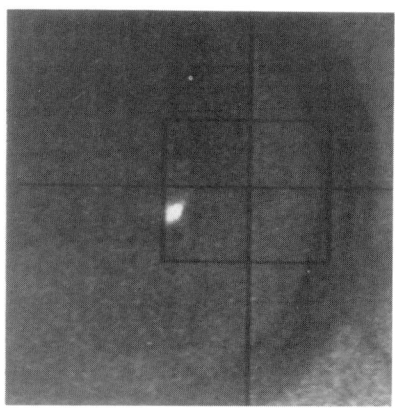

Fig. 5. Fig. 6. Minimum beam diameter.

The design of a higher resolution detector has been completed and the detector is now being assembled. With the use of higher magnification, higher extraction fields and a contrast diaphragm to limit the transverse energy of the transmitted secondaries, we expect to achieve an image resolutions better than 0.5 µm for a beam size of .05 mm or smaller.

The new detector is designed so that the detector body can be moved in and out of the beam path under computer control. We are also working on the design of a large area beam detector for beams of between 5 and 10 cm.

CONCLUSIONS

This imaging beam detector has an image resolution on the order of 5 µm, for a 1 mm particle beam diameter, a dynamic range of better than 10^7, and a beam transmission of better than 95%. The device has been very successful as a tuning aid since it allows the user to monitor the beam shape, in real time, while it is being adjusted. In addition, two detectors in series could be used to determine beam emittance and/or energy.

ACKNOWLEDGEMENTS

We gratefully acknowledge the support for this research provided by DOE Small Business Innovative Research (SBIR) Grant #DOE DE-AC03-88ER80666.

REFERENCES

1. P. Sigmund and S. Tougaard in "Inelastic Particle-Surface Collisions", edited by E. Taglauer, W. Heiland (Springer-Verlag, Berlin, 1981); H.J. Frischkorn and K.O. Groeneveld, Physica Scripta, T6, 89, 1983.

2. R.W. Odom, M.D. Strathman, S.E. Buttrill, S.M. Baumann, Nucl. Instr. Meth. B, 44, 405, 1990.

ANALYSIS OF THE BEAM POSITION MEASUREMENT WITH BUTTON-TYPE PICKUPS IN APS*

Y. Chung
Argonne National Laboratory, Argonne, IL 60439

ABSTRACT

The response of electrostatic button-type pickups for the measurement of the transverse position of charged particle beams was investigated and analytic formulae were obtained for the signal as a function of time t. The study was done for beam pipes of circular and elliptic cross sections, for rectangular and non-rectangular electrodes, and for several cases of longitudinal beam profiles. In particular, the error in the measurement of the beam position using circular electrodes as compared to rectangular ones was found to be less than 100 μm per 1 cm of beam excursion from the center of the beam pipe for the case of APS.

INTRODUCTION

For capacitive pickup devices, the position of the charged beam is measured through the difference between the electric potentials which develop on the electrodes. For slow, nonrelativistic beams, the image charge induced by the bunch of charged particles has considerably larger longitudinal dimension than the bunch itself. However, for highly relativistic beams, the image charge has the same longitudinal distribution as the beam, due to the Lorentz contraction of the longitudinal component of the electric field.

In this article, we will derive the response of the electrode as a function of time and the transverse position of the beam, assuming that both the angular and the longitudinal sizes of the electrodes are finite. The analytical model assumes a simple elliptic geometry for the beam chamber. The results are compared with those obtained numerically for the actual APS beam chamber, and they will be shown to agree quite well. This justifies the use of the analytical model rather than the time-consuming numerical methods to find the optimal position and size of the electrodes and to analyze how the shape of the electrodes affect the beam position measurement.

MONITOR RESPONSE

In order to get the signal from each electrode, we must first obtain the electric field inside the beam chamber. Consider an infinitely narrow beam moving along the longitudinal direction with the constant velocity

V, as shown in Fig. 1(a). Following the procedure by Cupérus,[1] instead of solving the full electromagnetic problem directly in the lab frame (unprimed), we will transform to the reference frame (primed) where the beam is at rest, obtain the field and then transform back to the lab frame.

The electric field $\vec{E}$ in the lab frame is then derived using the Lorentz transformation, which gives:

$$E_{\|} = E'_{\|} \quad \text{and} \quad \vec{E}_{\perp} = \gamma \vec{E}'_{\perp},$$

where $\gamma = \sqrt{1 - V^2/c^2}$. Therefore, the problem is reduced to that of an electrostatic case with linearly distributed charges.

Decomposing the electric field into Fourier components, we write

$$\Phi'(\vec{x}, t) = \int dk \, e^{ik(z-Vt)} \Phi'(\vec{x}_\perp, k),$$

where the linear dispersion relation $\omega = kV$ was assumed. The charge $q_p(t)$ induced on the electrode surface S is then

$$q_p(t) = \int dk \, e^{ik(z_1-Vt)} q_p(k),$$

$$q_p(k) = \gamma \varepsilon_0 \int dS \, e^{ik(z-z_1)} \Phi'(\vec{x}_\perp, k),$$

and the induced current $i_p(t)$ is given by

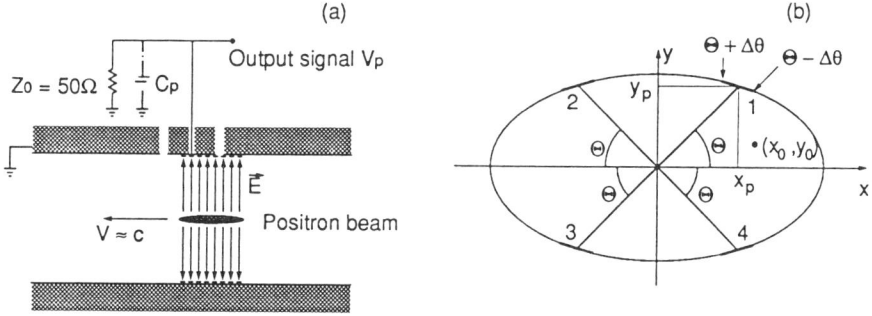

Fig. 1: (a) Schematic button-type pickup. (b) Geometry of the beam chamber and the pickup electrodes.

220 Beam Position Measurement

$$i_p(t) = \int dk\, e^{ik(z_1 - Vt)} i_p(k),$$

$$i_p(k) = i\gamma\varepsilon_0 kV \int dS\, e^{ik(z - z_1)} \frac{\partial \Phi'(\vec{x}_\perp, k)}{\partial n}. \tag{1}$$

The integration extends over the area of the electrode surface, and z_1 is the z-coordinate of a reference point, e.g., the center of the electrode z_p. $\vec{n}$ is the direction normal to the electrode surface. If the electrode is connected by a coaxial line of characteristic impedance Z_0 and if the capacitance between the electrode and the beam chamber is C_p, then the overall impedance $Z_p(k)$ for the electrode will be

$$Z_p(k) = \left(\frac{1}{Z_0} - ikVC_p\right)^{-1}. \tag{2}$$

If there is frequency filtering represented by F(k), the measured voltage $V_p(t)$ will be

$$V_p(t) = \int dk\, e^{ik(z_1 - Vt)} V_p(k),$$

$$V_p(k) = Z_p(k)\, i_p(k)\, F(k). \tag{3}$$

From Eqs. (1) and (3), it suffices to solve the Poisson equation for the 2-D static potential $\Phi'(\vec{x}_\perp, k)$ to obtain the electrode response $V_p(t)$. The equation is analytically solvable if the beam chamber geometry is somewhat simplified. In this work, elliptic coordinates will be used to approximate the APS beam chamber. We will consider highly relativistic beams only. Assuming a rectangular electrode flush with the interior surface, the Fourier component $i_p(k)$ can be expressed as[2]

$$i_p(k) = -i\Delta\theta \frac{Q_T V}{\pi^2} D(k) \sin k\Delta z \left\{ 1 + 2 \sum_{m=1}^{\infty} \frac{\sin m\Delta\theta}{m\Delta\theta} \times \right.$$
$$\left. \left(\frac{\cosh m\mu_0}{\cosh m\mu_p} \cos m\theta_0 \cos m\theta_p + \frac{\sinh m\mu_0}{\sinh m\mu_p} \sin m\theta_0 \sin m\theta_p \right) \right\}. \tag{4}$$

Here, $\Delta\theta$ and Δz are half the angular and the longitudinal sizes of the electrode. The subscripts 0 and p are for the bunch and the electrode, respectively. Q_T is the total charge in a single bunch and $D(k)$ is the Fourier transform of the longitudinal charge distribution $\rho(z)$.

By inserting Eq. (4) into Eq. (3), we find that $V_p(t)$ is separated into the time-dependent and the position-dependent factors for rectangular electrodes as

$$V_p(t) = T(t) \, P(\vec{x}_{\perp 0}), \qquad (5)$$

where

$$T(t) = \frac{Q_T \Delta \theta}{\pi^2 C_p} \int dk \, \frac{e^{ik(z_1 - Vt)}}{k + i\kappa} \sin k\Delta z \, D(k) \, F(k), \qquad \left(\kappa = \frac{1}{Z_0 V C_p}\right) \qquad (6)$$

and

$$P(\vec{x}_0) = P_0(\vec{x}_{\perp 0}) + \sum_{m=1}^{\infty} \frac{\sin m\Delta\theta}{m\Delta\theta} P_m(\vec{x}_{\perp 0}). \qquad (7)$$

For the elliptic case,

$$P_m(\vec{x}_{\perp 0}) = \frac{2}{1 + \delta_{m0}} \left(\frac{\cosh m\mu_0}{\cosh m\mu_p} \cos m\theta_0 \cos m\theta_p + \frac{\sinh m\mu_0}{\sinh m\mu_p} \sin m\theta_0 \sin m\theta_p \right). \qquad (8)$$

Once the longitudinal charge distribution $\rho(z)$ and the filtering function $F(k)$ are known, it is straightforward to calculate the electrode

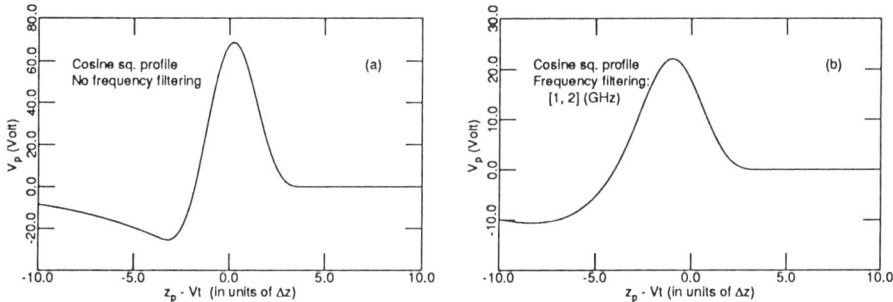

Fig. 2: The electrode signal V_p as a function of time t, (a) without frequency filtering, (b) Using a bandpass filter at [1,2] GHz. z_p is the electrode center coordinate and Δz is half the longitudinal size.

response $V_p(t)$. For certain cases of $\rho(z)$, it is possible to evaluate $V_p(t)$ analytically using the residue theorem of complex variables. In the discussion below, we will use the following parameters: $C_p = 2$ pF, $\Delta z = 0.5$ cm, $\sigma = 1$ cm, $Q_T = 3.5$ nC / mA (single bunch), $x_p = 1.38$ cm. Figure 2(a) shows the electrode signal $V_p(t)$ without frequency filtering and Fig. 2(b) using a bandpass filter at [1,2] GHz. If the electrodes are not rectangular, the results are similar but a bit more complicated.[3] One notable difference is that $V_p(t)$ for non-rectangular electrodes is not separable as in Eq. (5).

BEAM POSITION MEASUREMENT

As shown in Fig. 1(b), four button-type pickups will be installed for each unit. The quantities Δ_x, Δ_y, and Σ are defined as follows.

$$\Delta_x = V_{p1} + V_{p4} - V_{p2} - V_{p3},$$
$$\Delta_y = V_{p1} + V_{p2} - V_{p3} - V_{p4}, \qquad (9)$$
$$\Sigma = V_{p1} + V_{p2} + V_{p3} + V_{p4}.$$

The horizontal and the vertical positions of the beam are then determined from

$$X_0 = \frac{\Delta_x}{\Sigma} \approx S_x x_0 + R_x, \quad \text{and} \quad Y_0 = \frac{\Delta_y}{\Sigma} \approx S_y y_0 + R_y. \qquad (10)$$

y_0 (cm)	Analytical result	Numerical result	
	S_x (cm^{-1})	S_x (cm^{-1})	R_x
0.0	0.569	0.565	-0.0011
0.2	0.580	0.576	-0.0011
0.4	0.616	0.610	-0.0011
0.6	0.676	0.667	-0.0010
0.8	0.763	0.747	-0.0010
1.0	0.877	0.850	-0.0009
1.2	1.017	0.974	-0.0008
1.4	1.178	1.113	-0.0008

Table 1: Comparison between the analytical and the numerical results for the x direction. The offset R_x for the analytical case is zero.

The linear approximation is valid only when x_0 and y_0 are small. S_x and S_y are the sensitivity functions and R_x and R_y are the offset errors. Figure 3(a) shows X_0 as a function of the beam position x_0 for several cases of y_0, and Fig. 3(b) shows the contour plotting of both X_0 and Y_0. Table 1 lists the sensitivity function S_x obtained analytically for the elliptic beam chamber and numerically for the actual APS beam chamber. The two results agree quite well. The finite offset error R_x is due to broken symmetry in the x-direction due to the presence of the photon beam channel and the antechamber. The optimal position of the electrodes which gives the same sensitivity in both x and y directions was found to be $x_p = 1.32$ cm. However, this was shifted to $x_p = 1.38$ cm due to the mechanical constraint of the mounting flanges.

If the electrodes are not rectangular, X_0 and Y_0 are time-dependent, and the result of measurement will depend on when the data are taken.[3] This error due to the timing jitter will be larger for wide-band detection than for narrow-band detection at low frequency, say, a few hundred MHz. From Fig. 4(a), we find

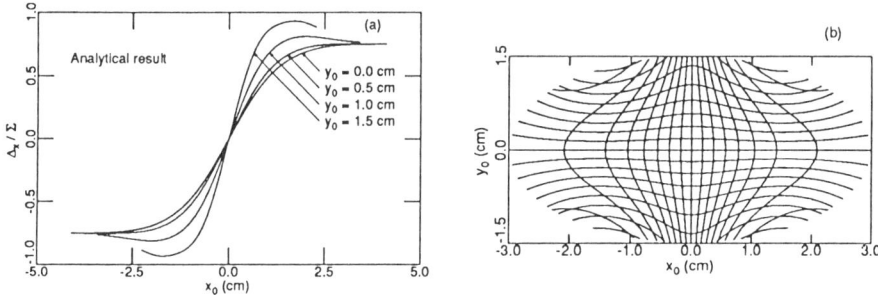

Fig. 3: (a) The ratio Δ_x/Σ as a function of the beam position x_0 for several cases of y_0. (b) The contour plotting for $X_0 = \Delta_x/\Sigma$ and $Y_0 = \Delta_y/\Sigma$ for the elliptic beam chamber.

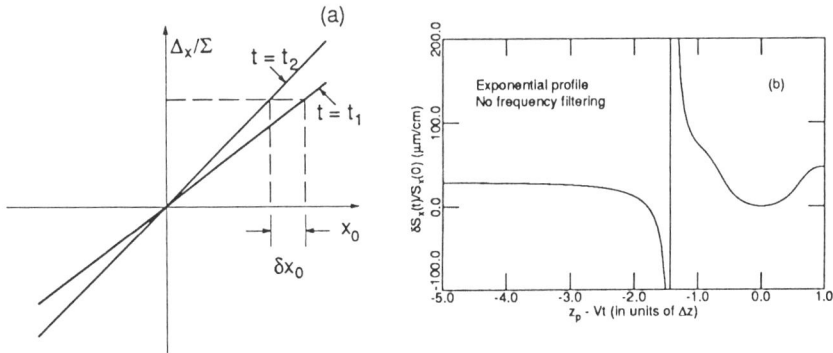

Fig. 4: (a) Measurement error due to timing error when the electrodes are not rectangular. (b) The relative change of the slope of X_0 at the center with no frequency filtering. The longitudinal beam profile is exponential.

$$\delta x_0 = -\frac{\delta S_x(t)}{S_x(t)} x_0.$$

$S_x(t)$ is the sensitivity as a function of time. Fig. 4(b) shows the plotting of the relative change of $S_x(t)$ as a function of time for the exponential profile ($\rho \sim e^{-|z|/\sigma}$) with no frequency filtering. Hexagonal electrodes were used in place of circular ones to facilitate analytic integration over the electrode surface. It is to be noticed that it diverges to infinity when Σ crosses zero while Δ_x does not. Typical error is 100 µm per 1 cm beam excursion from the beam chamber center. However, if the timing jitter is small (less than 10 ps) or if narrow band detection scheme at a few hundred MHz is used, this error will be reduced to a negligible level (less than 10 µm / cm).

CONCLUSION

The characteristic of the BPM system for the APS storage ring was studied analytically and numerically, and the results agree very well. This suggests that the presence of the photon beam channel and the antechamber has negligible effect on the BPM system. Using the analytical model, the optimal position of the electrodes was determined such that the sensitivity is as close as possible in x and y directions taking into account other mechanical constraints. A possible source of error in the measurement of the beam position using non-rectangular electrodes was analyzed. The error was found to be typically of the order of 100 µm per 1 cm of beam excursion from the center of the beam chamber and can be reduced significantly by employing proper timing schemes and signal processing.

REFERENCES

1. J. H. Cupérus, "Monitoring of particle beams at high frequencies," Nucl. Instr. and Meth. **145**, 219 (1977)
2. Y. Chung, Argonne National Laboratory, unpublished information (1989)
3. Y. Chung, Argonne National Laboratory, unpublished information (1990)

---

*Work supported by the U.S. Department of Energy, Office of Basic Energy Sciences, under Contract No. W-31-ENG-38.

A HIGH-FREQUENCY SCHOTTKY DETECTOR FOR USE IN THE TEVATRON*

D.A. Goldberg and G.R. Lambertson
Lawrence Berkeley Laboratory, Berkeley CA 94720

A vexing problem associated with detection of Schottky signals from a bunched beam is the presence of the coherent signal, which can be 10 or more orders of magnitude greater than the Schottky signal. To overcome this difficulty, we have constructed a Schottky detector for the Tevatron collider in the form of a high-Q ($\approx$5000) resonant cavity which operates at roughly 2 GHz, well above the frequency at which the single-bunch frequency spectrum begins to roll off ($\approx$200-300 MHz for the Tevatron). The detector is capable of sensing independently the vertical and horizontal particle motions. The 2 GHz Schottky signals are down-converted to frequencies below 100 kHz to permit relatively rapid high-resolution analysis using a FFT spectrum analyzer. The initial installation consists of a single cavity; a second detector will be built which employs a pair of phased cavities to permit discrimination between p's and $\bar{p}$'s. Details of the design of both the cavity and the associated electronics are presented. Spectra obtained from the detector show clearly observable Schottky betatron lines, free of coherent contaminants; also seen are the "common-mode" longitudinal signals due to the offset of the beam from the detector center. The latter signals indicate that at 2 GHz, the coherent single-bunch spectrum from the detector is reduced by >80 dB; therefore, in normal collider operation, the Schottky betatron lines are essentially entirely free of coherent contaminants. Experimental data will be presented showing how the detector spectra can be used to measure such properties as transverse emittance and synchrotron frequency.

1. INTRODUCTION

With the use of a suitable detector, one can sense fluctuations in the instantaneous number and/or positions of particles in a cyclic accelerator. The signals from such a detector, the so-called Schottky signals, have as their principal virtue that they can furnish a variety of diagnostic information about a particle beam, on a continuous, real-time basis, without perturbing the beam.

The nature of Schottky signals is described in detail in Ref. 1. Summarized briefly, the frequency spectrum of these signals consists of a set of bands occurring at integer multiples of the particle revolution frequency, and a second set which is shifted in frequency from the first set due to the particles' betatron motion; if the beam is bunched, the synchrotron motion splits these bands into a set of possibly overlapping satellite lines.

A vexing problem associated with detection of Schottky signals from a *bunched* beam is the presence of a coherent signal, which can be 10 or more orders of magnitude greater than the Schottky signal. To overcome this difficulty, we have constructed a Schottky detector for the Tevatron collider, sensitive to *transverse* motion in both planes, in the form of a high-Q ($\approx$5000) cavity whose resonant frequencies (made slightly different for the vertical and horizontal sensing modes) are roughly 2 GHz, well above the frequency at which the coherent single-bunch frequency spectrum begins to roll off ($\approx$200-300 MHz for the Tevatron), yet still below the cutoff frequency of the Tevatron beam pipe.

To achieve the frequency resolution necessary for measuring the widths of the central synchrotron satellite lines (< 3 Hz), and still provide reasonable data acquisition times, it is necessary to employ a FFT spectrum analyzer. This in turn necessitates heterodyning the 2 GHz cavity signal to something below 100 kHz,. To minimize spectral smearing due to phase noise from the beam, the heterodyning signal

---
*Work supported by the Director, Office of Energy Research, Office of High Energy and Nuclear Physics, High Energy Physics Division, U.S. D.O.E., under Contract No. DE-AC03-76SF00098.

© 1991 American Institute of Physics

is referenced to an RF signal from the Tevatron; because of the difficulties of making such a receiver tunable, it was decided to tune the cavity to the (fixed) receiver frequency, rather than vice versa.

The initial detector consists of a single cavity of rectangular cross section, which responds equally to p's and $\bar{p}$'s. Based on the success of this device, we are planning to build a second detector, which will employ a pair of phased cavities to permit discrimination between p's and $\bar{p}$'s. Our initial hope was that the detector be usable for both the fixed-target and collider modes of the Tevatron; however, in fixed target mode there appears to be a coherent signal from sources other than the bunch structure *per se*, (presumably intra-bunch coherent oscillation) of sufficient magnitude as to preclude observation of Schottky signals.

2. DETECTOR REQUIREMENTS

The principal requirements for the Schottky detector are summarized in Table 1. In the main, they arise from the considerations discussed above, in conjunction with the Tevatron parameters given in Table 2.

Table 1 Design Goals

S/N	>10 dB	
f_o (nominal cavity center frequency)	2044.5 MHz	
$\Delta f_{v,h}$ (vertical, horizontal frequency offsets)	±2 MHz	
Q_u	10,000	
allowed frequency variation	±50 kHz	(single cavity)
	±15 kHz	(double cavity)
allowed temperature variation	±1.0 °C	(single cavity)
	±.25 °C	(double cavity)

The above signal-to-noise ratio (S/N) is imposed on a machine operating to the specifications of Table 2; this makes it possible for the device to provide useful information for a machine operating at for example 1/10 of the stated intensity. Several factors governing the choice of operating frequency have already been discussed. In addition, to minimize the strength of the coherent signal in many-bunch operation, as well as to simplify the heterodyning scheme, we chose the cavity frequency to be a half-integer multiple of the rf frequency; the multiple of 38.5 gives a frequency which is sufficiently below the beam tube cutoff frequency that the single-detector response is relatively unaffected by the beam tube ports, and that it will provide adequate isolation between two-cavities separated by no more than 50 cm.

Table 2 Tevatron Operating Parameters (Collider Mode)

f_{rev} (beam revolution frequency)	47.71 kHz	
η (frequency/momentum dispersion factor)	.0028	
$\Delta p/p$ (fractional momentum spread)	3×10^{-4}	(full width @ 900 GeV)
N (number of particles)	6×10^{10}	(per bunch)
$\sigma_\perp$ (rms transverse beam size)	0.6 mm	(radius)

The dimensions of the rectangular cavity were chosen to displace the TM_{120} and TM_{210} frequencies by ±2 MHz from the above f_o. This splitting is low enough to permit reasonably narrow bandwidth in both the RF and 1st IF stages (see circuit discussion below), and yet high enough to avoid incidental coupling between the two modes; since the splitting frequency is also used as the frequency of the final IF stage, it should be high enough to permit signals from that stage to be transmitted over long distances without interference from the commercial AM band.

To translate the S/N requirement for the cavity into a Q-value we use the fact that the signal power (at frequency f) developed by a transverse pickup is given by

$$P = \langle (I_B x)^2 \rangle R_\perp T^2 (\pi f/c)^2 \tag{1}$$

where $\langle (I_B x) \rangle$ is the rms of the product of beam current and beam displacement, and $R_\perp T^2$ is the product of the transverse shunt impedance and transit-time factor. For a circular accelerator with particle revolution frequency f_{rev} and a circulating beam of N particles, the "source term" for the Schottky signal (per Schottky band) is given by

$$\langle (I_B x)^2 \rangle = N(e f^2_{rev}) \sigma_\perp^2 \tag{2}$$

Note that because Schottky signals result from fluctuations, this term is proportional to N (for many accelerators, N usually exceeds 10^{10}), rather than N^2 as would be the case for coherent signals (this accounts for the ability of signals due to coherent motion to completely overwhelm the Schottky signals); for the same reason, the signal is proportional to the r.m.s. displacement of the *individual particles* (i.e. the beam *size*) rather than that of the beam as a whole, as would be the case for a coherent signal.

When excited in the TM_{210} (or TM_{120}) mode, a closed rectangular cavity of length ℓ, an unloaded Q of Q_u, and a matched output, has shunt impedance *at resonance* given by

$$R_\perp T^2 = \frac{64}{25\pi^2} Z_o Q_u T^2 \frac{\pi f \ell}{c} \tag{3}$$

where Z_o is the impedance of free space, and T is the usual transit-time factor $\sin(\pi f \ell / c)/(\pi f \ell / c)$.

For the case of a cavity with beam ports at the end, Eq. 3 remains approximately correct as long as the port size remains small with respect to the wavelength. The product $T^2(\pi f \ell / c)$ has a broad maximum of ≈0.7 at $\pi f \ell / c \approx 1.17$, giving $R_\perp T^2/Q \approx 34$ ohms. Using the Tevatron operating parameters given in Table 1, and assuming a 3 dB noise figure for the electronics (amplifier noise plus cable attenuation), we find that an unloaded Q of 10000 gives us a S/N of >13 dB. Finally, a loaded Q of 5000 gives a response which is >8 Schottky bands wide (FWHM) so that the detector response is essentially flat over the span of the central Schottky band.

The final requirements on tuning and temperature regulation are linked: The temperature stability is simply what is required to maintain the tuning tolerance. For the double cavity, the tolerance is based on the requirement that gains and phases be sufficiently matched to give 30 dB directional rejection; for the single cavity the tolerance is somewhat more arbitrary, and is based on keeping the response constant to 5%. In either case, the required temperature tolerances exceed those of the available water systems at Fermilab, and we decided that the entire assembly be installed in a thermostatically controlled box to be maintained at 110 °F, roughly 10° higher than the warmest anticipated ambient temperature in the Tevatron tunnel.

3. EXPERIMENTAL APPARATUS

3.1 Detector

The single cavity detector is shown in Fig. 1; the interior of the device consists of a rectangular cavity with rounded corners. Attached to either end of the cavity are beam tubes which also serve to support the structure when it is installed in the Tevatron. The requirements of high Q, good vacuum properties, mechanical stability, and reliability and ease of fabrication led to a choice of 6061 aluminum for the cavity itself. Vacuum joints involving aluminum surfaces are made using a radially expanding metal (REM) seal, somewhat similar to a Mott seal; the REM seal also serves as an RF joint.

The tuning requirements for the cavity cannot be achieved by machining tolerances alone (the sensitivity to transverse cavity dimension is approximately 330 kHz/mil), and so a pair of micrometer-controlled tuning plungers, shown in the figure, must be employed. Each of the coupling antennas consists of an axial rod mounted on the end of an SMA fitting located roughly halfway between the cavity center and its outer wall. Both the tuning plungers and the antennas, as well as a third antenna for injecting test signals, come mounted on con-flat flanges. To minimize the number of REM seals, the cover plate containing these feed-throughs was fabricated from a plate made of 1/2" Al bonded to 3/8" stainless steel. The former metal provided the required low resistivity interior surface; the latter permitted the use of con-flat seals for all feed-throughs.

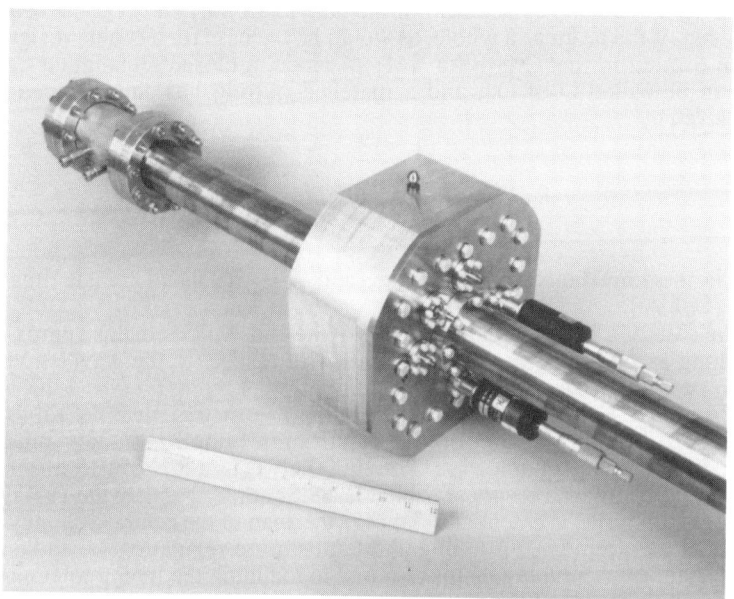

FIGURE 1 Assembled Schottky Detector

To tune and test the cavity, we first modified the antenna lengths so that they provided matched coupling to the external 50-ohm loads. The tuning plungers were then adjusted to give the proper "vertical" and "horizontal" frequencies (compensated for the system being at room temperature and not under vacuum). We then measured the unloaded Q-values and found them to be Q_v=9500 and Q_h=9200; the small difference is attributed to slight differences in the antennas and tuning probe positions. We then measured the detector response using the bead measurement technique described in the Ref. 2, and obtained an RT^2/Q of (29±1) ohms. The reduction from the calculated value of 34 is due principally to the reduction of the fields near the cavity ends due to the beam apertures; together with the slightly low Q values, this results in a reduction of S/N of just over 1 dB to a still acceptable value of 12 dB.

3.2 Electronics

The circuit for converting the 2 GHz Schottky signal to a signal at <100 kHz is shown in Fig. 2; selected data on the receiver and the various signal levels and frequencies is given in Table 5. The receiver's input rf stage can switch-select either the vertical or horizontal signals from either p's or (after the double cavity is installed) $\bar{p}$'s. To minimize S/N degradation, the switches are placed after the low-noise (< 1 dB NF) Miteq input amplifiers. The only component preceding the amplifiers is a band-pass filter needed to block out the strong coherent *longitudinal* signal due to excitation of the cavity's TM_{110} mode, which would otherwise drive the input stage into saturation; the S/N degradation resulting from these filters is only a few tenths of a dB.

Separate narrow-band filters for the vertical and horizontal signals in the first IF stage serve two purposes: They reduce noise power to the second mixer, thereby permitting additional gain in this stage; they also prevent the 3 dB degradation in S/N which results from the folding of the noise spectrum onto the signal following mixing. The band pass filters in the second IF stage also serve this latter purpose. (It was not possible to find a sufficiently narrow band filter to employ this technique in the rf stage as well.)

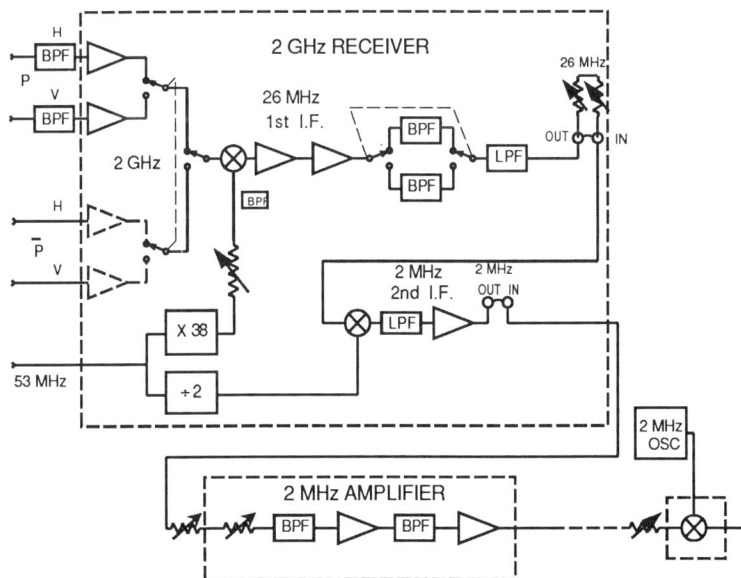

FIGURE 2 Frequency Converter Block Diagram

Table 5. Selected Receiver Data

Input Signal Power (includes input line losses):	
For $N = 10^{12}$, $\sigma_\perp = 1$ mm, 2800 Hz signal bandwidth:	
$P = -116$ dBm per Schottky betatron line	
$dP/df = -151$ dBm/Hz (averaged over betatron line)	
Input Noise Power Density ($T = 295$ K):	
$dP/df = -173$ dBm/Hz (includes 1 dB NF for amplifier)	
Receiver Conversion Gain: 86 dB (measured at output of 2 MHz amplifier)	
Receiver Operating Frequencies:	
Input rf 2.04 GHz (nominal) [Cavity frequencies: [f_h=2042.53, f_v=2046.53 MHz]	
1st IF 24.55 MHz, 28.55 MHz [f_h and f_v; $\Delta f \approx 300$ kHz]	
2nd IF 2 MHz; $\Delta f \approx 100$ kHz	
Baseband 0 - 100 kHz	
Phase Noise << 1 Hz	

As noted earlier, both the vertical and horizontal signals get down-converted to the same 2 MHz final IF. The final conversion stage is located on a separate chassis so that the 2 MHz signal can be used for transmitting from the detector location to the main control room, roughly 1 km away. The variable frequency oscillator in the final conversion stage permits selecting different Schottky bands, and positioning of the desired band within the 100 kHz frequency window of the spectrum analyzer.

To resolve the central synchrotron satellite lines, we need a frequency resolution small compared to the 38 Hz synchrotron frequency; to be able to measure the linewidths requires even better resolution. The principal limit to such resolution is phase noise introduced by either the beam or the measuring electronics. To minimize the effects of the former we have used a frequency conversion scheme referenced to the RF system; measurements on the beam (see below) indicated the system phase noise to be well below 1 Hz FWHM.

Finally, a word may be in order on the number of attenuators present. The original receiver design was based on the input power levels shown in Table 5. Initial experiments showed significantly higher input levels due to the presence of coherent longitudinal signals (see Sect. 4), and it was necessary to reduce the gain at the later stages to avoid saturation of the electronics.

3.3 Temperature control

To achieve the required temperature stability the detector assembly is enclosed in an insulated box having 2" styrofoam/plywood walls. The box temperature is maintained at 110 °F by means of a thermostatically controlled heater (in the form of a long-life 150 W light bulb) and a muffin fan to ensure uniform air temperature. The control unit is an Omega Model CN-2002-P2 controller, which has the capability of a programmably variable output to the heater. The temperature sensing element is a platinum RTD mounted on the detector. Initial tests indicated that the system can maintain the cavity temperature to $< \pm 0.2$ °C.

4. EXPERIMENTAL RESULTS

Data were acquired during two separate Tevatron collider runs. The first of these was at 273 GeV which utilized the "fixed-target" optics, for which the vertical and horizontal beta functions (beam sizes) at the detector location were approximately equal. The second run was at 900 GeV, and utilized the "low-β" optics, for which β_v is ≈ 8.5 β_h. A typical spectrum, obtained during the 273 GeV run, is shown in Figure 3, and illustrates most of the important features of the data (as explained in the Appendix, we "used" only the upper portion of the spectrum).

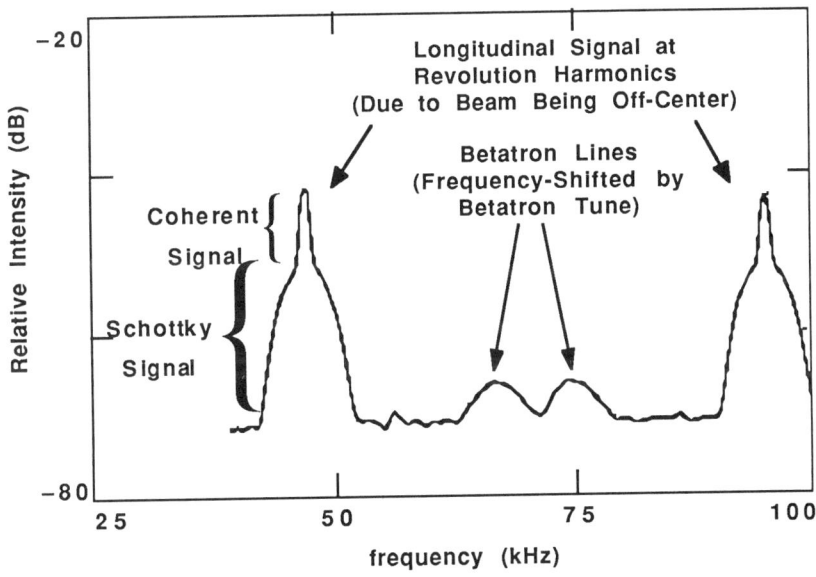

FIGURE 3 Spectrum of Vertical Output Signal from Schottky detector. E=273 GeV

The two large peaks at roughly 50 and 100 kHz are the longitudinal signals, variously known as revolution lines or common-mode lines. Their presence is due to the fact that the beam is not centered in the detector; in the spectrum coming directly from the detector (i.e., prior to heterodyning) they occur at integer multiples of the 47.7 kHz revolution frequency; both of the peaks in Fig. 3 are at frequencies corresponding to "weak" coherent lines (see Appendix). We see that these are compound peaks–a narrow, intense peak atop a broad, weaker one. The strong, narrow peak is the residual coherent signal at 2 GHz; the broad peak is the longitudinal Schottky signal.

The two peaks appearing between the two revolution lines are the betatron signals. The line widths are comparable to those of the broad peaks seen in the revolution lines, and there is no evidence of the narrow peak, i.e. they appear to be Schottky signals completely uncontaminated by any coherent signal!

The breadth of the Schottky line reflects the momentum variations of the individual particles within the beam, whereas that of the coherent line simply reflects the variation of the *centroid* of the beam, which is considerably less. The observed widths of the Schottky peaks are consistent with the expected momentum spreads of the beam. The identification of the broad and narrow peaks was inadvertently confirmed when the rf anode supply failed during one of the beam stores; the resulting debunching of the beam caused the coherent (narrow) peak to disappear immediately, where as the the Schottky signal was (initially) unaffected.

To assess the effectiveness of operating at high frequency to reduce the coherent signal, we make use of the expressions for the total power in the coherent and incoherent (Schottky) common mode lines.

$$P_c \propto \langle I_c^2 \rangle \cdot \langle x_d^2 \rangle \qquad\qquad P_S \propto \langle I_S^2 \rangle \cdot \langle x_d^2 \rangle \qquad\qquad (4a,b)$$

Since both signals are proportional to the mean-square beam displacement $\langle x^2_d \rangle$, the ratio of the mean-square Schottky and coherent currents is simply the ratio of the measured powers, i.e. of the areas under the respective peaks. As discussed in Sect. 2, at low frequencies this ratio is equal to N, the number of particles; hence dividing the above power ratio by N gives the effective coherent signal suppression. Over a number of runs at both 273 and 900 GeV at intensities of $\geq 5 \times 10^{11}$, we almost invariably found the strongest of the coherent peaks to be less than 35 dB greater than the Schottky peak,* meaning that at 2 GHz, the coherent signal is suppressed by more than 80 dB. (We also observed an unexpected [and as yet unexplained] result: The observed width of the coherent lines decreased by roughly a factor of 2 over the first few hours of a beam store [the width of the Schottky lines remained essentially unchanged during the same period], accompanied by the appearance and disappearance of gross structure in the line shape; there did not appear to be any correlated change in the amplitude of either the coherent or the Schottky peaks.)

We can now estimate the degree to which the Schottky betatron lines are free of coherent contaminants. Expressions for the total power in the coherent and Schottky betatron peaks are given by

$$P_c^b \propto \langle I_c^2 \rangle \cdot \langle x^2_{\beta c} \rangle \qquad\qquad P_S^b \propto \langle I_S^2 \rangle \cdot \langle x^2_{\beta i} \rangle \qquad\qquad (5a,b)$$

where $\langle x^2_{\beta c} \rangle$ and $\langle x^2_{\beta i} \rangle$ are the mean-square coherent and incoherent betatron amplitudes, respectively. Hence in addition to the ratio of mean-square coherent to Schottky currents, the relative powers depend on the ratio $\langle x^2_{\beta c} \rangle / \langle x^2_{\beta i} \rangle$. We know from experiments done with the Tevatron tune detector[3] that $\langle x^2_{\beta c} \rangle / \langle x^2_{\beta i} \rangle \leq 10^{-8}$, so that this additional suppression of 80 dB means that the coherent contaminant in the betatron lines is some 45 dB below the Schottky signal (even for the case where the contaminant is produced by a *strong* coherent betatron line), consistent with the apparent total absence of a coherent betatron peak.

The longitudinal lines in Fig. 3 are not in detail smooth curves but consist of a set of so-called synchrotron sidebands or satellites which result from the modulation of the particles' revolution frequency as a result of synchrotron oscillation. A greatly expanded view of the center of a revolution line (from the 900 GeV data) is shown in Fig. 4, which shows the central line (equivalent to the "carrier frequency") as well as the first two satellites on either side. The separation of the satellites is equal to the frequency at which the revolution frequency is modulated, i.e. the synchrotron frequency; from the data in Fig. 4 we determine that to be 37.6 ± 0.2 Hz.

As we noted earlier, the ability to obtain resolution of better than 1 Hz on signals which are originally at 2 GHz results from using the signal from the Tevatron rf system as the source of the heterodyning frequency; the sharpness of the peaks also indicates that the receiver has very low phase noise. (The central satellite, theoretically infinitely narrow, has a width consistent with instrument resolution when viewed at resolution bandwidth as low as .05 Hz!) Hence one is able to observe the considerable fine structure seen in Fig. 2. The fact that the structure is mirror-symmetric about the central satellite suggests that it is actually signal on the beam, and not either noise or an artifact of the heterodyning process. While the source(s) of the structure have

---

*As discussed in the appendix, the relative strengths of the various coherent peaks in a given spectrum can vary considerably, depending on the uniformity of the bunch populations. For the case shown in Fig. 3, the relative areas under the coherent and Schottky peaks differ by little more than a factor of 2; however the ratio of the strong coherent line (not shown) to the weak ones was nearly 30 dB.

not been identified, it is consistent with a group of particles within an annulus in the longitudinal phase space of the beam quite near to the outer edge of the rf bucket (not necessarily the same bucket as that containing the main portion of the beam); several individuals at Fermilab have independently suggested that such an annulus could be populated at the time the individual beam bunches are coalesced in the Main Ring.

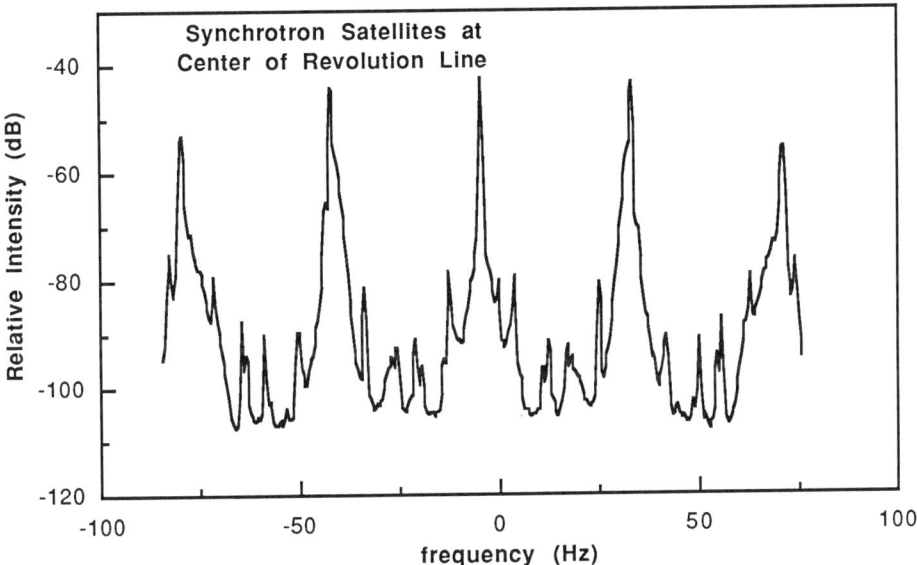

FIGURE 4 Central synchrotron satellites observed during operation at =900 GeV.

One also expects to see satellite structure on the betatron peaks. When only protons are present we do indeed see them, with the same frequency spacing at which they are observed in the longitudinal spectrum. The lines (including the central peak) are much broader, about 11 Hz FWHM. The most likely cause of this is the variation of the tune with betatron amplitude; the present technique appears to be a particularly convenient way of measuring this quantity. When p̄'s are introduced along with the protons, the satellite structure disappears, presumably smeared out due to the beam-beam tune shift which, in addition to broadening the peaks, shifts those for p's and p̄'s by different amounts. Were one to detect signals from p's and p̄'s separately, it is possible that satellite structure could be recovered.

The final item for discussion is the betatron spectrum itself. The total power in the individual betatron lines, conveniently measured by viewing them with broad resolution, should provide a measure of the beam emittance. To see if this could be usefully done, we looked at the S/N and found it to be within 1 dB of the calculated value (based on the measured emittance and beam current). We should note that due to the asymmetry of the beta functions in the "low-β" mode, the S/N observed in the vertical spectrum is enhanced by roughly 3 dB over what it would be if one were using the more symmetric fixed-target optics (the obverse of that result is that the S/N for *horizontal* betatron line is degraded in this mode, having a S/N value of roughly 8 dB lower than that of the vertical signal for the same operating conditions). However, we are advised that a new set of "low-β" optics has been developed for the next collider run which has more nearly balanced beta functions, and should yield comfortable S/N ratios for signals from both planes.

ACKNOWLEDGEMENTS

We have benefitted greatly from the advice and counsel of Ferd Voelker, who did several feasibility studies on the initial idea of a resonant cavity detector. Design and construction of the cavity was done with the invaluable assistance of Tom Henderson and John Meneghetti. Similar assistance in the design and construction of the electronic circuits and the temperature control apparatus was provided by Walter Barry and Jim Wise. We are grateful for the encouragement and counsel of numerous Fermilab personnel throughout this project, most notably to Jim Crisp for his continual cooperation in his role as our Fermilab liaison, and to Gerry Jackson for his help in acquiring and digitally recording the data from the 900 GeV run. In what is no doubt a departure from the usual procedure, we feel we must acknowledge the Hewlett-Packard Corporation for developing the microwave instrumentation without the aid of which this entire undertaking would have been impossible. Finally, we would like to acknowledge Jim Hinkson for numerous helpful suggestions on how to translate the jargon of beam physicists into a form intelligible to instrumentation engineers.

APPENDIX

A NOTE ON THE PERIODICITY OF THE SCHOTTKY SPECTRA

When the Tevatron collider is operated with three beam bunches (of p's and/or $\bar{p}$'s), one expects strong coherent lines at intervals of $3f_{rev}$; because of the non-uniform population of the individual bunches, weaker signals are also expected at the intermediate multiples of f_{rev}. With six bunches, the same situation obtains, because the operating rf harmonic (1113) is not an integer multiple of 6. The cavity is normally tuned so that its peak responses (in both planes) lie midway between two of the $3f_{rev}$ lines; over that ≈ 150 kHz span the cavity's response is flat to $\approx \pm 0.3$ dB.

Since the FFT analyzer can span 100 kHz, and the Schottky band spacing is only 47.7 kHz, the observed spectrum normally includes three revolution lines. However, due to an artifact of the final heterodyning stage, the lower half of the spectrum is generally distorted; consequently we analyzed only those data lying between (and including) the upper two revolution lines in any of the spectra. In the 273 GeV data, the final frequency conversion was normally done so as to make these both weak lines; for the 900 GeV runs the lower of the two was usually a strong line.

REFERENCES

1. S. Chattopadhyay, "Some Fundamental Aspects of Fluctuations and Coherence in Charged-Particle Beams in Storage Rings," CERN Report 84-11 (unpublished)
2. W. Barry and G.R. Lambertson, "Perturbation Method for the Measurement of Longitudinal and Transverse Beam Impedance,"*et al*, in *Proc. 1987 IEEE Particle Accelerator Conf.*, p.1602.
3. G.P. Jackson, private communication.

A WALL CURRENT BEAM POSITION MONITOR BUILT ON A CERAMIC CHAMBER

YAN YIN
TRIUMF, 4004 Wesbrook Mall, Vancouver, B.C. V6T 2A3, Canada

Abstract

A wall current beam position monitor (BPM) using metal strips pasted on a ceramic chamber is proposed for the KAON Factory Booster and Driver ring. The original purpose of using a ceramic chamber and metal strips is to reduce eddy currents while providing the wall current with a path. Since the strips are already there, making them into a position monitor will save a lot of space and money, and will avoid extra parts in the ring which may increase longitudinal impedance. It is like a regular wall current monitor which has wide band frequency response, so can be also used to observe high mode components of the beam. An analysis, calculations and the results of all tests and measurements are presented here.

1.Introduction

When beam goes through a beam pipe, an image current of the beam flows along the wall of the pipe. When the beam is off the center of the pipe, the distribution of the wall current is changed and related to the beam position. A regular wall current BPM consists of 4 groups of resistors around a ceramic ring between two metal beam pipes. The wall current BPM proposed here is made by cutting several strips in the middle of each side into two sections, then bridging them with resistors. As same as for the regular wall current BPM, the voltage across the resistors is formed by wall current, and therefore related to beam position. The difference between the voltages on the resistors on opposite sides normalized by the sum signal is used as a beam position indication.

Three models of the BPM have been made and tested. The first model, made of perforated board proved the feasibility of the idea. The second one is made of G-10 boards, with most strips on the inner surface; the few strips on the outer surface of the "chamber" are used to sense beam position. The third one is also made of G-10 boards, but with all strips on the inner surface. The pick-up strips are in the center of each side. The signals are led out with feedthroughs which are pins going through the walls.

2. Wall Current Distribution

236 A Wall Current Beam Position Monitor

Since the Booster and Driver rings will be cycling in a strong magnetic field, especially the Booster will have 50 Hz cycling frequency, a ceramic chamber has been chosen as a beam pipe. To provide a path for the wall current, silver strips are pasted on the inner surface longitudinally. The strips are separated from each other, so eddy currents are substantially reduced. The BPM proposed here makes use of those strips.

The beam pipe used for KAON Factory will be a rectangular ceramic chamber with strips pasted on the inner wall. A calculation of wall current distribution has been done for a metal rectangular pipe. A measurement of the wall current distribution has also been done with the model 1, see Fig. 5.

Since the beam is ultrarelativistic, the charges induced on the wall will have the same longitudinal intensity modulation as the beam due to Lorentz contraction of the longitudinal components. Assume the wall is at a uniform potential and the beam is very thin. The frequencies above the critical frequency, f_c, which can propagate freely, are not considered, because only the properties of the beam in the vicinity of the pick-up is interested. The wavelenth of the critical frequency for TM10 mode is 20 cm for a pipe of 10 cm by 4 cm, so we only discuss the wavelengths longer than 20 cm.

Expand the charge of the beam in a Fourier series and study the field of the each component separately. According to J.H.cuperus[1]:

The surface charge density, induced on the wall y=-b (Fig.1) is approximately:

$$\sigma_{y=-b}(x,z) = -D_\lambda \cos[\frac{2\pi}{\lambda}(Vt+z_\lambda-z)]\sum_{m=1}^{\infty}\frac{\sin[\frac{m\pi(x_0+a)}{2a}]sh[\alpha_m(b-y_0)]}{a\,sh(2b\alpha_m)}\sin[\frac{m\pi(x+a)}{2a}]$$

$$\alpha_m = \sqrt{[(\frac{2\pi}{\gamma\lambda})^2+(\frac{m\pi}{2a})^2]}$$

In the formula, a is half width of the chamber, b is the half height of the chamber. V is the speed of the particle, D_λ is the amplitude, λ is the wavelength of the Fourier component.

By exchanging $-y_0$ for y_0, we get the charge on the wall y=b and, by exchanging (x,x_0,a) for (y,y_0,b), we get the charge on the other two walls.

A program was written to calculate the sum of the series. Assuming the pick-up strips are in the middle of each side of the chamber, and the beam is located at x_0, y_0, it calculates the distribution of the wall current. The distribution plots are for beam at x=0, y=0. The distribution is a function of the beam position. The program also calculates the difference of the signals of the two pick-up points at y=-b and y=b, normalizing them by their sum. By changing beam position along y direction the curve of non-linearity can be found.

The plot shows 3 curves for a chamber 10x8 cm (Fig.2). Curve 3 is for pick-up strip at the center of the side of the chamber. This curve shows for this situation how the non-linearity is only a function of beam position at x direction. Curve 1 and 2 show when the pick-up strips are off center 5% of the chamber size at x=-0.1 scale, the non-linearity of y direction is related to beam position at x direction. Curve 1 is for antenna at x=-0.7, curve 2, at x=0.7. There are 4 curves in the plot for chamber 10x4 cm. The curve 4 is measured non-linearity of the model 1 prototype (Fig.3).

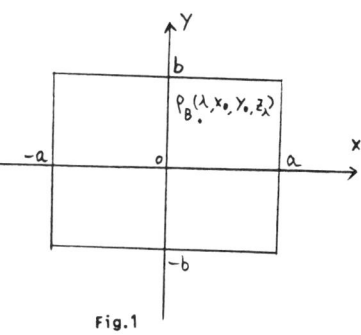

Fig.1 Cross section of a rectangular Chamber for calculation of the wall current distribution

The plots of wall current distribution and non-linearity are made for scaled chamber size.

The calculation shows:

1). The distribution is related to the wavelength observed, but the dominant factor is the chamber geometrical size and beam position. If the ratio of chamber height and width changes, the distribution changes. The strips on the side closer to the beam will pick up a larger signal. Fig.4 is for 10x8 cm and 10x4 cm chamber for a beam at $x_0=0$, $y_0=0$.

2). The BPM has intrinsic non-linearity. But if the pick-up points are at symmetric line of each side, the non-linearity is a function only of beam position in one dimension. Which means the non-linearity at y direction is independent with x direction. But if the pick-up points have an offset from the symmetric line of the side, the non-linearity will be related to the other direction. This means that a well designed BPM requires a precise mechanical process to put pick-up strips in the middle of the side, and since one can not make it perfectly symmetric, a calibration by mapping the area within the BPM is very important.

3). From the distribution curve, one can determine the number of strips to be used for picking up signal. If the strips are uniformly pasted, 1/3 of the strips at one side will pick up 43% wall current of that side, i.e. 11% of the total charge of the beam, if the beam is not very far off the center. This is consistant with the experimental measurement (see description in the last paragraph of the section). From the bench tests of model 2, a 20 ns signal of peak voltage 2.7 v was sent in, the equivalent peak current going through the antenna is 54 mA. The picked up signal is 28 mV at 5 ohm load, equivalent to 5.6 mA, which is 10.3% of the beam peak current. Actually the capacitance plays important role if the frequency goes up to GHz, because the signal is not only picked up by the resistors, but also picked up by the capacitance parallelly bridged on the strips together with resistors. The calculation indicates the signal reduces as the frequency increases. The wall current monitor shows much wider frequency response, however, as the pick-up is not purely resistive.

4) The calculation has been done for different beam energies and wavelengths. The non-linearity is only a function of beam position which means one can use a frequency signal which is different from beam signal to calibrate the BPM. This can make calibration easier.

The measurement of wall current distribution shows that the distribution is more close to Gaussian than the calculation. This means that the equation can only give a rough estimation of the wall distribution, which is convenient especially when the pick-ups are in the middle of the sides. But for more accurate analysis, it is not enough.

The first model is made of perforated board which is normally used for electronics circuits. The chamber size is 100 mm wide, 40 mm high and 150 mm long. The second model was made corresponding more closely to the physical dimensions of the Booster beam pipe. It is made from 4mm thick G-10 board, 106 mm wide, 80 mm high and 700 mm long. On the inner surface of the chamber 5 mm wide copper foil strips were longitudinally pasted, leaving the 25 mm wide band in the center of each plate no strips. On the outer surface of the chamber corresponding to the middle unpasted space, several copper strips are pasted. As for the first model, resistors bridge a 5 mm gap outside the chamber. Four 10 ohm resistors are soldered to these strips; the total resistance is 2.5 ohm, for X, -X, Y, -Y faces separately.

At both ends, a simple flange is mounted. An antenna goes through to map the BPM. The calibration curve for model 1 is shown in Fig.6.

The third model were made of the same size but with all strips on the inner surface of the chamber. The signals are led through out of the chamber by "feedthroughs" which are only two pins.

With these two prototypes a study has been done to investigate the longitudinal impedance of the structures.

3. Sensitivity and Frequency Response

From a circuit point of view the image current (wall current) is as close as you can get to an ideal current source--the image current will flow through any impedance placed in its path.

The pictures below show the signals going into the antenna and the signal out of the BPM.

The pick-up strips with loads which consist of resistors and capacitance can be analyzed as a circuit with a current source driving the parallel impedance. V_{bpm} is the voltage produced across the load. The longitudinal sensitivities is defined as $S = V_{bpm}/I_{beam}$ [2]. For example, as the second module, when the "beam" is in the center of the BPM, the load is 2.5 ohm:

$I_{beam} = .030$ A, $V_{bpm} = 20$ mV
$S = 0.67$ V/A   for 20 ns pulses

The wall current BPM also has a good position sensitivity as well. Take model 1 as an example(see Fig.7,8). The position sensitivity is defined as

$K=dx_0/dx_{bpm}$. From the plot, $x_0=30(p_2-p_4)/(p_2+p_4)$, $K=30$ mm, same as a button type BPM which has 90 mm beam pipe with 25 mm buttons.

Like all wall current monitors, this is a broad band high frequency monitor. The frequency response can accordingly be estimated. From the equivalent circuit:

$$V(\omega)=I_{beam}(\omega)\frac{1}{j\omega C+\frac{1}{r}+\frac{1}{j\omega L}}$$

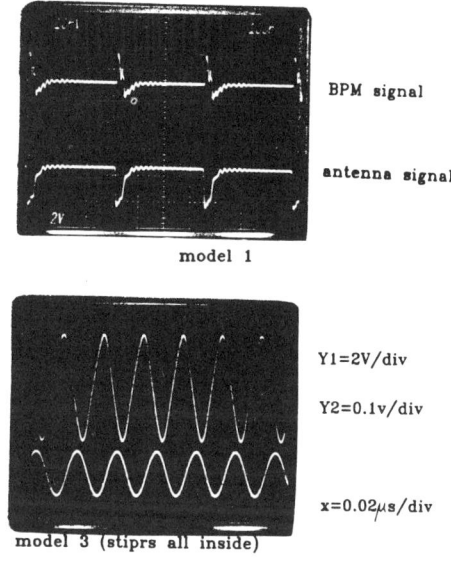

BPM signal

antenna signal

model 1

Y1=2V/div

Y2=0.1v/div

x=0.02μs/div

model 3 (stiprs all inside)

Upper:antenna signal, lower: BPM signal

BPM output signals

Assume the capacitance formed by the slot is 20 pf. For two parallel conductors, the inductance is:
~$2\mu x_t /\pi$, $\mu=4\pi \cdot 10^{-7}$, $x_t=0.1$ m (length of model 1).

For the first model, load is 5 ohm, since 4 strips were connected together, the inductance is reduced to 40 nH. This gives a lower corner frequency of R/L =20 MHz, upper corner of 1/RC of 1.5 GHz. This is consistent with Fig.8. The frequency response can be improved by reducing resistance and increasing inductance. As the BPM discussed here, as mentioned above since the resistor is 5 ohm, the inductance is very small, so the low frequency responce is not good, but it is good when the frequency is above 20 MHz.

A network analyzer was used to measure the frequency response of this prototype wall current monitor. Even for such a simple BPM model, the frequency spectrum shows that it is a wide band monitor, much wider than a button type BPM or a short stripline BPM.

The frequency response of model 2 is shown in Fig.9 which is measured with time domain method.

The longitudinal coupling impedance of this BPM model has been studied and measured both in the frequency domain by TSD (Through-Short-Delay) method and Time-domain method[4,5]. The measurement shows that the coupling impedance of the wall current BPM as designed here is within the requirement. The total impedance caused by 40 BPM of the ring is around 0.1 Ohm for resistance and 0.12 Ohm for reactance.

A ceramic chamber BPM is being made, see Fig.10.

Acknowledgement
   The author gives her thanks to Sigi Turke for making the wall current BPM mechanical drawings.

Reference
1.Monitoring of particle beams at high frequencies.
   J.H.Cuperus, CERN, 1977, NIM V145 p219

2.Bunched beam diagnostics. H.Siemannn Cornell Univ.

3.A 1.5 GHz wide-band beam-position and intensity monitor for the electron-positron accumulator (EPA)
   G.C.Schneider, CERN

4.Transmission line impedance measurements for an advanced Hadron Facility
   Linda Walling, Los Alamos,LA-UR-88-3533

5.Characteristics of directional coupler beam position monitor
   Robert E. Shafer, IEEE NS-32 No.5, p1933, Oct. 1985

6.The measurement of impedance for KAON beam pipe structures by means of the Through-short-delay calibration method and Time domain method.
C.Oram,Y.Yin, N.Ilinsky, P.Reinhardt-Nikulin, TRI-DN-90-K142.

7.BPM calibration and its set-up. Y.Yin, TRI-DN-90-K143

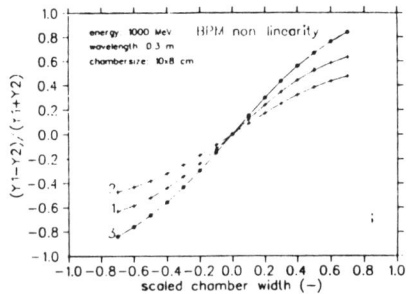
Fig.2 Wall current BPM non-linearity (Chamber:10x8 cm)

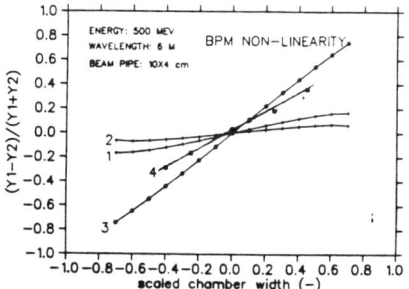
Fig.3 Wall current BPM non-linearity (Chamber:10x4 cm)

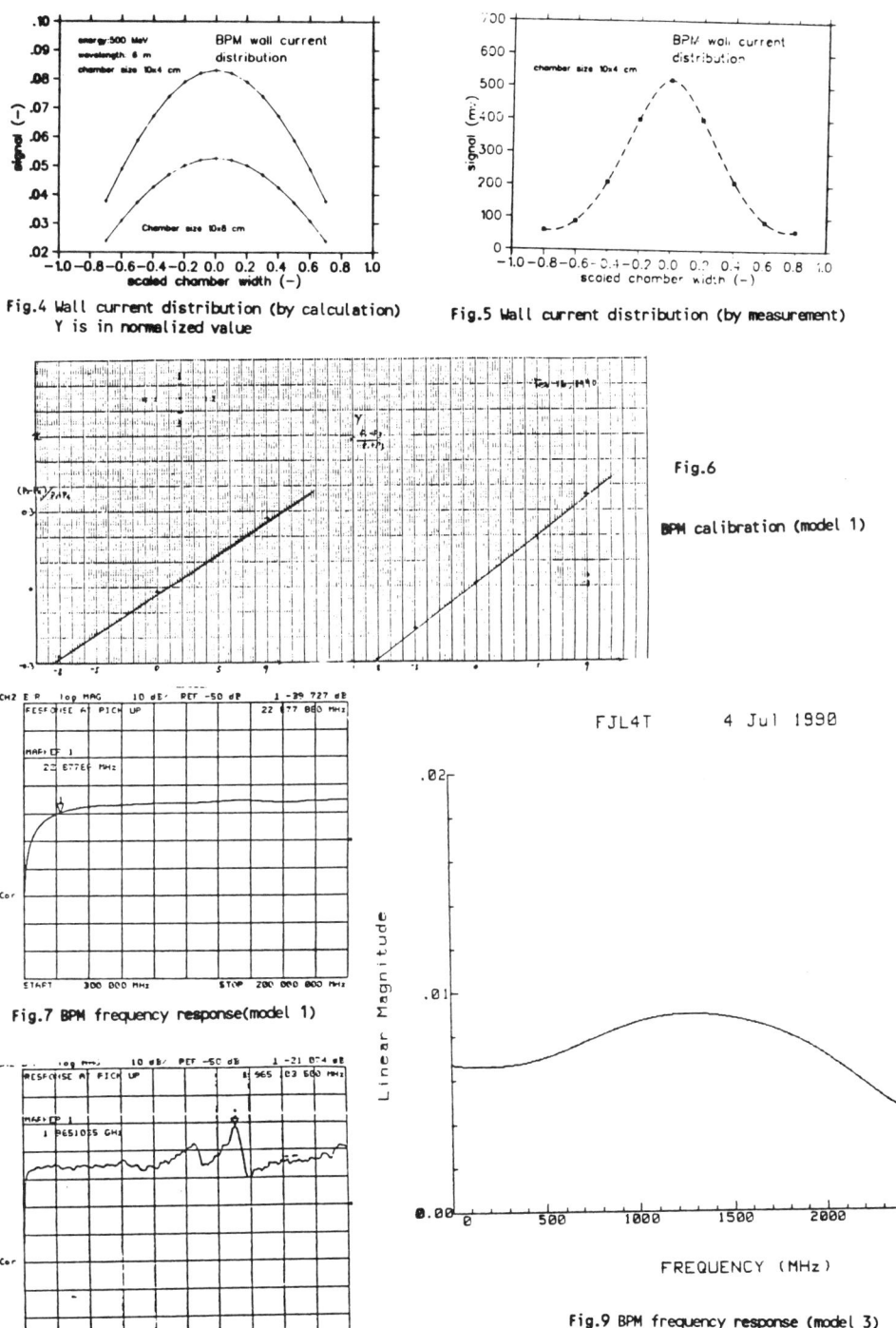

Fig.4 Wall current distribution (by calculation) Y is in normalized value

Fig.5 Wall current distribution (by measurement)

Fig.6 BPM calibration (model 1)

Fig.7 BPM frequency response(model 1)

Fig.8 BPM frequency response(model 1)

Fig.9 BPM frequency response (model 3)

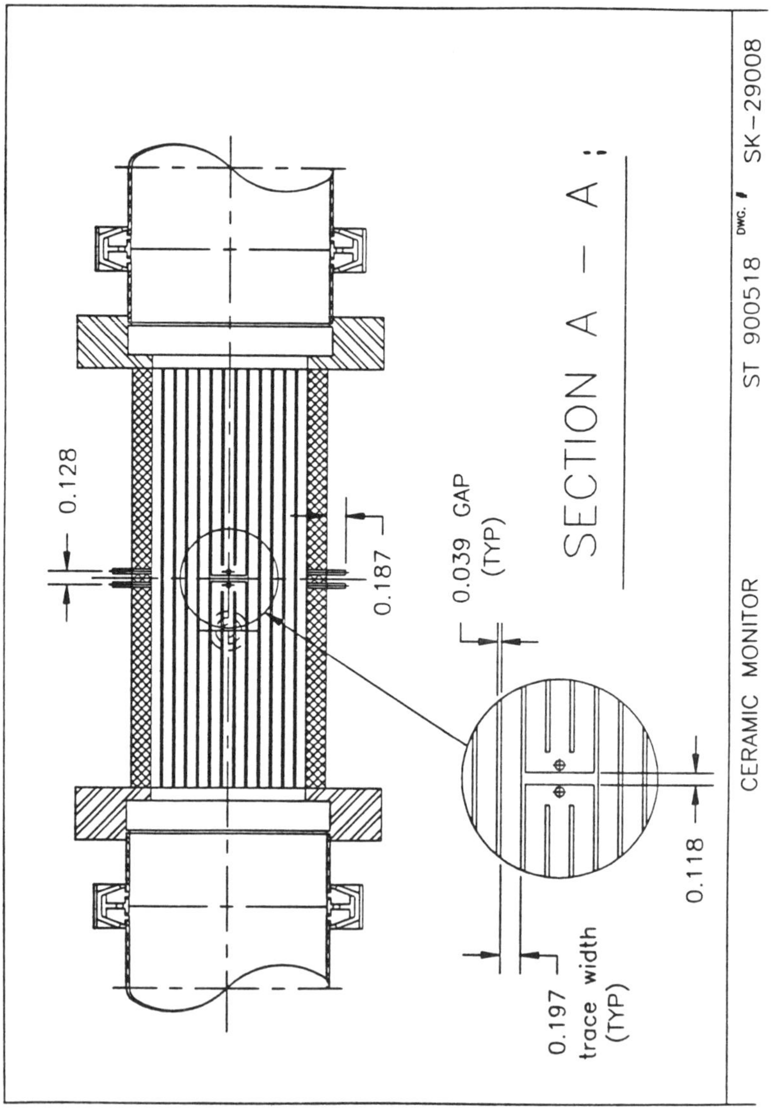

242 A Wall Current Beam Position Monitor

Fig. 10 A wall current BPM with a ceramic chamber

A Fast Beam Loss Monitor System for the KEK Proton Synchrotron Complex

J. A. Holt, J. Kishiro, D. Arakawa and S. Hiramatsu

National Laboratory for High Energy Physics
Oho 1-1, Tsukuba, Ibaraki, 305, Japan

Efforts to increase the intensity of the KEK proton synchrotron (PS) have led to the need for a new fast response beam loss monitor system. The design and some preliminary test results of a new beam loss monitor system were presented. The same detector, a beryllium oxide cathode electron multiplier tube, is used throughout the accelerator complex. The time resolution of the detector is good enough to see the bunch structure of the PS. This is useful not only for tuning but also for the study of beam-loss dynamics. The detector signal is digitized, read into a computer and displayed after each acceleration cycle, enabling quick visual feedback of parameter tuning. Spatial characteristics of the beam loss and the ability to track the loss of a single bunch through most of the acceleration cycle are among several modes of display

TUNED-ANTENNA DRIVER FOR MICROSTRIP PROBE SENSITIVITY TESTING

R. B. Shurter and J. D. Gilpatrick
Los Alamos National Laboratory, Los Alamos, NM, 87545

A tuned-antenna system addresses several needs when used in the microstrip-probe sensitivity-characterization system. As the antenna is swept across the probe's aperture, the power level (as seen by a single lobe) changes by as much as 30 dB for a large diameter probe. To keep the pickup signal well above the noise level and within the power sensor's range (-70 dBm to -20 dBm for the HP 8484A), the signal generator's output power must be efficiently coupled to the drive antenna. Whereas previous setups relied on resistive matching, the tuned approach is more efficient. This approach allows for better sensing of the probe aperture over the dc resistance method that we used earlier. The older method used an ohm meter to measure the resistance between the wire and the aperture ring. Surface contamination at the point of contact made the resistance reading fluctuate, thereby introducing positional uncertainty.

The new method is based on sensing reflected power. A directional coupler is inserted in the line from the signal generator to the tuned antenna. As the antenna approaches the aperture ring, its resonance changes and sharply increases the amount of reflected power measured by a power meter at the coupled port of the directional coupler. Not only is there an indication of increasing close proximity of the wire to the ring, but the actual contact is much better defined. There is also the advantage of not having to reconfigure the system between dc resistance measurements and rf measurements.

A unique aspect of the new antenna design is the method of locating and tensioning the wire. Rather than stretching a wire between two plates with the probe inserted between them, the wire is attached to a coaxial connector on one end and a ball bearing on the other with the tensioning force supplied by a strong magnet. This precludes the need for precise separation of the plates and resultant wire breakage if not done correctly, as well as disassembly of the fixture to insert or remove the probe. The antenna/ball assembly is simply pulled out of the fixture, the probe installed and re-inserted in the antenna. The magnet repositions the ball repeatedly within ten-thousandths of an inch.

Another aspect of this design is the way the signal return path is implemented. A copper spool encloses the antenna with one end electrically anchored to the probe. The other end is capacitively coupled through the fixture top plate to the outer shield of the signal-generator coaxial feed.

The capacitance of a parallel plate capacitor is given by:
$$C = K \varepsilon_0 A/d .$$

With a dielectric constant K=1 for air, permittivity of free space ε_0=8.854 × 10^{-12} $C^2/(Nm^2)$, area of the plate A = 6.627 in^2, and the distance between plates d=0.015 in, the capacitance is found to be approximately 0.004 μF.

Calculating the capacitive reactance by the formula $X_c = 1/\omega C$, an impedance of approximately 0.1 Ω is obtained; thus the capacitive coupling at 425 MHz presents a low impedance return path, without necessitating contact fingers.

A sketch of our apparatus is given in the Figure.

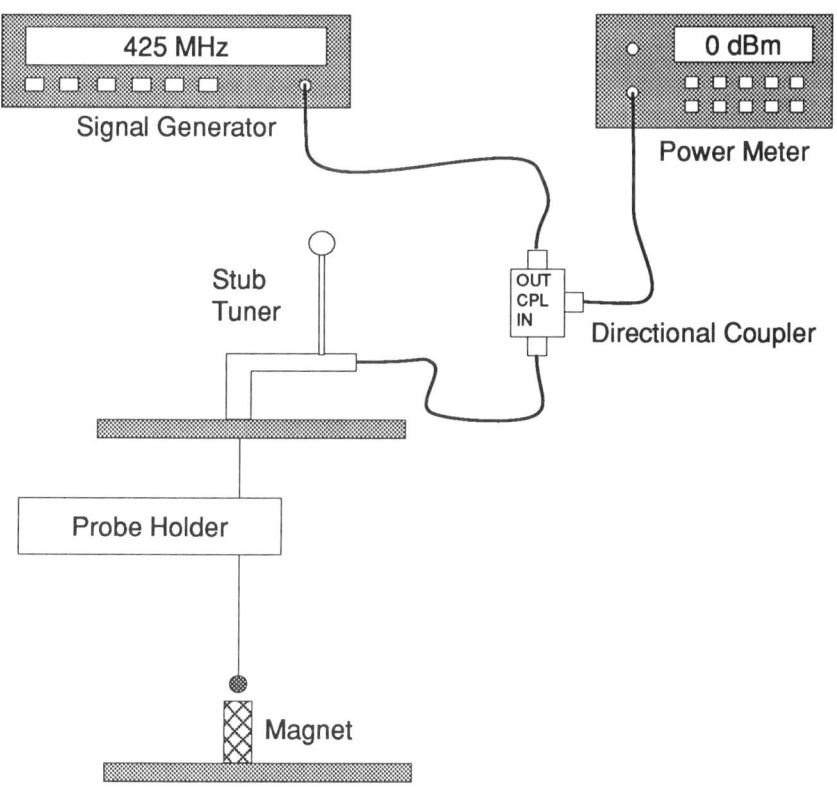

A Controlled Master Frequency Oscillator For The SSC Low Energy Booster

L.K. Mestha
SSC Laboratory*, Dallas, Texas 75237

Abstract

A variable frequency oscillator is normally used to generate the master frequency for an accelerator. For the Low Energy Booster at SSC the prospects of using a Direct Digital Synthesizer (DDS) to vary the frequency from 47MHz to 59MHz are considered. A Digital Signal Processor (DSP) which can be loaded with the desired frequency values is interfaced to the DDS. These values are launched on the DDS at known intervals to ramp the frequency smoothly. In this paper the observations made while ramping the frequency are discussed. The design principle of the digital synchronization of Low Energy Booster frequency source with the Medium Energy Booster frequency source is also briefly discussed. When developed it can be used to synchronize the two machines.

1. Introduction

In this paper, a scheme for generating a controlled accelerating frequency for the SSC Low Energy Booster is outlined. The fully developed system will be capable of delivering the sinusoidal low level rf signal with feedback paths to synchronize the LEB and MEB frequency sources by controlling the phase (not just the frequency) of the LEB throughout the LEB acceleration. This will help in extracting the beam from the LEB at a known MEB turn. Description of the hardware to establish the 'proof of principle' is shown.

2. Master frequency

It is well known that the desired master frequency can be obtained from the relationship shown in reference 1 when the magnetic field profile is known with respect to time. For the Low Energy Booster the frequency curve is shown in Figure 1. The LEB cycle has frequency of 10Hz and the momentum has a pure sine-wave dependency with the LEB cycle frequency. The injection and extraction momenta are 1.219Gev/c and 12Gev/c respectively. The corresponding magnetic fields are 0.12157Tesla and 1.1968Tesla[2]. For these parameters the LEB rf sweeps between 47.518MHz and 59.776MHz. It is clear from the df/dt curve in Figure 1 that a maximum rate of change of frequency of 1036.8MHz/sec takes place at about 5.7ms into the LEB cycle. A digital system should be able to follow the ramp rate and while doing so it should not produce large frequency steps that would cause unwanted beam excursions. A sampling interval of about a microsecond is thought to be adequate; the system under development will be capable of stepping as fast as 0.18μsec interval. This means, the highest step in the revolution frequency will be 1.728Hz (assuming a harmonic number of 108).

---

*Operated by Universities Research Association Inc., under contract with U.S. Department of Energy, Contract No. DE-AC02-89ER 40486

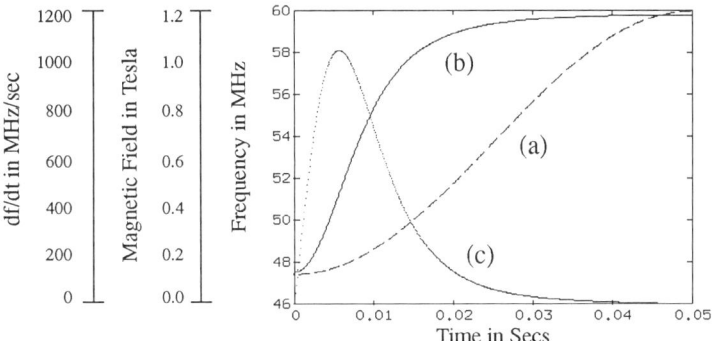

Figure 1.: (a) Magnetic field (b) Frequency, and (c) Rate of change of frequency as a function of time

3. Synchronization Principle

When the field ramps up in the LEB, acceleration of the beam occurs. As the extraction time looms near, the phase of the LEB rf must be adjusted to match the phase of the next accelerator, in this case the Medium Energy Booster (MEB). During transfer the MEB rf is essentially running exactly at the LEB extraction frequency. In the literature one can see two schemes for achieving synchronous transfer of bunched beams from one accelerator to the other[3,4] namely the phase-locking and the phase-slippage schemes. In the phase-locking scheme, at a known time before injection to the next accelerator the phase and frequencies of the accelerating machine are locked to the extraction frequency. The time required to phase lock the two frequencies depends on the lattice parameters of the accelerating machine[1,2]. In the phase-slippage scheme, at a predetermined time before transfer, the frequency of the accelerating machine is offset a few cycles relative to the extraction frequency. As a result, the phase of the accelerating frequency slips relative to the fixed frequency. A consequence of this is several phase coincidence points of the reference wave which occur at a beat frequency equal to the offset frequency. One of them is used to trigger the synchronous transfer of the beam.

A more general approach for synchronization is to consider the 'locking' of the phase of the variable frequency source with the phase of the extraction frequency source throughout the acceleration. Since the fundamental concept of this approach consists of maintaining control of the phase with a suitable feedback mechanism it would be more appropriate to call it 'phase-control' scheme. In this scheme the rf wave (also called the reference rf wave) is forced to follow a preprogrammed 'trip-plan' so that at the instant of transfer it will have the same phase as the MEB rf wave (except for constant phase offset). The feedback loop corrects for any deviation from the 'trip-plan'. The 'trip-plan' has at least one point at flat B-field region where the machines have exact phase for the purpose of synchronous beam transfer. With this scheme it is possible to synchronize two RF sources starting from a random phase difference. When there is no error in the field, we can calculate the time at which the two reference rf waves match for beam transfer. Field deviations make this not valid. The phase of the LEB reference rf wave from a given time during acceleration is equal to

$$\psi_{LEB} = \frac{2\pi}{h} \int_0^\tau Rf(t)\, dt \qquad (1)$$

248 A Controlled Master Frequency Oscillator

Similarly the phase of the MEB reference rf wave can be written as:

$$\psi_{MEB} = \frac{2\pi R'}{h'} f' \tau \qquad (2)$$

The notations R and R', the radii of the orbits, f and f', the rf frequencies, and h and h', the harmonic numbers of the LEB and the MEB, are used respectively. τ is the time interval over which the integration is carried out. The difference of equations 2 and 3 is the 'synchronizing phase', ψ. That is,

$$\psi = |\psi_{MEB} - \psi_{LEB}| \qquad (3)$$

For synchronous transfer, the synchronizing phase is equal to zero (ignoring the transfer line delays). Since the present discussion is for two circular machines with non-integer circumference ratios (22/3), at the end of one MEB turn, the LEB reference rf wave would have completed several full turns and a semi-turn. If the two frequencies are the same, then the LEB reference rf wave completes 7 full turns and a semi-turn equal to one-third the LEB circumference, which is the case with flat B-field. The semi-turn is not one-third the circumference ratio when the frequencies are different. Equation 3 is plotted in Figure 2 for each MEB turn. The '+' marks represent the position of the LEB reference rf wave for each MEB turn away from the transfer point. The phase values are shown from 0 to 540 metres to cover the full circumference of the LEB. Path length representing full turns completed by the LEB reference wave is ignored, since it is of no significance for synchronization. In Figure 3 the position of two rf reference waves are shown for the first four turns followed by the last four turns of the MEB. The transfer line delays are ignored for simplicity. Numbers with extension 'm' represent the path length to be completed by the LEB reference rf wave in metres to reach the transfer point after each MEB trun. In Figure 2 three curves approach constancy at 50 msec; this is due to the fraction "1/3" in the circumference ratio. The decay is due to the fact that the difference in frequencies is narrowing as the time approaches the transfer time at which the magnetic field is maximum. In this region every third point is on the same curve. This means the relative phasing

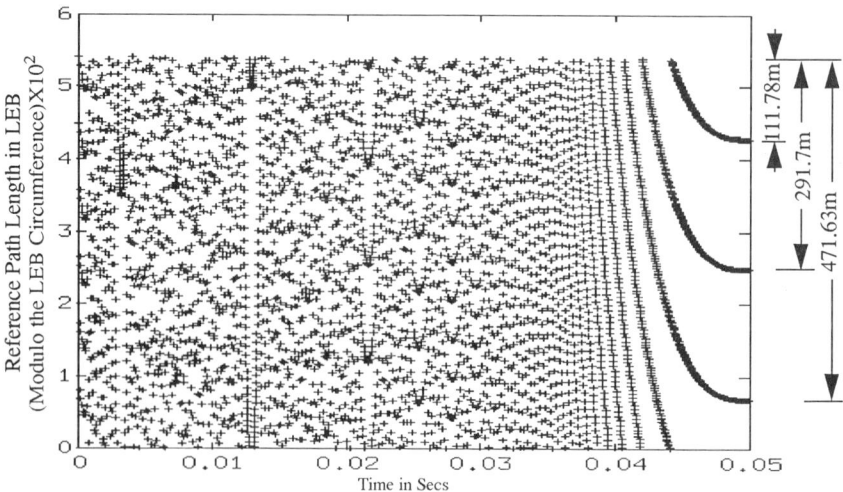

Figure 2.: Synchronizing phase during acceleration in the LEB;Phase not adjusted

between the two reference waves is constant every time the MEB reference wave completes three turns. This is clear from Figure 3 by comparing turn numbers 3771 and 3774.

The synchronizing phase as seen from Figures 2 and 3 is not zero for transfer purpose. Also, if the LEB reference wave is controlled to the 'trip-plan' shown in Figure 2 there may be synchronization with the MEB reference wave, but transfer cannot be matched. Instead if the 'trip-plan' is offset by 111.78 metres right at the beginning and an arbitrary rf wave is arranged to follow the new synchronizing phase, then there will be several transfer points. The new 'trip-plan' is shown in Figure 4 for last ten milliseconds. Figure 5 is same as Figure 3, but with the new LEB phase of Figure 4. From Figure 5 it is clear that the turn numbers 3771 and 3774 have zero phase.

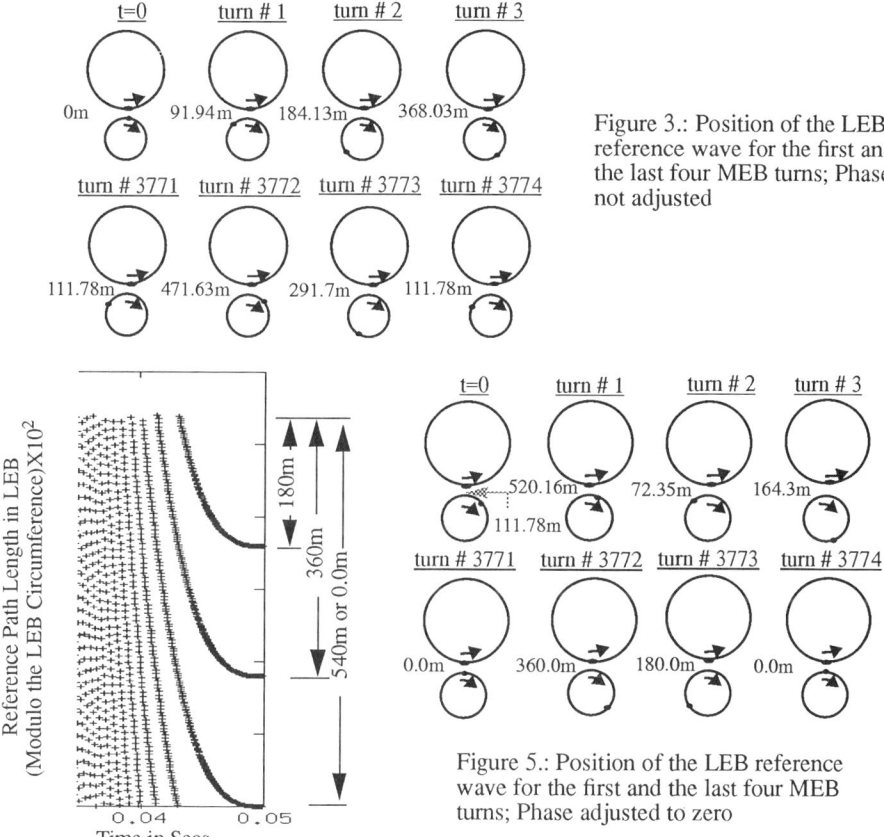

Figure 3.: Position of the LEB reference wave for the first and the last four MEB turns; Phase not adjusted

Figure 5.: Position of the LEB reference wave for the first and the last four MEB turns; Phase adjusted to zero

Figure 4.: LEB 'Trip-Plan'; Phase adjusted to zero at transfer time

4. System Model

The system model can be obtained by introducing error terms for the beam orbit and the frequency in equation 1. Let $\delta R(t)$ be the deviation in the radius of the orbit and $\delta f(t)$ be the error in the master frequency, then the phase equation becomes:

$$\psi'(t) = \frac{2\pi}{h} \int (R + \delta R(t))(f(t) + \delta f(t)) \, dt \qquad (4)$$

The sign of $\delta R(t)$ depends on which side of transition the machine is operating. Since the control philosophy is to maintain the 'trip-plan' shown in Figure 4 by modulating the LEB frequency, an equation for the error in synchronizing phase in terms of the frequency is required. This phase error is equal to the offset in the path length covered by the LEB reference wave from the ideal path. Hence it is given by:

$$\delta\psi(t) = \psi'(t) - \psi_{LEB}(t) \tag{5}$$

Substituting equations 1 and 4 in equation 5 and taking the first derivative with respect to time, the following differential equation is obtained.

$$\frac{d\delta\psi(t)}{dt} = \frac{2\pi}{h}\left(\delta R(t)f(t) + R\delta f(t) + \delta R(t)\delta f(t)\right) \tag{6}$$

In equation 6, in the absence of B-field error (i.e., $\delta B(t)=0$), $\delta R(t)$ can be related to frequency in the usual way as follows:

$$\frac{\delta f(t)}{f(t)} = -\frac{\delta R(t)}{R}\left(1 - \frac{\gamma_T^2}{\gamma^2(t)}\right) \tag{7}$$

where $\gamma(t)$ is a function of frequency. Substituting for $\delta R(t)$ from equation 7 in equation 6, a first order non-linear differential equation is obtained.

$$\frac{d\delta\psi(t)}{dt} = \frac{2\pi R}{h}\left(\frac{\gamma_T^2}{\gamma_T^2 - \gamma^2(t)}\right)\delta f(t) + \frac{2\pi R}{hf(t)}\left(\frac{\gamma^2}{\gamma_T^2 - \gamma^2(t)}\right)\delta f^2(t) \tag{8}$$

Equation 8 shows that the first term is time dependent due to the ramping in frequency. The second order terms are due to the errors in the radial position. If the phase error is known, then a feedback loop can be arranged to control $\delta\psi(t)$ to be near zero.

A scheme to calculate the phase error by measuring the LEB phase is described below. By knowing this quantity it is clearly possible to set up a feedback control loop which can modulate the LEB frequency to reduce the error. A feedback controller based on Sliding-Mode techniques which accommodates the time-varying nature of the system is being considered. Since the loop gain of such a system is infinite, the system is not susceptible to parameter variation and external disturbance within the limits set by the controller [5] constants. Detailed design associated with the controller is discussed in reference 5. In this paper the equations which calculate the frequency shift are quoted. At first, a sliding line is calculated by knowing the measured phase error, which is equal to

$$S(t) = \delta\psi(t) + c\int \delta\psi(t)\,dt \tag{9}$$

where 'c' is the eigen value which is responsible for fixing the response time of the loop. A control function is obtained by monitoring the sign of the function $S(t)$, given by

$$u(t) = -(k_1|\delta\psi(t)| + k_0)\,\text{sgn}\,S(t) \tag{10}$$

where $\text{sgn}\,S(t)$ is +1 when $S(t)$ is positive and -1 when $S(t)$ is negative. and $|\delta\psi(t)|$ is the absolute value of the phase error. k_0 & k_1 are positive constants and must be set so as to make the loop stable and more robust. After this the frequency shift, $\delta f(t)$ is calculated from the following equation.

$$\delta f(t) = \frac{h}{2\pi} u(t) - (\frac{h}{2\pi})^2 \frac{1}{f(t)} (\frac{\gamma(t)}{\gamma_T})^2 u^2(t) \qquad (11)$$

The coefficient of $u^2(t)$ is constant for a given MEB turn. These are stored in the DSP at each MEB turn or however many times the corrections are needed. Values of the constants c, k_0 and k_1 are also stored in the DSP. The control functions satisfy the global stability of the loop by satisfying the conditions set by Lyapunov Stability Theory. In Figure 6 the synchronizing phase error is plotted against time for $c=2000$, $k_1=25$ and $k_0=0.5$ by assuming an initial phase error of 5 metres. It can be seen that the phase error becomes zero within the first 10 milliseconds. Loop response can be made slow by suitably selecting the controller constants. In Figure 7 the frequency shift, $\delta f(t)$ is shown with respect to time.

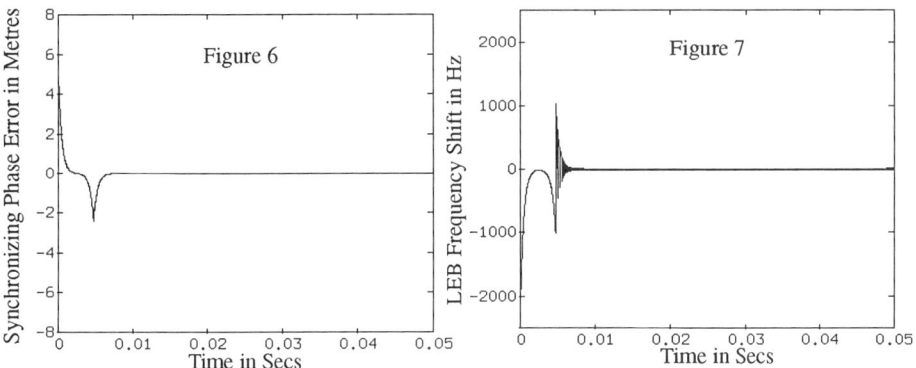

Figure 6 & 7.: Decay of the synchronizing phase error, $\delta \psi(t)$ and the frequency shift, $\delta f(t)$ as a function of time when the feedback loop is closed.

5. Phase Measurement

Equation 8 above has given an indication of the system to be controlled. Using this system equation the controller functions shown by equations 9, 10 and 11 were obtained to correct for errors in the synchronizing phase. An accurate detection of the phase error is important to guarantee the controller operation. The phase error can be detected for one or two MEB turns depending on the magnitude of the error.

Waiting for much longer than one MEB turn to measure the LEB phase could lead to large error. The consequences of this will be greater frequency shift generated by the feedback controller, which may lead to beam oscillations. Since the MEB completes about 3774 turns before it is ready to accept the beam from the LEB, if the phase is measured at every MEB turn, then there will be 3774 points in the LEB cycle where the frequency shift has to take place. This is possible if the hardware can extract the phase error from the measured LEB phase and then solve equations 9, 10 and 11 within the available time. One of the ways with which the LEB phase can be measured approximately is by time-tagging the arrival time of the LEB reference wave from the instant the MEB reference wave completes the turn. In the development system considered here, a Time to Digital Converter (TDC) will be used to record the arrival time of the LEB reference wave. Using this time of flight information, the phase of the LEB reference wave is calculated in terms of the path length by using the following equation,

$$\psi'_{measured}(t) = 2\pi(R + \delta R(t))\frac{f(t) + \delta f(t)}{h}\tau' \quad (12)$$

where τ' is the arrival time of the LEB reference wave, $\delta f(t)$ is the frequency shift of the LEB frequency in the previous MEB turn, and $\delta R(t)$ the average radial position offset during the time-tagging period. Equation 12 above assumes no frequency shift while measuring the arrival time of the LEB reference wave. The phase error is calculated by subtracting the measured LEB phase values with those shown in Figure 4 for a given MEB turn. This will be done inside the DSP in real time.

6. Hardware Description

The choice of hardware was driven by the need to demonstrate the basic principle established above to synchronize the variable frequency source of the LEB with the fixed frequency source of the MEB. Three stages of development are planned. The first stage will be to prove that the phase of the LEB can be controlled to the inexorable 'trip-plan' to achieve synchronization at the designated MEB turn, in the absence of field errors. The second stage will be to insert errors in the B-field and study the effect of the control loop. In the third stage the feedback controller will be redesigned to include the radial position correction. When all these three stages work together, then the system will be delivering controlled master frequency to the accelerator cavities.

A schematic representation of the system block diagram for the first stage is shown in Figure 8. Two 32 bit floating point Digital Signal Processors, TMS320C30, from Texas Instruments were used as the processing units. The configuration of the C30 DSP board is made by

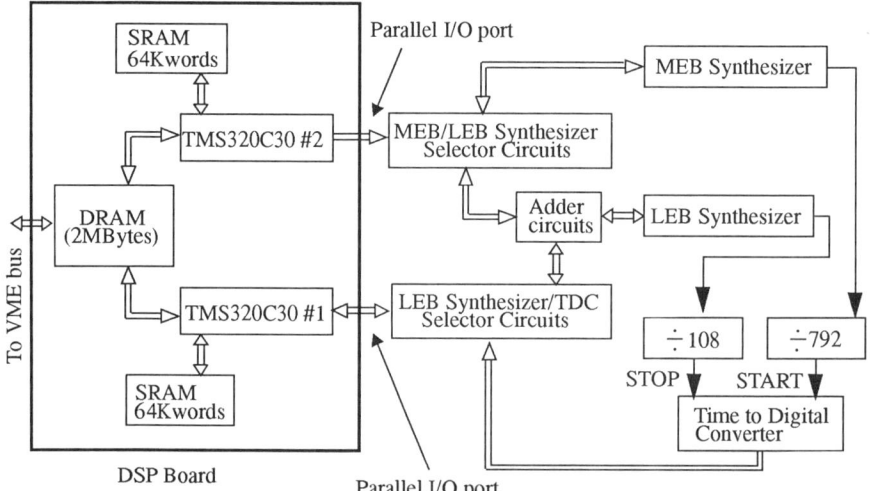

Figure 8.: Schematic representation of the System Block Diagram

SKY Computers. The DSP board has 2Mbyte of Dynamic RAM onboard and the usual control registers to interface to the VME bus. All the communication to the UNIX machine is done through the VME interface bus. The Node RAM in each C30 processor is a static memory and is 64Kwords long. The first C30 is used to generate the master frequency by reading the frequency data stored in the SRAM and then launching it to the LEB frequency synthesizer at the fine time intervals mentioned above. The MEB frequency synthesizer is running at the LEB extraction frequency. The counters, dividing by 108 and 792, represent the number of rf waves in one revolution in the LEB and the MEB respectively. The TDC measures the arrival time of the first LEB reference pulse after every MEB reference pulse. The data is then read into the second C30 to solve equations 9, 10, 11 and 12 by reading appropriate values from the global memory. The frequency shift $\delta f(t)$ is fed to the adder circuits to correct the master frequency. The operation of the feedback loop is repeated every MEB turn. The feedback loop must execute the following functions within one MEB turn which is about 13.249µs long. (a) 'Time of arrival' measurement using the TDC (b) Handshaking between the TDC and the DSP for data acquisition (c) LEB phase from equation 12 (d) Phase error by subtracting the results of (c) from the 'trip-plan' (e) Frequency shift by solving equations 9, 10 and 11 (f) Addition of the frequency shift with the master frequency. Due to the simple control algorithm shown above and with the aid of a C30 processor, it is possible to compute all the functions within the available time.

6. Conclusions

In this paper the principle behind generating the controlled master frequency is shown. It is important to note at this stage that the benefit of this approach is subjected to how well the system works on the real accelerator. The greatest advantage of the control scheme is the ability to select any bunch in the Low Energy Booster and then transfer to the Medium Energy Booster. This helps to include gaps in the Low Energy Booster to prevent radiation on the extraction beam line due to finite kicker rise time. Present day technology, has provided Direct Digital Synthesizers which can sweep the LEB frequency at more than a 5MHz rate, combined with 32 bit frequency resolution and RF to 400MHz. The phase discontinuity for all bit combinations needs to be tested before they can be used on the accelerator. From the feedback point of view it is believed that the DSP, TDC and DDS combination can produce the required control function. Also, since a DSP is used in the system, it will be easier to implement the control function when an additional feedback loop is needed to contain the transverse beam orbit within the specified limits.

7. Acknowledgments

Many people are involved in the development of this system. The author wishes to thank John Mangino and Jim Santana for their direct involvement in the engineering aspect of the project. The author appreciates Richard Talman and Bob Webber for their insight and timely suggestions. Technical discussions with Prof. Kai Yeung on the controller design are greatfully acknowledged.

8. References

1. D.J. Martin, et al.: ' Early Instrumentation Projects at the SSC', International Industrial Symposium on the Super Collider (IISSC), Miami (USA), March 14-16, 1990
2. Site-Specific Conceptual Design of the SSC, SSCL-SR-1056, July 1990.
3. Y. Kimura, et al.: 'Synchronous transfer of beam from the booster to the main ring in the KEK proton synchrotron', IEEE Transactions on Nuclear Science, Vol. NS-24, No. 3, June 1977.
4. S.R. Koscielniak.: 'RF Synchronization during transfer of batches from Booster to Collector', TRIUMF Design Note, TRI-DN-89-K74, October 1989.
5. L.K. Mestha.: 'Phase-Control Scheme for Synchronous beam transfer from the Low Energy Booster to the Medium Energy Booster', SSC Laboratory Report, SSCL-340, 1990.

A NEW SCHEME FOR MEASURING THE LENGTH OF VERY SHORT BUNCHES AT CEBAF*

C. G. Yao
Continuous Electron Beam Accelerator Facility
12000 Jefferson Avenue, Newport News, VA 23606

ABSTRACT

The CEBAF injector is designed to bunch a 100 keV beam chopped to 60° to a very short bunch of less than 1° at 5 MeV energy. Presented are a new scheme[1] for measuring the phase distribution in the bunch that gives information about bunch length, and preliminary experimental results that show a resolution of better than 0.1° in phase at 1.5 GHz (or 0.2 ps in time). The advantages of the scheme compared with existing methods for bunch phase width measurement, such as using streak cameras, measuring the transverse spot size of the bunch by deflecting the beam[2], measuring the energy spread by bunch traversing a section of accelerating structure at zero cross phase, or observing transition radiation[3], are: high resolution is obtained; it is suitable for any level of energy; and it is inexpensive. The scheme is not easy to apply to an intense beam where the space charge effect has an important impact on bunching.

INTRODUCTION

A superconducting accelerator at CEBAF will provide a high quality CW beam of energy from 0.5 GeV to 4.0 GeV. The relative energy spread (95% of particles) of the machine will be about 1×10^{-4}. This requires that the injector must supply a well-bunched beam to the main accelerator of phase width less than 1° at 1.5 GHz. Obviously, a method to measure the length of the very short bunch is of importance. There are several different methods available for measuring bunch length, including 1) a streak camera, 2) measuring the transverse spot size of the bunches after traversing a deflecting structure, 3) measuring the energy spread with a spectrometer after beam passage through a section of accelerating structure at zero phase, and 4) measuring the transition radiation emitted by the beam. After careful investigation we conclude that these methods are expensive and that the resolution of these methods is not good enough to measure the bunch length of the well bunched beam from the CEBAF injector (PARMELA shows that the anticipated phase width of a 5 MeV bunch could be as small as 0.5° at the entrance to the first full cryogenic unit if all parameters are set optimally). Also, some of these methods do not work well in the lower energy regions of the injector. The method proposed here is

---

* This work was supported by the U.S. Department of Energy under contract DE-AC05-84ER40150.

to measure the phase distribution within the bunch by detecting a phase shift of fields induced by a series of sub-bunches in a cavity while sweeping the phase of the chopper system with respect to the rest of the injector. The sub-bunches are obtained by having a narrow slit at the chopping aperture.

PROPOSED SCHEME

A layout of the CEBAF injector and the scheme for measuring bunch length in this system is shown in figure 1. The injector is comprised of a 100 keV gun, two identical square pillbox chopper cavities, a buncher, a side-coupled capture section, two superconducting five cell cavities (at this point beam energy is 5 MeV), and two full cryogenic units (each of them consists of eight cavities). The output energy is expected to be 45 MeV. Two identical dipole modes polarized perpendicular to each other are excited in the chopper cavities. A CW beam passing by the first chopper cavity, where the two modes have equal amplitudes and 90° phase difference, will project onto a circle with a diameter of $\approx$ 1 cm on a slit plane which is about 0.5 m downstream from the first chopper cavity. A 60° radial slit allows 60° out of every 360° to pass the chopper system. There are lenses on both sides of the slit. The second chopper cavity deflects the beam back to the axis if its RF phase is correct with respect to the first cavity. In the experiment a cavity which is used for picking up beam fourth harmonic signal is located in the 5 MeV region (wherever the bunch length is to be measured) about two meters downstream from the output of the second superconducting cavity.

The CEBAF accelerator operates at 1.5 GHz. The cavity is a simple cylindrical one without nose cones, operating in a TM_{020} mode at 6 GHz to detect the fourth harmonic of the beam induced fields for higher sensitivity. It is made of stainless steel, and has a radius of $\approx$ 4.51 cm and a length of 2.5 cm. Its loaded Q value is about 1000. The radius of the beam pipe is 1.2 cm. The R/Q of the cavity is 24.6 Ω. It is expected that for a 10 μA average current of the sub-bunches, the output signal from the cavity is about -26 dB.

The slit system consists of two slits, one of them has 60° radial open angle for normal beam operation, and the other has about 10° open angle for measuring the bunch length. During normal operation the 60° slit is used. The injector is tuned by minimizing the energy spectrum of the output beam with help of a spectrometer. Then the 10° slit is moved to the position where the 60° slit used to be. A sawtooth waveform signal is applied to the x terminal of the oscilloscope and to an electronic phase shifter, which sweeps the phase of both chopping cavities around the set point, i.e., with respect to the rest of the injector. The signal from the double balanced mixer is connected to the y terminal of the oscilloscope. As a result, the phase of the centroid of the sub-bunch with respect to initial phase is shown on the screen of the oscilloscope. The bunch length is determined by measuring the vertical width of the curve after calibration.

256 Measuring Length of Very Short Bunch at CEBAF

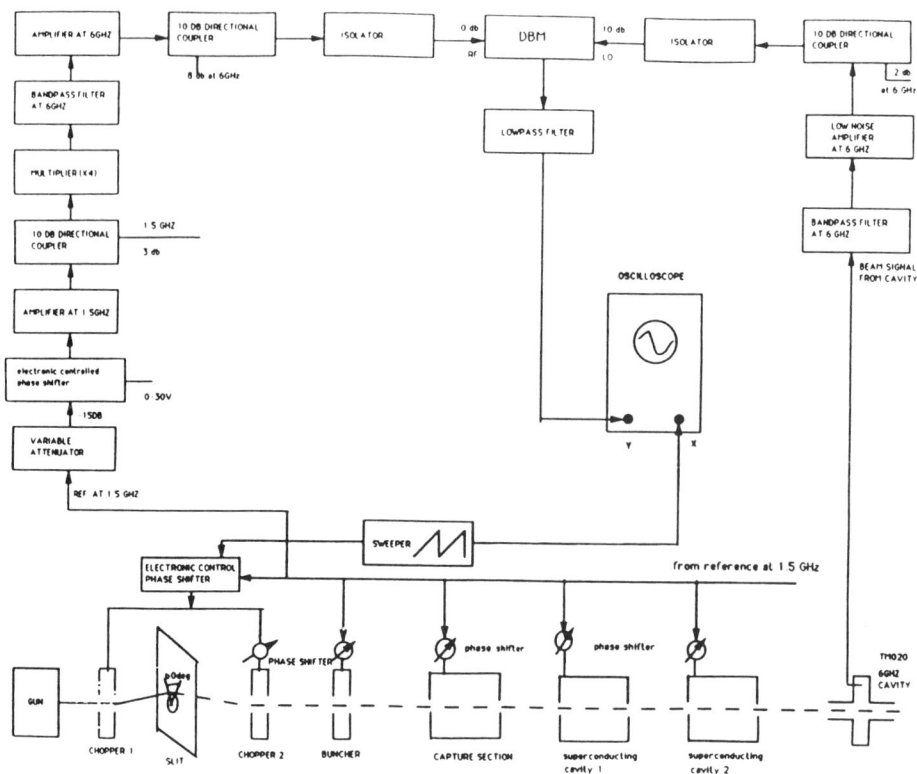

Figure 1. Bunch length measurement.

SEVERAL CONSIDERATIONS

First, the scheme is not easily applied to cases where space charge effects play an important role. Basically, this scheme is good for CW accelerators and, perhaps, for high duty factor accelerators also.

Second, because the beam current is constant for each sub-bunch series, the beam loading effect during phase sweeping is ignorable to the first order, as long as sweep phase ranges within the phase capture region of the injector. The beam loading is very small in the room temperature components of the CEBAF injector. By contrast, the superconducting cavities are highly beam loaded. The filling time for the room temperature devices is of order a microsecond and for the superconducting cavities is of order a millisecond. If the sweeping speed is slow enough, the feedback system that stabilizes the RF amplitude and phase in each device will eliminate the second order beam loading effect that is caused by the slightly different energy gain of each sub-bunch in the device during the measurement. In the experiments the phase sweeping frequency is 2 Hz.

Though the initial phase width of each sub-bunch is a constant and determined by the narrowed slit open angle, the final phase width of each sub-bunch varies during phase sweeping. By Fourier analyses, the amplitude of the fourth harmonic of beam-induced field is

$$a_4 = 2a_0 \frac{\sin 2\phi}{2\phi} \tag{1}$$

where a_0 is the average current which is constant, and ϕ is the final sub-bunch phase width. It is seen from eq. (1) that even if the final phase width of the sub-bunch varies from 0.1° to 2°, the relative variation in the amplitude of the fourth harmonic is under 0.1%.

INITIAL EXPERIMENTAL RESULTS

Figure 2 shows a signal of beam-induced field from the fourth harmonic cavity. Figure 3 gives experimental results with different phase sweeping ranges. The horizontal axis shows the time in units of 100 ms per division. As mentioned above, the sweeping frequency is 2 Hz. The vertical axis shows the detected phase of the centroid of the 10° sub-bunch. The sensitivity is calibrated to 1.7 mV per degree at 6 GHz. Figure 3(a) is a case without phase sweeping. It shows the bunch phase jitter. The peak to peak is ≈ 1.7 mV, or 1° at 6 GHz, yielding 0.25° bunch phase jitter at 1.5 GHz. This result is consistent with what we expect from the specifications on variations in RF amplitude and phase in each of the upstream components. Averaging gives a straight line parallel to the x axis. Figure 3(b) shows a case with 25° phase sweeping. The peak to peak is 2.1 mV, or 0.31° at 1.5 GHz, and the variation on average is 1.2 mV, or 0.18° at 1.5 GHz. Figures 3(c) and 3(d) show results when the phase is swept through 50° and 100°, respectively. The peak to peak phase fluctuations are 0.78° and 1.6°, and on average the phase variations are 0.5° and 1.5° at 1.5 GHz, respectively. The measurements are summarized in figure 4. Curve 1 shows the intrinsic bunching performance of the injector. Curve 2 shows the bunch length including time-dependent errors. It is possible that time-independent phase error, for example from energy spread caused by the buncher or chopper cavities, sets a limitation to the performance of the injector. The measurements may not be sensitive to the effects of time-independent errors. However, such effects should not normally have an obvious impact on the bunching characteristic of the injector. The measurements show that the phase jitter is 0.2°, and they are consistent with a total bunch length of 0.9° at 1.5 GHz.

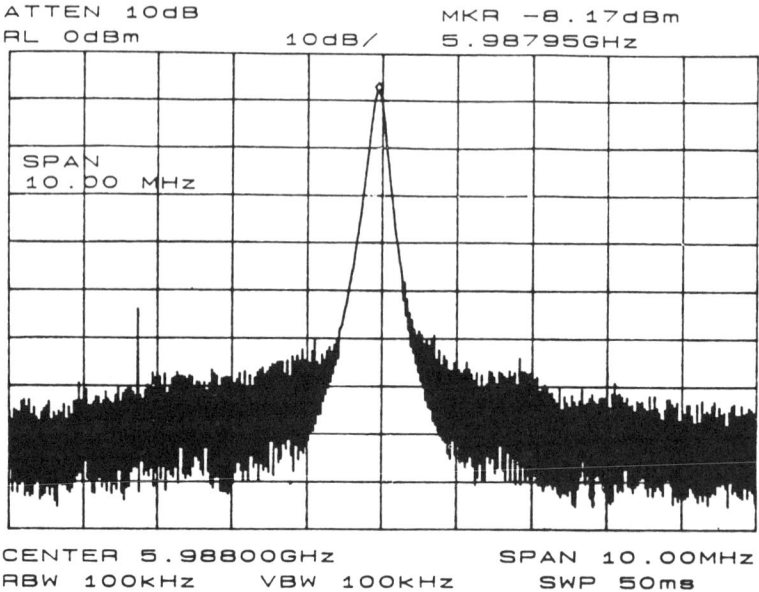

Figure 2. Signal of beam-induced fourth harmonic from the cavity.

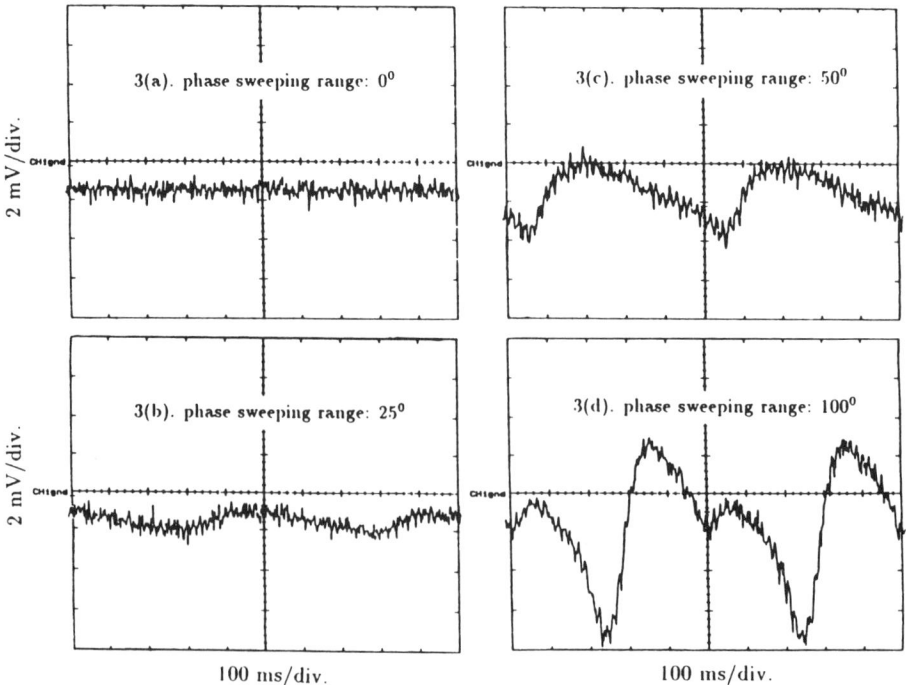

Figure 3. Preliminary experimental results of bunch length measurements.

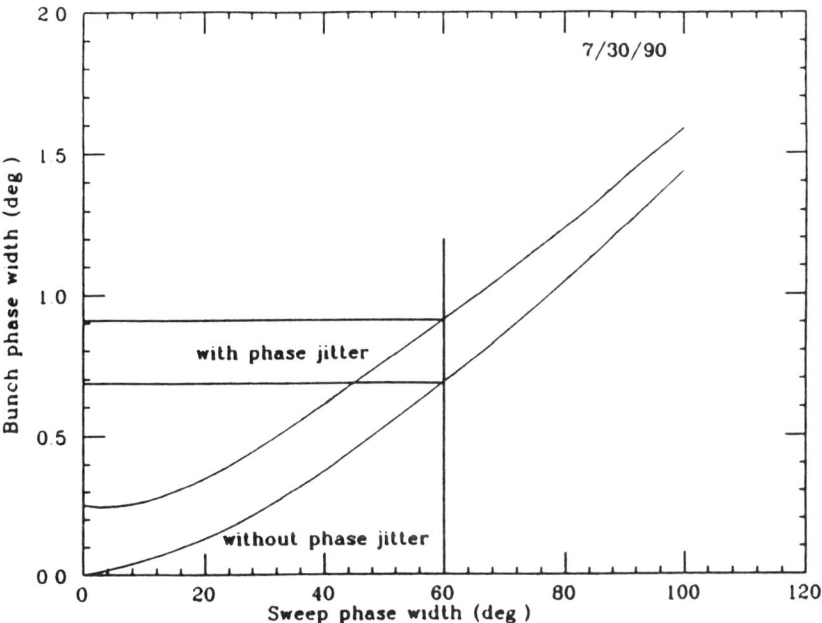

Figure 4. Final bunch width vs. initial phase width.

ACKNOWLEDGMENTS

The author would like to thank C. Leemann and C. Sinclair for their consistent support of this work, and R. York and G. Krafft for helpful discussions.

REFERENCES

1. C. G. Yao, CEBAF TN-131, June 1989.
2. G. Loew, et. al., 5th International Conf. on High Energy Accelerators, Frascati, p. 551 (1965).
3. R. B. Fiorito, et. al., Proc. European Particle Accel. Conf., Rome, Vol. 2, p. 1078 (1988).

Instrumentation at the Bates Linear Accelerator Center

J.B. Flanz, E.E. Ilhoff, K.D. Jacobs, T. Russ
MIT Bates Linear Accelerator Center
P.O. Box 846
Middleton, MA 01949

Introduction

The Bates Accelerator Center is a small laboratory for nuclear physics operated by MIT. It consists of a 500 MeV electron linac and a magnetic beam recirculation system to allow energies up to 1000 MeV to be achieved. In addition there are 3 experimental halls and a total of 5 beam lines. Also a 1 GeV storage ring is under construction. The nuclear physics program demands high quality beams of all energies in the range between 50 MeV to 1 GeV. The beam emittance is $10\pi/\gamma$ mm-mr with an energy spread of about 0.3%. A schematic layout of the facility is shown in Figure 1. Over the past few years the instrumentation and diagnostics capabilities installed in the machine have been improved to allow more efficient and reproducible tuning and operation.

The beam diagnostic systems we are developing are primarily designed to help set the accelerator/recirculator and beam line magnetics to achieve to appropriate magnetic optics and beam phase space. Some of the systems we have installed will be described. The small laboratory staff, ongoing experimental program and ring construction have contributed to stretching out the implementation of these systems for normal operation. However we have already made progress with encouraging results.

Beam Profile Measurements

A consequence of the low emittance is that weak focussing is generally used in single pass beam operation. Also there is a fairly wide band pass in energy for a given focussing solution. However focussing two beams of different energy, as is done when recirculating the electron beam to achieve high energy, is more difficult. The focussing strength must be stronger to transmit the second pass beam through the accelerator through apertures as small as 12mm. However that strength must not be excessive for the lower energy single pass beam.

It is important to provide an operational procedure for the determination of the focussing solution. We have done so by implementing a system of wire scanners for beam profile measurements in the linac. The system is interfaced with a computer display which is extremely important for operator interaction. Each scanner consists of a horizontal and vertical 20 μm wire driven across the beam by a stepper motor which moves a specified number of steps after each beam burst (up to 1KHz). The system is capable of steps as fine as 0.5 mils. An accumulation of many

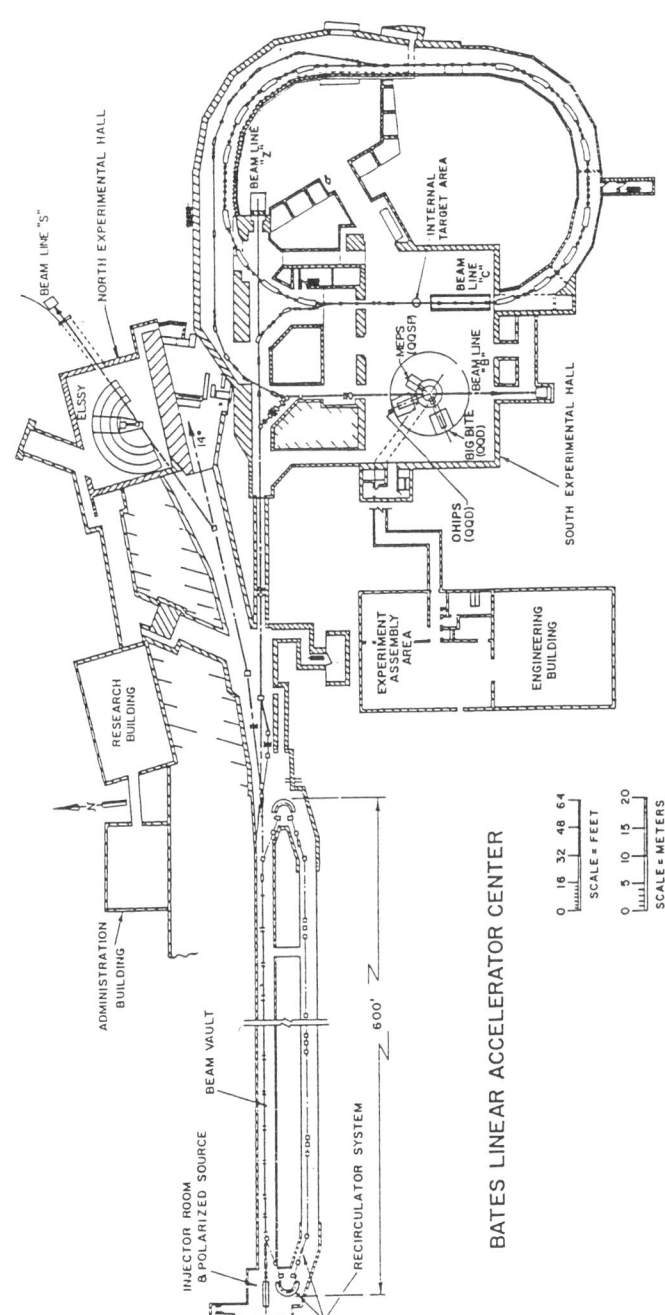

Figure 1. Schematic Layout of Bates Accelerator Center

beam profile measurements indicates a combination of the wire scanner and beam reproducibility. The tests indicate reproducibility and/or resolution to 10 mils.

There are 13 wire scanners installed in the accelerator allowing for a more or less complete beam profile measurement through the accelerator. Figure 2 shows the raw data of 5 wire scanners in a case when there are two beams in the accelerator during recirculation. Figure 3 shows the beam envelope formed from the beam profile measurements for a single pass beam. Another display we use is shown in figure 4. It compares the beam outline to the known apertures at various points in the accelerator.

Beam Phase Space Measurement

The above described wire scanners are also installed at the entrance to the beam switchyard. The beam phase space can then be measured. This is particularly useful since different beam lines require different input beam phase space orientations for proper matching. At this time the beam phase space is determined by measuring the beam profile at three points in a drift space. We use a separation of 10 m between lutes and expect to be able to measure the emittance of a .01mm-mrad beam to within 10% assuming profile monitor resolution of 25 um.[1] Figure 5 shows an example of the on-line computer display presenting the results of a phase space measurement.

Magnetic Optics Measurements

The beam recirculation system consists of 10 dipoles, 15 quadrupoles and is over 200m long. This system is required to deliver a highly achromatic beam to the accelerator, essentially a long collimator, with an energy band pass of several percent. The system is designed to third order and requires almost spectrometer quality optics. Due to uncertainties in the reproducibility of the magnetics, the best method to tune this beam line is to measure the first order optics automatically. This is done by toggling steering coils at the input to the system in order to vary the input angle and input position separately. The resulting position variations at critical points in the system are measured and the resulting magnetic optics matrix elements are compared to the ideal values. Figure 6 shows the computer display of a "typical" result.

Before the implementation of this system it would take a day to make the proper measurements and perform the analysis. Now several iterations can be accomplished in 15 minutes. This allows easier tuning, and decoupling from beam fluctuations. Further implementation of this type is system in the switchyards and in the storage ring under construction is envisioned.

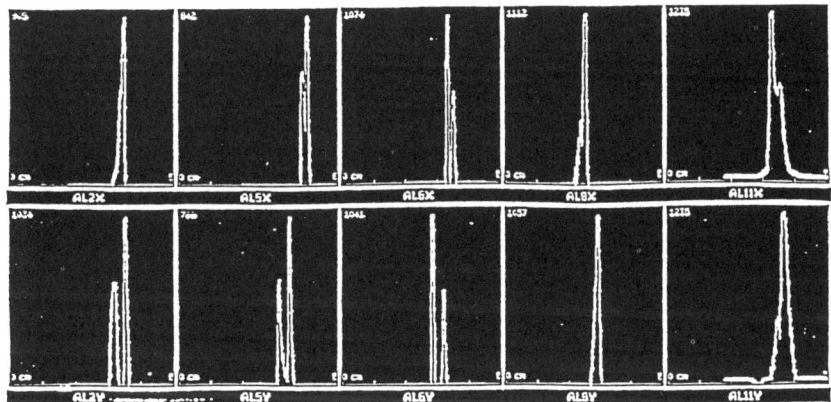

Fig. 2 : Beam profile measurement with two beams in the accelerator. The top row shows horizontal profiles, and the bottom row shows vertical profiles, at different locations in the accelerator.

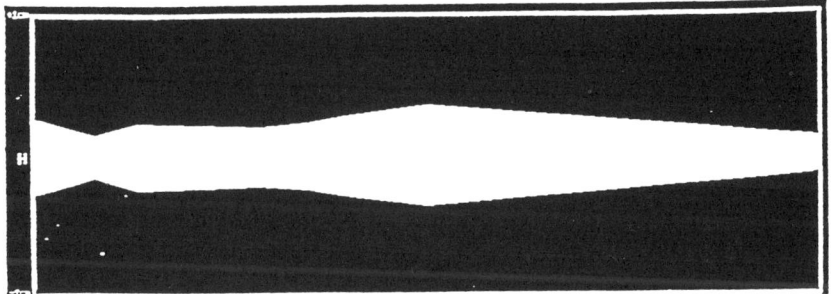

Fig. 3 : Measured Horizontal beam envelope in the accelerators.

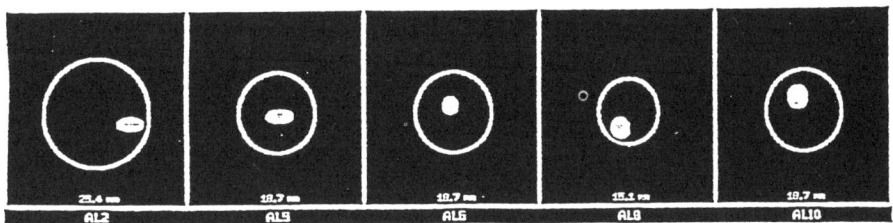

Fig. 4 : Horizontal and vertical beam profile information compared with acccelerator apertures at different locations in the accelerator.

Rf Phase and Amplitude Monitor

An important contribution to beam stability and maintenance, is the monitoring of the Rf system. We have implemented SLAC phase and amplitude monitors on our linac in order to monitor and identify Rf drifts and time dependent ramps. Our implementation is depicted in figure 7. In particular we use beam monitors to compare the phase of the beam to the phase of the Rf input to the accelerator section. Using the phase of the beam as a clock provides the most accurate comparison. The system is in the process of being debugged as time allows.

South Hall Ring

We hope to implement these systems with a more powerful control system for commissioning the new ring under construction. Automatic beam phase space matching and optics measurements should prove to be important. This will be in addition to the beam position information required for tune measurements and closed orbit corrections etc.

Conclusions

The implementation of automatic computer controlled beam diagnostics has proven very valuable in understanding the beam and magnet systems. It has been a significant help in developing tuning procedures and in diagnosing problems. Of particular importance is the operator displays and user interface in order to ensure that these systems can be used properly by operations staff. Such systems, however, require a fair number of personnel to install and debug properly and limited manpower in a small laboratory can lead to an extended period of time for such debugging. We are sufficiently encouraged by the results to continue the program of upgrading controls and instrumentation, especially preparing for the commissioning of the new ring.

References

1. K.D. Jacobs et. al., *Emittance measurements at the Bates Linac*, proceedings of the 1989 Particle Accelerator Conference, (1990) p1526.

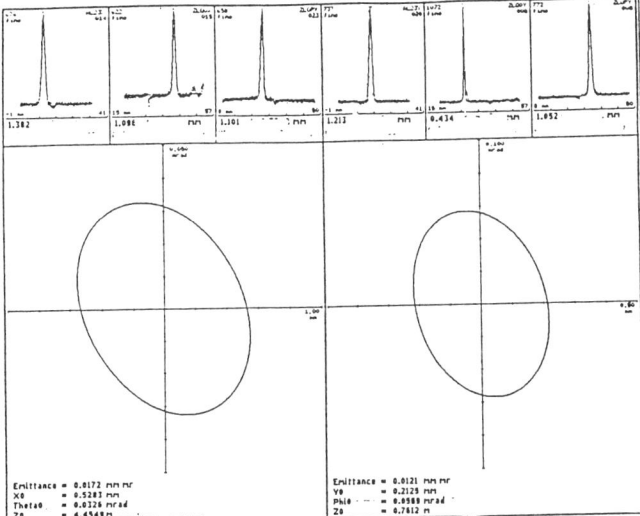

Figure 5.
Computer Display of Beam Phase Space Measurement

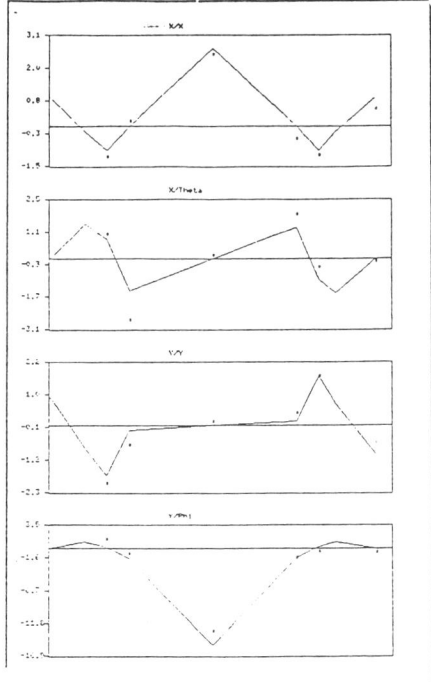

Figure 6.
Computer Display of Magnetic Optics Measurement

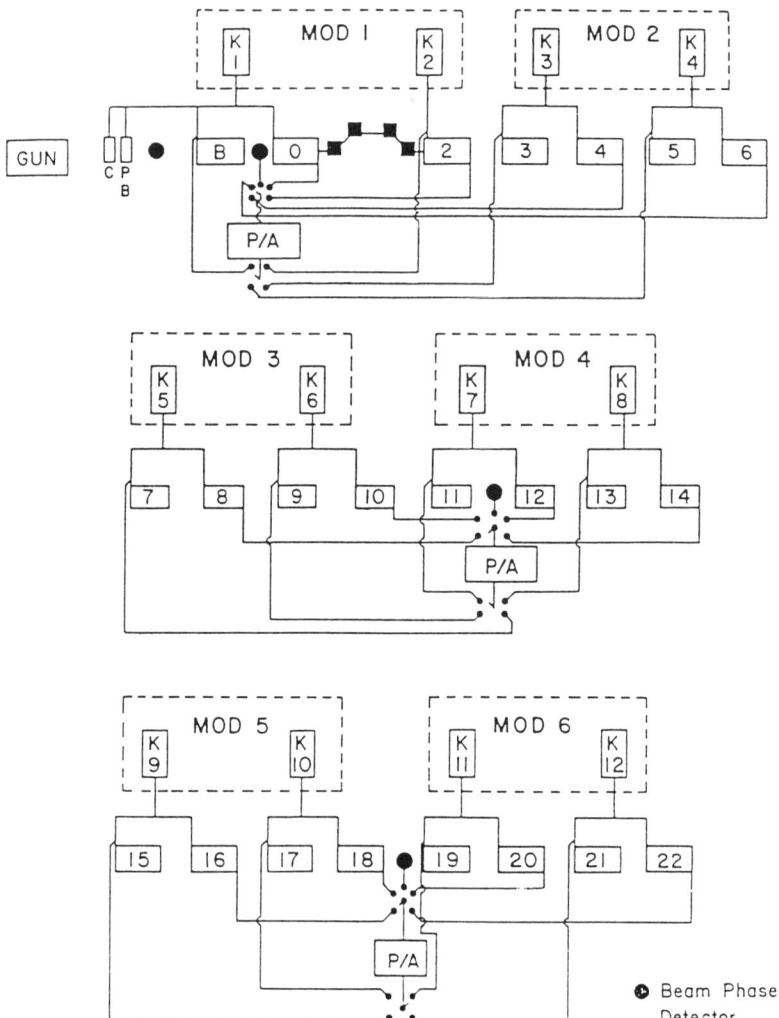

Figure 7. Arrangement of phase and amplitude monitors.

CORNELL SYNCHROTRON TUNE CORRECTION

C. Dunnam, J. Byrd, R. Meller

Laboratory of Nuclear Studies
Wilson Laboratory, Cornell University, Ithaca, N.Y. 14853

ABSTRACT

Methods are described for measurement and stabilization of energy dependent transverse tunes observed in the Cornell synchrotron. For this accelerator, operating over an energy range of 0.15 to 5.5 GeV, tune correction is required to compensate the low energy offset contribution of synchrotron magnetic field errors. Without dynamic correction, synchrotron transport losses and beam instability limit injection rates into the CESR collider and may induce long-term damage in the CLEO detector. In the system described, a real-time digitization technique ["MIRABILE"] has been applied for analysis of tune spectra over the energy ramp cycle. FTM (frequency-time-magnitude) plots derived from rapidly-digitized time domain data are analyzed to determine appropriate correction waveforms for excitation of vertical and horizontal correction quadrupoles distributed about the synchrotron ring. A description of the synchrotron tune stabilization system and related operational data are presented.

INTRODUCTION

In recent years, peak Cornell Electron-positron Storage Ring [CESR] luminosities of 9×10^{31} have been achieved through a series of accelerator upgrades, machine studies and refinement of operating procedure[1]. Improvement of integrated luminosity for HEP has been similarly addressed by a program to significantly increase the HEP duty cycle of the CESR collider. This has been accomplished through successive increases of charge injection rates into the storage ring. One facet of this upgrade program has been the investigation of synchrotron transport efficiency improvement with active tune modulation

The CESR electron/positron injection system presently consists of a common 30 meter 160 MeV linac, separate transport lines for electrons and positrons, a 5.5 GeV synchrotron, and discrete e^+/e^- transfer lines for conduction of particles into the CESR storage ring proper. Originally configured for e^- acceleration from 300 MeV to 12 GeV, the synchrotron is presently tuned to accommodate a reduced 160 MeV linac injection energy for both electrons and positrons.

Over an acceleration cycle, the magnetic field of the CESR injection synchrotron varies by a ratio of up to 40. Because of this, the synchrotron lattice functions vary substantially, with maximum effect ($\approx$12% of fractional tune) at the lowest energies. Synchrotron transport efficiency is reduced when particles are lost as the lattice-dependent synchrotron tunes cross inherent resonances. Although the Cornell synchrotron was originally designed with two families of d.c. correction quadrupoles, static excitation of these elements no longer assures accurate correction. Relatively large remanent field contributions through the low energy portion of the ramp cycle guarantee that at least one resonant line will be encountered.

Project conducted under National Science Foundation grant PHY-8713986

MIRABILE

Enhanced synchrotron transport efficiency requires application of tune stabilization over a portion of the synchrotron ramp cycle. Study of stabilization techniques is, in general, greatly facilitated by the ability to make one-shot records of time-dependent tunes. Such records are then used to evaluate lattice correction functions. "MIRABILE" [MIllion Revolution Accelerator Beam Instrument for Logging and Evaluation] is a system which combines fast digitization hardware [Figure 1] and a FFT post-processing program to create files from which

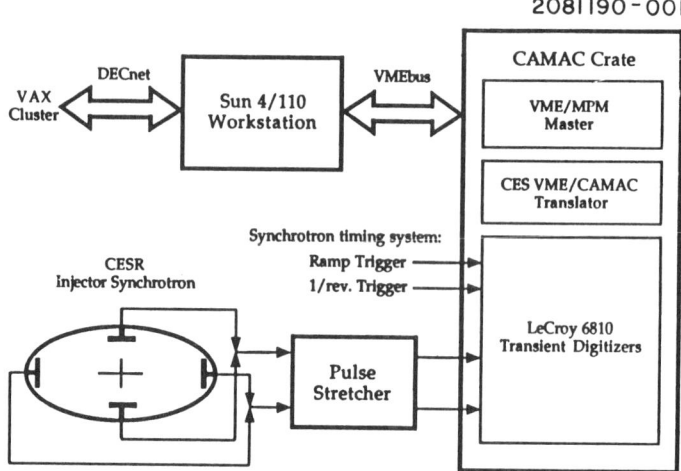

Figure 1. CESR MIRABILE system configured for synchrotron tune analysis.

frequency-time-magnitude [FTM] plots may be readily formulated. MIRABILE has evolved from its earlier incarnation as a data logging system for the Tevatron E778 beam dynamics experiment[2]. In the CESR MIRABILE system as configured for synchrotron record-taking, 100 picosecond pulses from two synchrotron beam position pickup button pairs are stretched in fast differential detectors. Over a selected injection cycle in the synchrotron, time-varying analog channels corresponding to horizontal and vertical displacements are sampled at the CESR revolution frequency by two 5 MHz, 12 bit LeCroy model 6810 transient digitizers. Each digitizer has sufficient memory to store the 32 kiloword sequence required for a complete record of one synchrotron ramp cycle. The digitizers are CAMAC standard, and are controlled by a Sun 4/110 workstation via a CES CAMAC 8210 Branch Driver. This configuration allows the digitizer memory to be mapped directly into the memory of the Sun 4/110, facilitating rapid data transfers and simplifying the interface software.

Cached data corresponding to one "snapshot" are analyzed in 256 turn overlapping segments. The segment length is chosen as the minimum which permits observation of the horizontal and vertical signals, while the choice of segment overlap depends on desired time resolution. Each segment is FFT'd to determine the magnitude of the power spectrum; typical processing time for a single ramp cycle is approximately 15 seconds. Graphical output may then be displayed at the workstation for immediate analysis, or plot routines can be run from the host VAX system following DECNET transfer of ASCII files.

SYNCHROTRON TUNE STATIC CORRECTION

Transport efficiency through the synchrotron is strongly influenced by the unavoidable particle loss which occurs when the accelerator tunes cross resonant lines. Figure 2 is a

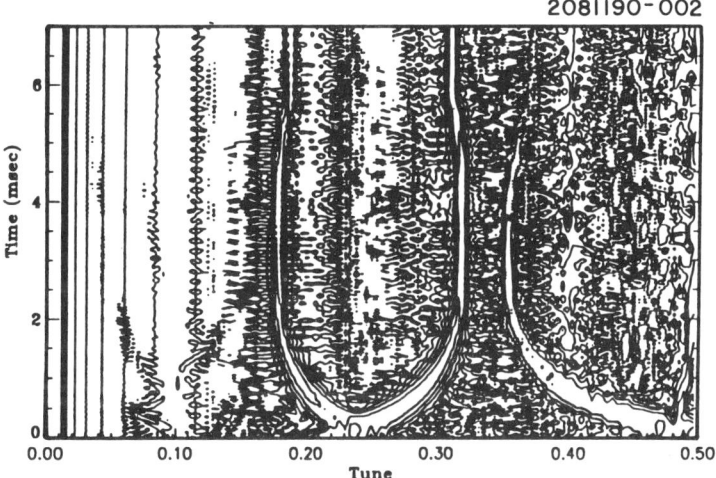

Figure 2. FTM plot of synchrotron tunes with nominal d.c. lattice correction.

MIRABILE plot of synchrotron tunes under conditions of horizontal and vertical quadrupole d.c. bias optimized for CESR injection. An operational dilemma is clearly evident in the disruptive interactions of the synchrotron tunes with two naturally occurring resonance lines, $Q_{0.25}$ and $Q_{0.33}$. Particle loss during the $Q_{0.33}$ interaction interval is acceptable; however, a large percentage of linac-injected charge has already been lost due to interaction with the $Q_{0.25}$ line. Adjustment of the correction quadrupole bias in either direction reduces injector efficiency since the tune locus is effectively boxed in by resonant barriers. Synchrotron tune behavior

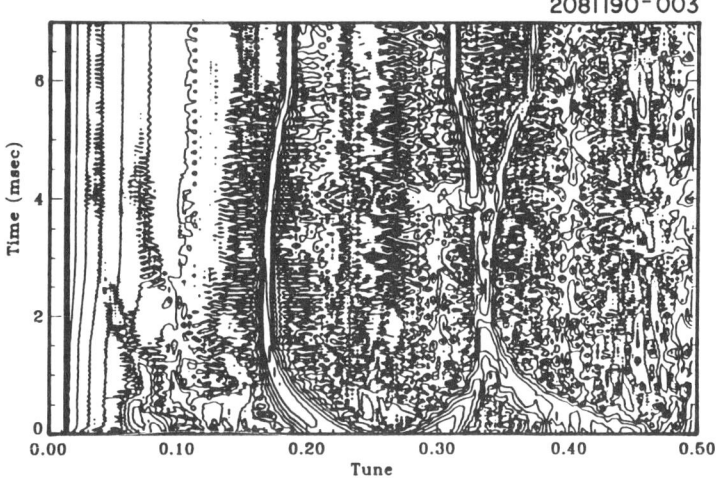

Figure 3. FTM synchrotron tune plot for d.c. quadrupole overcompensation.

under conditions of d.c. overcompensation is plotted in Figure 3 to illustrate the inherent restriction on acceptable d.c. adjustment range. As shown, the horizontal and vertical low energy tunes have been successfully separated from each other and properly distanced from the $Q_{0.25}$ resonance. Unfortunately, this also forces the horizontal tune well into the $Q_{0.33}$ interaction region, which results in unacceptably high particle loss. Clearly, avoiding both resonant lines requires dynamic tune parameter correction over the ramp cycle.

TUNE CORRECTION HARDWARE

Six sets of correction quadrupoles provide magnetic focussing for tune compensation. A block diagram of the active synchrotron tune correction (STC) system is illustrated in Figure 4.

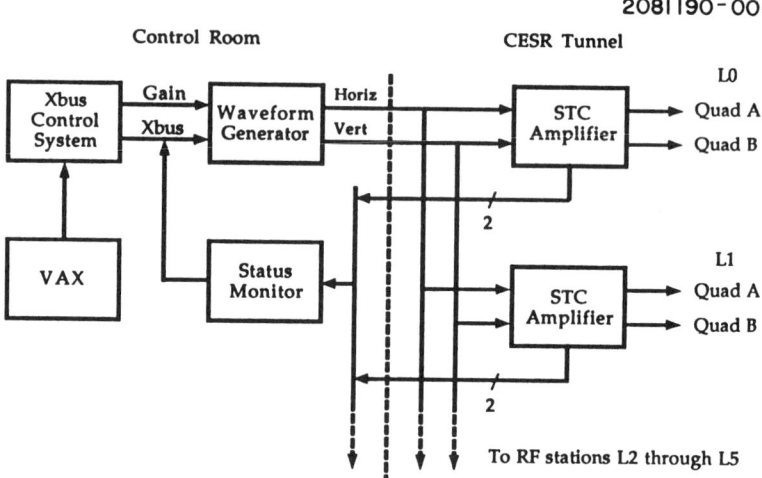

Figure 4. Synchrotron tune correction (STC) system.

Correction elements are driven by dual power amplifiers which are located at the six quadrupole sites around the CESR tunnel. Each driver chassis [Figure 5] includes a dual current-mode power amplifier, high-current switchmode power supply and thermostatically controlled forced-air cooling system. The amplifiers' output capability is a hefty ±10 amperes at ±30 volts$_{pk}$. Bridged high power monolithic operational amplifiers permit inclusion of two bipolar channels and heatsink on one 6 x 4 inch removable card. Output current slew rate, constrained by supply voltage and quadrupole coil inductance to approximately 4 amperes per millisecond, exceeds the bandwidth required for synchrotron tune correction.. Status and fault detection circuitry are included on the amplifier card in addition to full overvoltage and overcurrent protection. Collocation of the amplifiers with quadrupole pairs minimizes voltage drop and crosstalk from associated power cabling.

A proprietary waveform generator card in the control room outputs horizontal and vertical drive signals which are daisy-chained to each of the amplifier chassis. New waveform parameters are loaded into the generator card's RAM via the injector system control bus for intentional value changes or a system boot sequence. Amplitudes of the STC drive signals scale with an analog control voltage from the operator interface and are the only parameter controls available to casual users. Status logic levels from the distributed amplifier monitor circuits are displayed in the control room and are also available for direct polling by the control system.

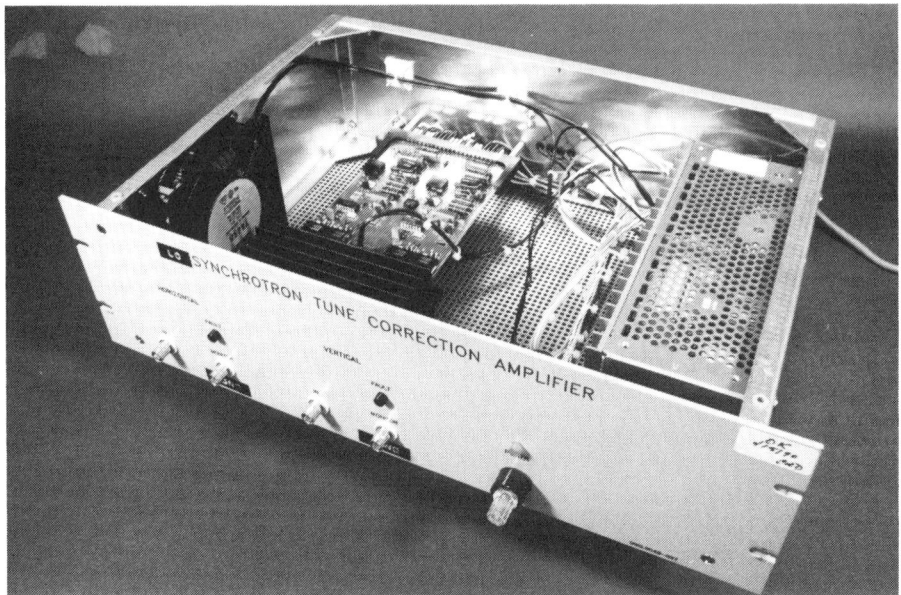

Figure 5. STC driver chassis.

SYNCHROTRON TUNE DYNAMIC CORRECTION

Optimum tune correction for synchrotron operation with CESR positron injection conditions is shown in the Figure 6 FTM contour plot. Here, dynamic correction for the low energy remnant field successfully pushes the horizontal tune far from the $Q_{0.25}$ resonant line. Unlike the d.c. excitation case of Figure 3, however, this lattice correction does not induce an unwanted excitation of the $Q_{0.33}$ resonance.

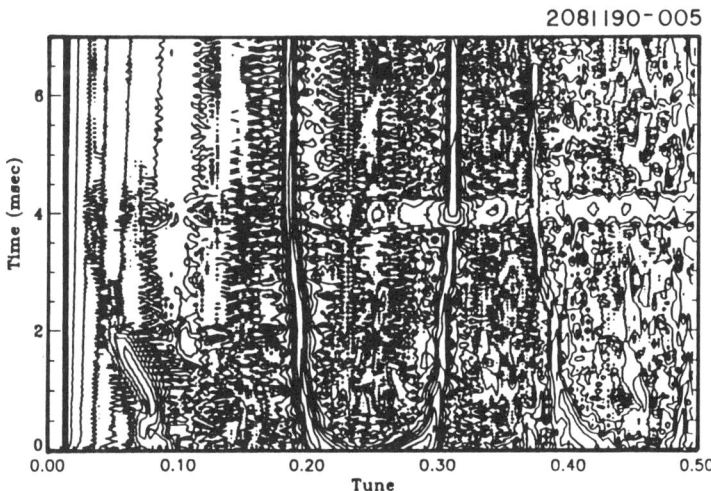

Figure 6. Synchrotron tunes with optimum dynamic correction.

To implement dynamic lattice compensation, the synchrotron tune FTM plots are analyzed and a saveset is created which defines a time-varying waveform to be applied to the correction quadrupoles. Values for waveform increments of 100 μsec are initially derived from a survey of MIRABILE plots at various d.c. bias levels. Dynamic correction waveforms for both horizontal and vertical quadrupoles are then established through an iterative procedure of FTM plot analysis and incremental compensation adjustment.

To illustrate the range of dynamic correction attainable with the STC system, Figure 7

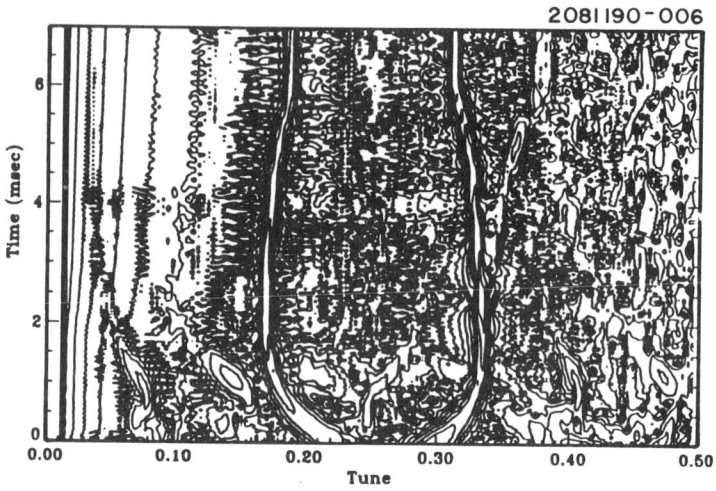

Figure 7. Synchrotron tunes with dynamic overcompensation.

demonstrates the effect of a moderately overdriven condition in the horizontal plane. As the plot reveals, bowing of the tune locus causes severe interaction with the $Q_{0.33}$ resonance. When observing turn-to-turn charge transport in real time, a corresponding particle loss is readily evident. Also of interest in this example is the STC system's ability to strongly modulate synchrotron tunes to the peak ramp energy of 5.5 GeV.

CONCLUSION

MIRABILE capture and transform of data has proven remarkably useful as an adjunct to development and refinement of a tune correction system for the Cornell synchrotron. Graphical data from MIRABILE are readily analyzed for insight into time-dependent corrections applicable to the synchrotron lattice function. In spite of the iterative character of the process, convergence on a suitable correction waveform has proven rapid and unambiguous. FTM plots developed in MIRABILE clearly demonstrate dynamic correction of the synchrotron tunes to be a substantial improvement over the d.c. lattice correction scheme previously employed. Tangible benefits of the dynamic tune correction we have applied to the Cornell synchrotron include both improved charge transport and reduced background during CESR "fill" intervals.

---

[1] D. Rice, et al, "High Luminosity Operation of the Cornell Electron Storage Ring", Proc. IEEE Part. Accel. Conf., 1989.

[2] S. Peggs, C. Saltmarsh, R. Talman, "Million Revolution Beam Instrument for Logging and Evaluation" SSC-169 (Internal Note).

A BEAM POSITION MONITORING SYSTEM FOR BROOKHAVEN'S LINAC TO BOOSTER TRANSFER LINE[*]

T. J. Shea, C. M. Degen, D. M. Gassner, V. LoDestro
Brookhaven National Laboratory
Upton, NY 11973

Abstract

A beam position monitor system has been developed for Brookhaven's Linac to Booster transfer line. Beginning in early 1991, this line will transport a chopped, RF modulated H⁻ beam from the 200 MeV Linac to the AGS Booster. Over a 15dB dynamic range in beam current, the position monitor system will provide a real-time, normalized position signal accurate to ±0.5mm. Directional coupler style pickups have already been installed in the line and one prototype has been tested in beam. Analog processing electronics will be located in the tunnel and will incorporate the amplitude modulation to phase modulation normalization technique. To avoid interference from the 200 MHz linac RF system, processing will be performed at 400 MHz. In bench tests, prototypes of the electronics meet the above requirements for accuracy and dynamic range. This paper will provide a system overview and a detailed description of the analog processing electronics.

The Transfer Line and the Beam

As its name suggests, the Linac To Booster (LTB) transfer line carries beam from the 200MeV linac to the booster. The line is approximately 30 meters long and executes a 135° bend. The seven beam position monitors are distributed as shown in Figure 1. In addition, the entrance and exit are each equipped with a current transformer and a harp profile monitor.

Because the line immediately follows the linac, the beam at the entrance still exhibits the micro-structure imposed by the 200 MHz cavities. The micro-bunches at this point

---

[*] Work performed under the auspices of the U. S. Department of Energy

274 Brookhaven's Linac

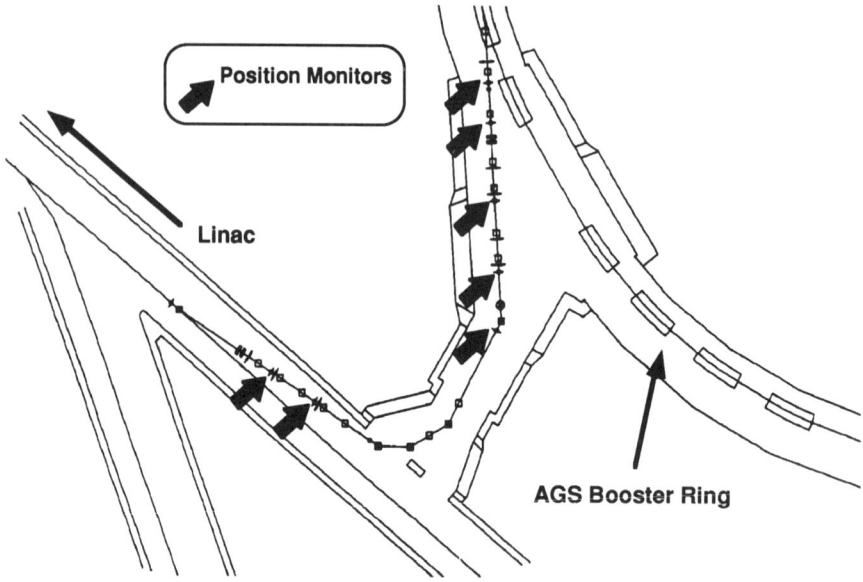

Figure 1. Layout of the Brookhaven Linac to Booster Transfer line. The line is about 30 meters long.

are less than 150 picoseconds long at the base. As these bunches drift through the line, they gradually lengthen, and reach a length of 720 picoseconds by the exit. Therefore, at any point in the transfer line, the Fourier spectrum of this beam's longitudinal current density will contain strong components at high harmonics of the bunching frequency. This allows processing at the first harmonic of the 200 MHz bunching frequency, thereby avoiding 200 MHz pickup from sources other than the beam. In fact, the amplitude of this 400 MHz component suffers a mere 5% decrease as the beam travels from the first to the last monitor. Other beam parameters can be summarized as follows:

- full linac pulse length   10 to 400 μs
- chopped pulse length   100 to 300 ns
- chopping frequency   2.5 MHz
- bunching frequency   201.25 MHz
- current inside chopped pulse   10 to 25 mA

The Position Monitors

For machine commissioning, operation, and studies the position monitor system for the LTB line must fulfill the following requirements:

- Measure the position of H⁻ beam over a 5 to 25 mA range
- Provide a normalized, real-time position signal
- Operate reliably in high EMI environment
- Provide ±0.5 mm accuracy and resolution
- Provide a ±30 mm aperture

A cross section of the monitor itself is shown in figure 2. The striplines are each 13.5cm long in order to provide a frequency response peak at 402.5 MHz when excited by a beam travelling at .566 times the speed of light. These striplines are mechanically

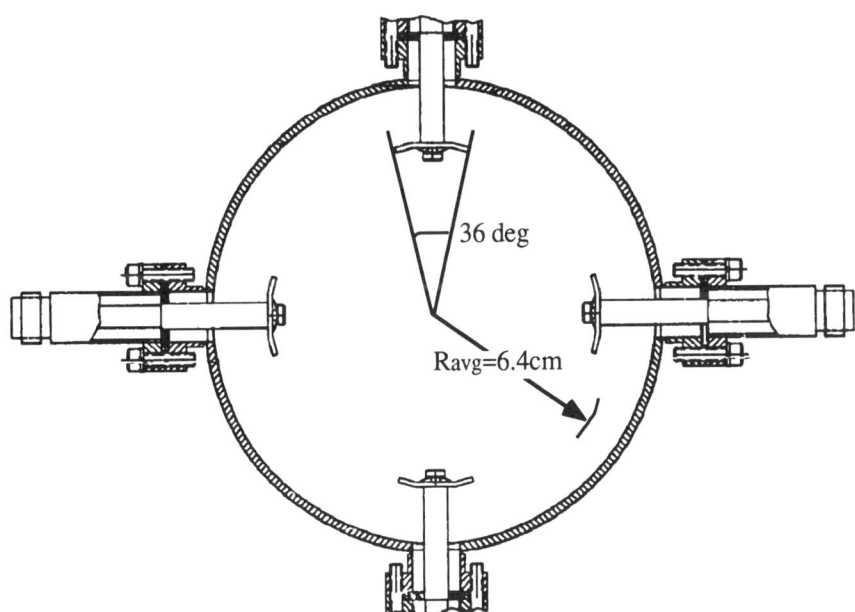

Figure 2. End view of the LTB position monitor. The horizontal and vertical striplines are actually offset from each other in the longitudinal direction.

276 Brookhaven's Linac

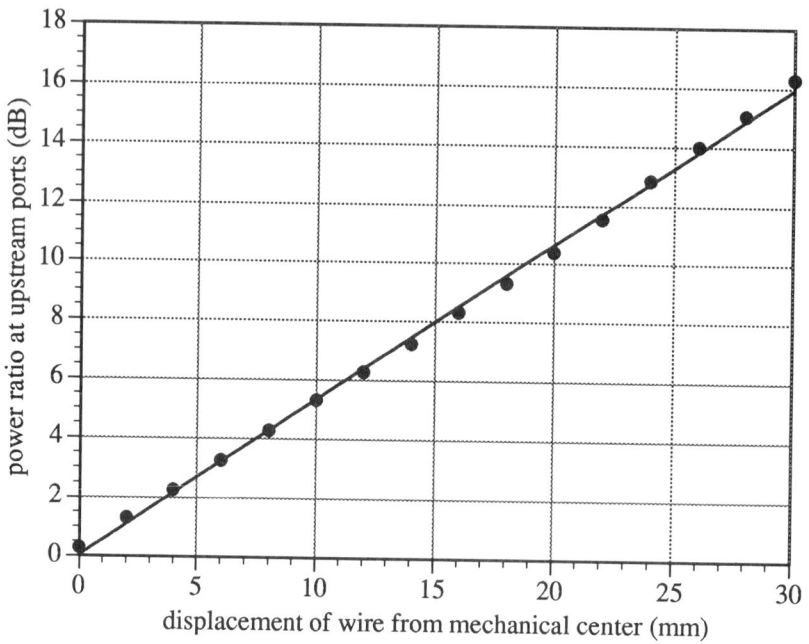

Figure 3. Results of a scanning wire measurement depicting the .53 dB per mm sensitivity in one plane.

aligned and terminated with precision resistors so that the mechanical center is located to within 0.5mm of the electrical center. A few monitors were measured with the same scanning wire system used to calibrate the AGS Booster ring position monitors. With a technique similar to that performed at DESY for calibration of HERA position monitors[1], data from these wire scans were used to find a sensitivity slope for small beam offsets and additional higher-order coefficients for larger displacements. As shown in figure 3, the monitor's sensitivity is nearly linear over a 30 millimeter aperture. The line shown in figure 3 is plotted from the following approximation for position monitor sensitivity[2]:

$$S_x = \frac{160}{\ln 10} \frac{\sin\left(\frac{\phi_0}{2}\right)}{\phi_0 b}$$

where S_x =.53 dB/mm is the sensitivity, $\phi_0 = 36°$ is the angle subtended by the stripline, and b = 64mm is the average of the stripline radius and the container radius.

The Front End Electronics

The front end analog processing electronics incorporates amplitude modulation to phase modulation (AM/PM) normalization without down-conversion in a manner similar to that of the Fermilab Tevatron system[3]. Other systems incorporate down-conversion to improve performance[4], but due LTB's chopped beam and relaxed dynamic range requirements, down-conversion is neither desirable nor necessary.

A block diagram of the front end electronics is shown in figure 4. Signals from the monitors travel in Heliax to the modules. Amplifiers provides isolation from the 402.5 MHz bandpass filters for broadband matching. The filters have a 3 dB bandwidth of approximately 40 MHz and this defines the bandwidth of both the position and intensity signals. In order to obtain an intensity signal, two-way power splitters route half of each input signal to a four-way combiner followed by a self-synchronous detector. For the position information, AM/PM conversion (with a slope of 6.6 °/dB for small input differences) takes place in the combination of a 90° coax delay followed by the 180° hybrid. A cascade of four Plessey limiters follows each output of the hybrid and provides a constant amplitude signal to the phase detector.

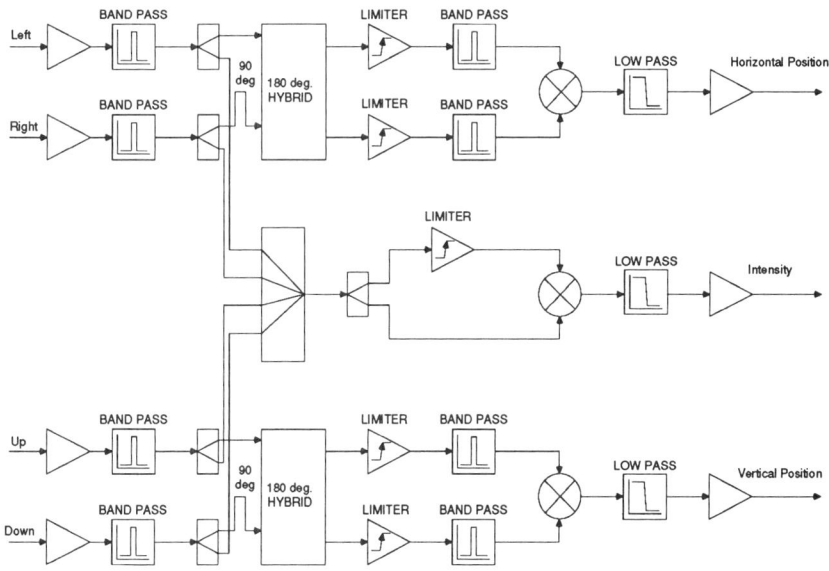

Figure 4. Block diagram of the front end electronics. The bandpass filters are tuned to 402.5 MHz and the limiters each consist of four Plessey 532C limiters.

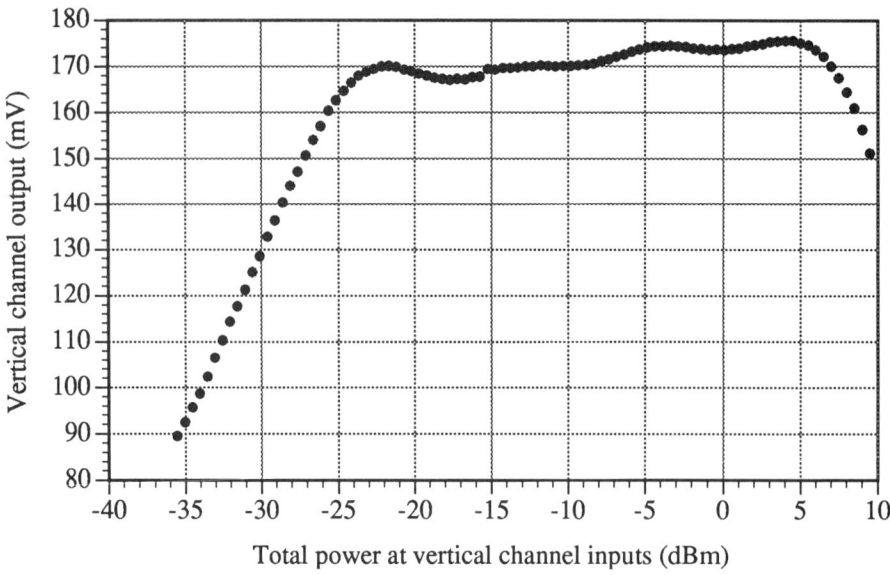

Figure 5. Variation of vertical position output with change in input power sum. Power ratio between the up and down inputs is held to 3dB. Note that the zero is suppressed on the vertical axis.

Because the limiters are usually the Achilles heel of AM/PM systems, their implementation here deserves detailed discussion. Before installation, 128 limiters were matched in pairs from a pool of 168. The system used for matching incorporated an HP8753C vector network analyzer controlled by a Macintosh computer running National Instruments' LabVIEW software. A curve of phase verses input power at 402.5 MHz was recorded for each limiter and fitted to a high order polynomial. The zeroth order term was then dropped and the result was used to match limiters in pairs using peak phase deviation as the figure of merit. This process, performed automatically under LabVIEW, produced pairs with a worst case phase deviation of 180 millidegrees. When installed in a prototype front end module, equivalent position error over the required dynamic range was within the ±0.5mm specification. An example from the dynamic range test is shown in figure 5. If necessary, better results could be obtained by gain matching each of the four stages.

The prototype was also tested in the time domain by providing the inputs with 100ns long bursts of 402.5 MHz. Signals measured after the 50Ω output buffer show the clean leading and trailing edges characteristic of a 40 MHz bandwidth with no noise on the output between bursts. This performance should be adequate for future machine studies. Before installation, each module will also be fully characterized in an automated test system. Data from these tests may then be used by the control system to correct for the nonlinearities that affect measurements of large beam offsets.

Acknowledgments

The authors gratefully acknowledge the help of the following people who contributed to the development of this system: R. Witkover for conceptual design and many helpful suggestions, R. Bossart for monitor design and conceptual help, J. Brodowski for mechanical implementation of the position monitors, D. Ciardullo for suggestions on the front end electronics, and R. Thomas for the scanning wire measurements.

References

1. W. Shutte, "Results of measurements on the HERA Proton Beam Position Monitors", Proceedings of the 1989 IEEE Particle Accelerator Conference, Chicago 1989.

2. R. Shafer, "Characteristics of Directional Coupler Beam Position Monitors", IEEE Transactions on Nuclear Science, NS-32, No. 5, October 1985.

3. S. P. Jachim, R. C. Webber, R. E. Shafer, "An RF Beam Position Measurement Module for the Fermilab Energy Doubler", IEEE Transactions on Nuclear Science, NS-28, No. 3, June 1981.

4. F. D. Wells and S. P. Jachim, "A Technique for Improving the Accuracy and Dynamic Range of Beam Position Detection Equipment", Proceedings of the 1989 IEEE Particle Accelerator Conference, Chicago 1989.

A REAL TIME BUNCH LENGTH MONITOR IN THE TRISTAN AR

T. Ieiri
KEK, Oho 1-1, Tsukuba-shi, Ibaraki-ken, 305 Japan

ABSTRACT

The bunch length can be measured by the comparison of two Fourier components of the beam. This technique has been applied to an electron bunch in the TRISTAN Accumulation Ring (AR). Since the rms bunch length of the AR is in a range from 1 cm to 4 cm, detected frequencies are in the Gigaherzt region. However, the detected frequency is restricted by the cutoff frequency of the vacuum pipe, which imposes high resolution requirements on a detector. Preliminary experiments with a spectrum analyzer show that it is possible to measure the bunch length in real time with a stripline electrode.

INTRODUCTION

Bunch lengthening is one of the most important issues in electron/positron rings from the aspect of beam dynamics. In the AR, bunch lengthening actually occurs [1]. Anomalous bunch lengthening phenomena are also observed when two bunches or more are stored [2]. It is expected that the bunch length should be measured dynamically to understand these phenomena. The bunch length has been measured using the synchrotron light by a streak camera, whereby an rms bunch length is obtained from a stored longitudinal profile [3]. This method has an advantage of detecting a charge distribution. However, it is difficult to follow dynamical changes of the bunch length, for instance, when bunch length oscillations occur.

In order to measure the bunch length more easily in real time, a bunch length monitor has been proposed in the AR using the method of detecting Fourier components of the beam [4,5]. A monitor for real time measurement of the bunch length is now being constructed with a stripline electrode of 30 cm in length. This note describes a design of the monitor, detector electronics and preliminary results obtained with a spectrum analyzer.

DESIGN OF THE MONITOR

The AR is an injector for the TRISTAN Main Ring, an electron and positron collider, and also an electron storage ring for use of synchrotron radiation. Main parameters related to the monitor are summarized below:

1. RF Frequency f_{RF} = 508.58 MHz
2. Revolution Frequency $\omega_0/2\pi$ = 794.6 kHz
3. Number of Bunches   1
4. Natural rms Bunch Length σ_l = 1.0 - 2.0 cm
5. Synchrotron Frequency f_s = 20 - 40 kHz
6. Beam Current I_b ~ 50 mA max.

Assuming a Gaussian longitudinal charge distribution of rms bunch length σ_t, the Fourier component of the n-th harmonic is given by

$$V(n\omega_0) = 2V_0 \cdot \exp[-n^2\omega_0^2\sigma_t^2/2] \quad . \tag{1}$$

Here n is integer, $V(n\omega_0)$ is induced voltage on an ideal beam pickup and V_0 is the DC component. The rms bunch length is obtained from the ratio between two Fourier components as

$$\sigma_t = \frac{1}{\omega_0}\sqrt{\frac{2}{(n_2^2 - n_1^2)} \cdot \ln\left\{\frac{V_1(n_1\omega_0)}{V_2(n_2\omega_0)}\right\}} \quad , \tag{2}$$

where V_1 and V_2 are Fourier components at n_1-th and n_2-th harmonics of the revolution frequency and $n_2 > n_1$.

When designing this monitor, we must consider two things. The first one is what kind of pickups we should use to detect the beam signal. It is desirable for the pickup to have wide and flat frequency response up to a few Gigaherzt. The pickups already installed in the AR, striplines, button electrodes and current tranformers (CT) are considered. The button has high-pass characteristic and the CT has low-pass characteristics. The stripline installed at A042 seems to be satisfactory.

The second one is which harmonics we detect to measure the expected bunch length. The frequency response of the stripline of 30 cm in length has high sensitivity to the beam around frequencies of 250 MHz, 750 MHz, 1250 MHz, 1750 MHz and so on.

Detected frequencies are chosen among them. We select the 250 MHz component as the lower frequency component, because it is hardly affected by the bunch length. Though we want to use a frequency component above 2 GHz as the upper frequency to detect the minimum bunch length of 1 cm, it is limited by the cutoff frequency, 1.63 GHz, of the vacuum pipe. Therefore, the upper frequency is chosen to be 1.62 GHz. The calculated amplitude ratio between 250 MHz and 1.62 GHz components is shown in Fig.1 as a function of the rms bunch length. The expected bunch length is in the range from 1 cm to 4 cm. When the bunch length is 1 cm, the ratio is 0.5 dB, which will require a detector with resolution of 0.1 dB or 1%.

The stripline is also sensitive to transverse beam position. However, position dependence will be cancelled out by the ratio between 250 MHz and 1.62 GHz components if the dependence is independent of the picked up frequencies.

Fig. 1
The calculated amplitude ratio between 250 MHz and 1.62 GHz components as a function of rms bunch length

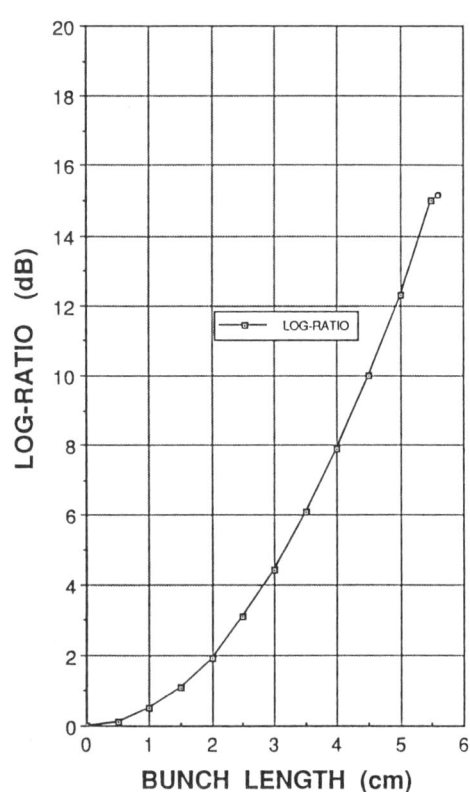

DETECTOR ELECTRONICS

A block diagram of the detector is shown in Fig.2. The beam signal from the stripline is divided into two paths. One goes through the Band Pass Filter (BPF) of 250 MHz and the other of 1.62 GHz. The detector is composed of two identical heterodyne and synchronous receivers/detectors, and an analog calculator unit (ACU). The heterodyning lowers the input frequencies to 10.7 MHz with two stages of mixers. Local oscillator signals of 320 MHz and 1690 MHz are produced from multiplied RF frequency and crystal oscillators, and fed to the mixers. The ACU calculates a square root of log-ratio to get a signal propctional to the bunch length.

In order to measure the bunch length in the AR, the following specifications for the detector are required.

1. Dynamic Range   60 dB
2. Accuracy between two heterodyne receivers   0.1 dB
3. Accuracy of ACU   1 %
4. Stability   less than 0.5 %
5. Frequency Response   100 kHz

The dynamic range of 60 dB is covered by using a remotely controlled attenuator (ATT.1). The accuracy of 0.1 dB is necessary to measure the minimum bunch length of 1 cm with 20 % accuracy. A gain difference between the two receivers will produce an offset at the output of the detector. The ATT.2 is used for coarse balance. Precise balance tuning with a calibration signal is required between the two receivers. A total frequency response of 100 kHz is necessary to detect quadrupole bunch oscillations; this is possible by using AD538s by Analog Devices Inc.

PRELIMINARY EXPERIMENTS

In order to check validity of this application, the bunch length is measured using a spectrum analyzer (Tectronix-492PGM). A beam spectrum picked up by the stripline is shown in Fig.3 as an example. One can see notches at every 500 MHz, which is due to resonances at the half wavelength and its harmonics. One may also observe disturbances in the spectrum over 1.63 GHz because of waveguide modes. Since these raw data include signal attenuation through a coaxial cable (Hitachi-HF20D, 35 m), the

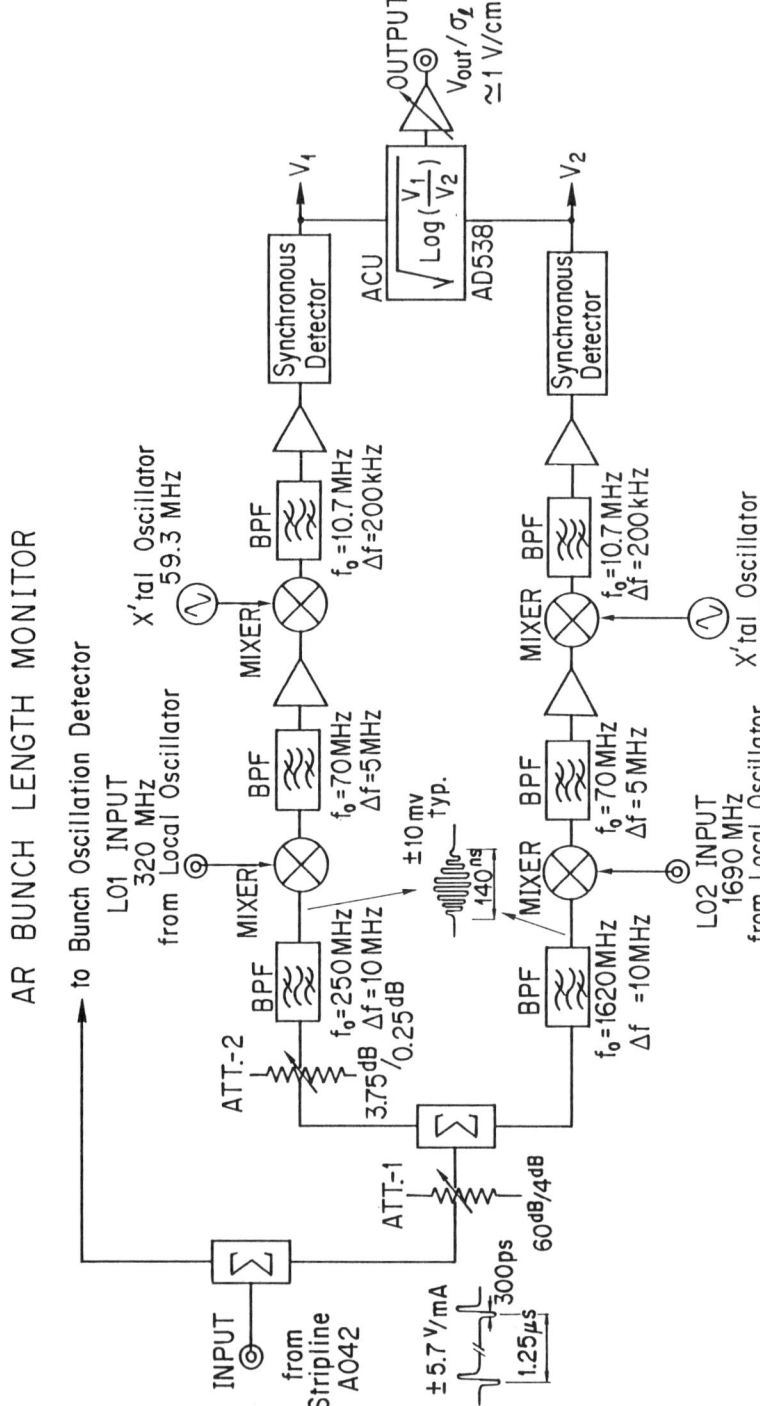

Fig.2. Block Diagram of Detector

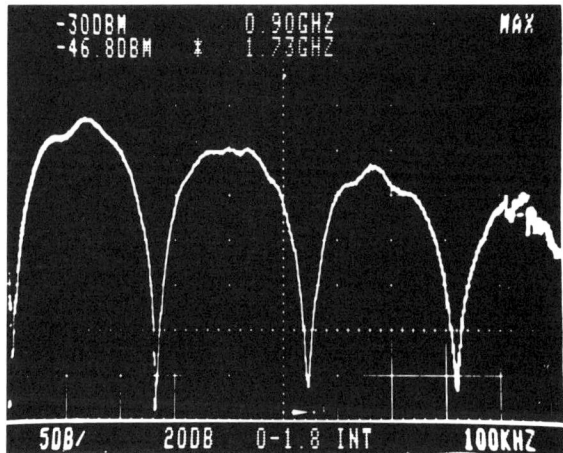

Fig. 3 Beam Spectrum picked up by the stripline at beam energy of E= 2.5 GeV and beam current of I_b= 19.8 mA. Horizontal scale is 180 MHz/div. and Vertical is 5 dB/div.

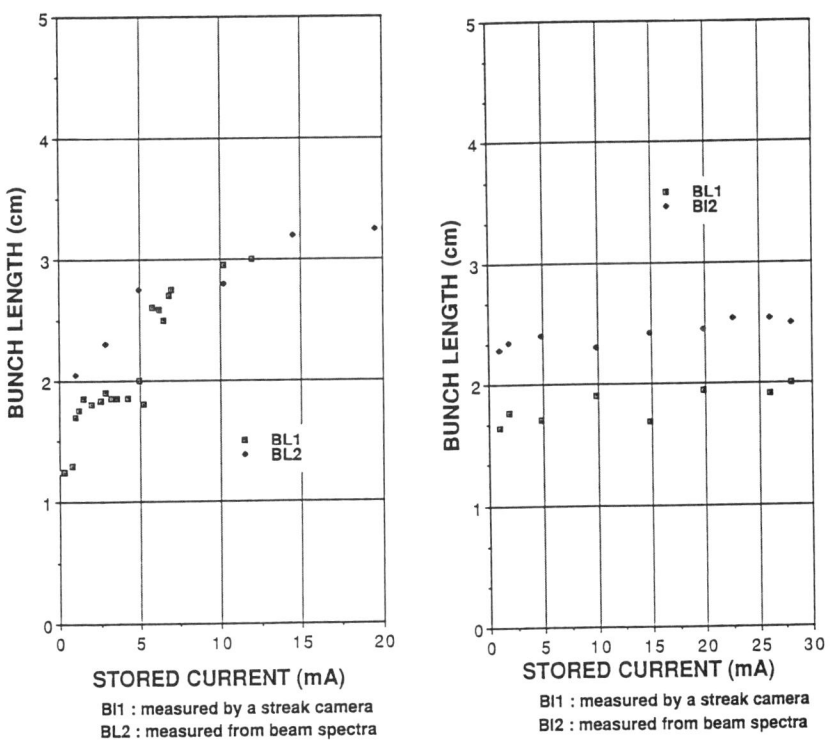

a) E=2.5 GeV   b) E=6.5 GeV

Fig. 4 Measured rms bunch length as a function of beam current.

cable loss must be compensated. A difference of the cable loss between 250 MHz and 1.62 GHz is 3.5 dB from a measurement with a network analyzer. The bunch length is obtained from the modified data of the beam spectrum and using the calculated curve in Fig.1.

The bunch length measurements were done at the injection energy of 2.5 GeV and the storage energy of 6.5 GeV. Experimental data are shown in Fig.4 together with data obtained by a streak camera. They are consistent. A precise calibration is necessary to get more precise absolute values.

In conclusion, it is possible to measure the bunch length in real time with the stripline in the AR. Although the minimum detectable bunch length depends on resolution and stability of the detector, the monitor will be able to measure a bunch length of 1 cm.

ACKNOWLEDGEMENTS

The author would like to thank Prof. Y. Mizumachi for continuous encouragement.

REFERENCES

1. K. Nakajima et al., "Bunch Lengthening in TRISTAN AR", Proc. of 5th Symp. on Accelerator Science and Technology, KEK, (1984) p.297.
2. Y. Funakoshi et al., "Observation of Multibunch Phenomena in TRISTAN Accumulation Ring", Particle Accelerator vol 27, (1990) p.89.
3. A. Ogata and Y. Mizumachi, "Optical Beam Diagnostics in TRISTAN Accumulation Ring", Proc. of 5th Symp. on Accelerator Science and Technology, KEK, (1984) p.151.
4. G. Jackson and T. Ieiri, "Stimulated Longitudinal Emittance Growth in the Main Ring", Proc. of IEEE PAC, Chicago, (1989) p.863.
5. T. Ieiri et al., "A Bunch Length Monitor by Detecting Two Frequency Components of the Beam", Proc. of 7th Symp. on Accelerator Science and Technology, Osaka, (1989) p.367.

Discharge phenomena in the button electrodes of the beam position monitors of TRISTAN MR

M.Tejima,T.Ieiri,H.Ishii,T.Shintake,K.Mori and Y.Mizumachi
KEK, National Laboratory for High Energy Physics
Oho, Tsukuba-shi, Ibaraki-ken, 305, Japan

Abstract

In the TRISTAN Main Ring, when bunch intensity exceeds about 1mA, abnormal burst signals appear in the position monitor electrodes. The burst pulses appear after about 10 nsec from the bunch signal. The polarity of the pulses are mostly negative. They occur mainly in each position-monitor-electrode of RF cavity regions. We found that we can suppress them by a biasing voltage of about 290 volts. Moreover, we verified that they don't appear in the new button-electrode which has a wide gap between the button and the chamber. We think that the cause of the abnormal signal is "Multipacting discharge" excited by higher order modes of the RF cavity.

1 Introduction

When the operation of the TRISTAN Main Ring started, our beam position monitor (BPM) system worked well at correcting the closed-orbit-distortion. The reliability was due partly to the rejection of erroneous data in the position measurement procedure. The log of the errors rejected suggests that maintenance of a bad monitor is necessary.

Two years ago, we noticed an increase of errors occurring in the radio frequency(RF) straight sections. Moreover, we noticed that the error occurred whenever the beam intensity was high. We found that the measurement errors were caused by an abnormal burst signal which appears after about 10 nsec after the beam pulse, as shown Fig. 1a.

We observed the beam signal of every BPM, and obtained the distribution of abnormal output monitors, as shown in Table 1.

We noticed that abnormal signals also appeared at special locations in the bending section.

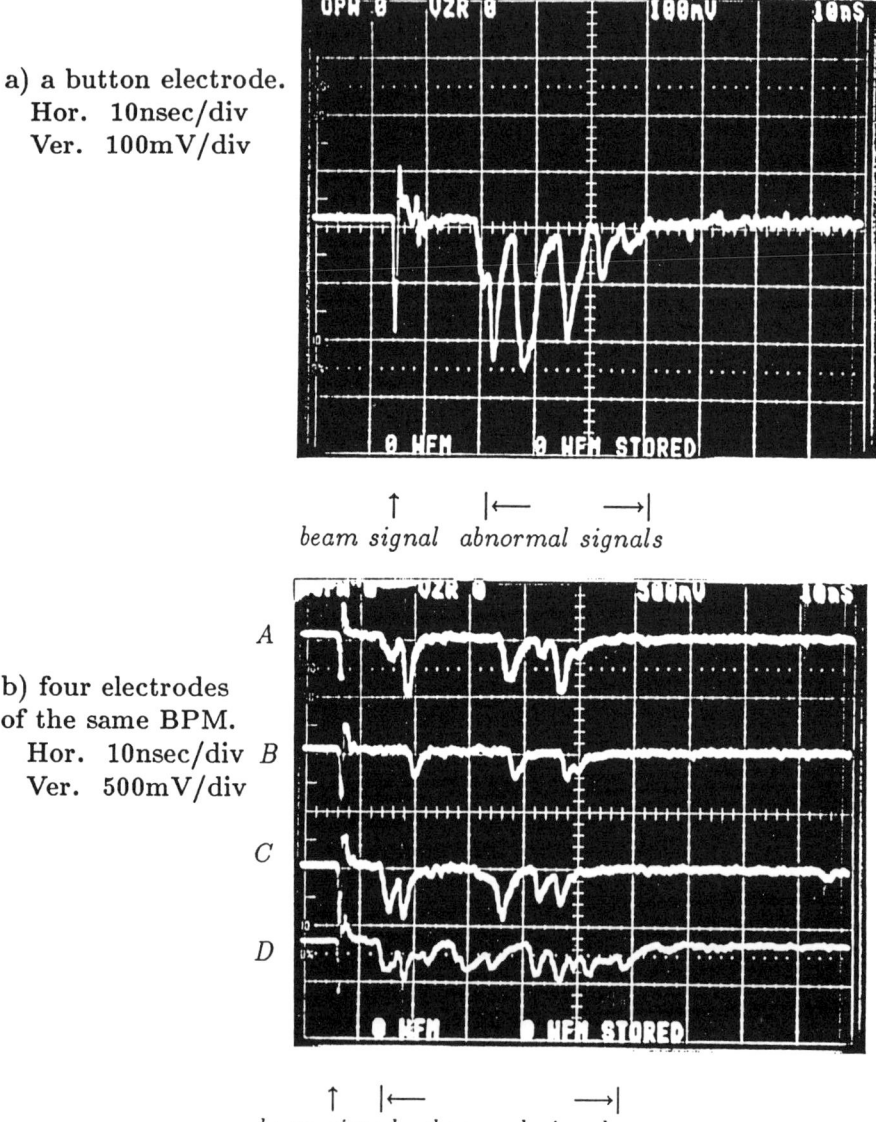

a) a button electrode.
   Hor. 10nsec/div
   Ver. 100mV/div

b) four electrodes of the same BPM.
   Hor. 10nsec/div
   Ver. 500mV/div

Fig. 1 Waveform of beam signals and abnormal signals in BPM in the straight section near RF cavity.

Table 1 Distribution of abnormal output in the MR beam position monitors.

Position Monitor	Ring Octant								
	Tsukuba		Oho		Fuji		Nikko		
	L	R	L	R	L	R	L	R	
QCS									
QC1		N	N	N	N	N	N	N	
QC2		N	N	N	N	N	N	N	
QC3	N	+	+	N	N	N	N	N	DC Separators
QC4	N	N	N	N	N	N	N	N	
QC5	−	−	−	−	−	−	−	−	··· The electrodes were taken out and examined. (see experiment 3)
QC6	−	−	−	−	−	−	−	−	
QRD1	−	−	−	−	−	−	−	−	
QRF2	−	−	−	−	−	−	−	−	
QRD3	−	−	−	−	−	−	−	−	RF Section
QRF4	−	−	−	−	−	−	−	−	
QRD5	−	−	−	−	N	N	N	−	
QRF6	−	−	−	−	N	N	N	−	
QRD7	−	−	−	−	N	N	−	−	
QRF8	N	N	N	N	N	N	−	N	
QS1	N	N	N	N	N	−	N	N	
QS2	N	N	N	N	N	N	N	N	
QS3	N	N	N	N	N	N	N	N	
QS4	N	N	N	N	N	N	−	N	
QS5	−	−	−	N	N	N	−	N	
QS6	N	N	N	−	N	N	N	N	
QS7	N	N	N	N	N	N	N	N	
QF1	N	N	N	N	N	N	N	N	
QD2	N	N	N	N	N	N	N	N	
QD3	N	N	N	N	N	N	N	N	
QD4	N	N	N	N	N	N	N	N	
QD5	N	N	N	N	N	N	N	N	
QD6	N	N	N	N	N	N	N	N	
QD7	N	N	N	N	N	N	N	N	
QD8	N	N	N	N	N	N	N	N	
QF9	N	N	N	N	N	N	N	N	
QD10	N	N	−	N	N	N	−	N	Bending Section
QF11	N	N	−	−	N	−	N	−	
QD12	N	N	N	N	N	N	N	N	
QD13	N	N	N	N	N	N	N	N	
QD14	N	N	N	N	N	N	N	N	
QF15	N	N	N	N	N	N	N	N	
QD16	N	N	N	N	N	N	N	N	
QF17	N	N	N	N	N	N	N	N	
QD18	N	N	N	N	N	N	N	N	
QF19	N	N	N	N	N	−	N	−	
QD20	N	N	N	N	N	N	N	N	
QF21	N	N	N	N	N	N	N	N	
QD22	N	N	N	N	N	N	N	N	
QF23	N	N	N	N	N	N	N	N	
QD24	N	N	N	N	N	N	N	N	
QW1	N	N	N	N	N	N	N	N	
QW2	−	−	N	N	N	N	N	N	Wiggler Section
QW3	N	N	−	N	N	N	N	N	
QW4	N	N	N	N	−	N	N	N	

Experimental building name
L=Left side, R=Right side

N: Normal Waveform
−: Abnormal Waveform of Negative Porality
+: Abnormal Waveform of Positive Porality
⌐: Gate valve

2 Observations of abnormal signals

1. They appear when the bunch current exceeds about 1mA.

2. They appear about 10 nsec after the beam, and then they repeat several times as shown in Fig. 1a.

3. Their polarities are negative for both e^- and e^+ beams.

4. They appear mainly in the straight sections, more specifically, near RF cavities.

5. Their magnitude seams to be sensitive to the bunch shape; it increases as the bunch length decreases. The abnormal signal becomes very small or disappears when two beams collide at the maximum energy.

6. Their waveforms are not the same for e^- and e^+ beams.

7. Their waveforms in the four electrodes of one monitor are not the same, as shown in Fig. 1b.

8. There are a few abnormal monitors in the bending sections. They seem to gather around several particular places in the lattice (QS5, QF11, etc.), namely, in the neighborhood of the gate valve for the vacuum duct (Fig. 2).

Hor. 5nsec/div
Ver. 100mV/div

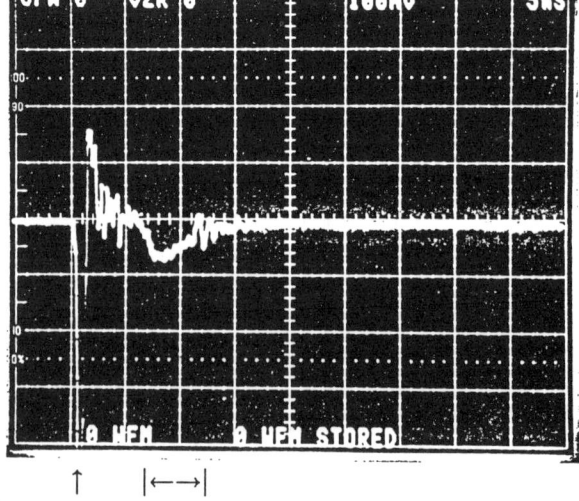

↑ |←→|
beam signal abnormal signals

Fig. 2 Waveform of beam signal and abnormal signals in a button electrode at the bending section near the gate valve of the vacuum duct.

3 Experiments

In order to find out the cause of the abnormal waveform, we performed the following experiments:

1. We applied a bias voltage on the electrode as shown in Fig. 3. Fig. 4 shows the change of electrode current with bias voltage. When the electrode is short-circuited to the ground (V=0), negative DC current flows. When the bias voltage reaches + or − 290V, the DC current disappears and, at the same time, the abnormal signal disappears.

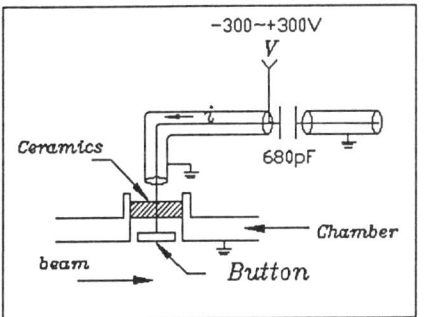

Fig. 3 Measurement scheme using a bias voltage.

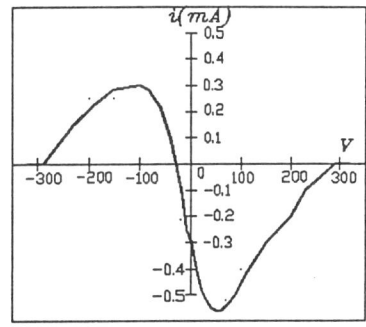

Fig. 4 Effect of bias voltage on electrode current.

2. We installed a new button electrode as shown in Fig. 5 in the RF section. As compared with the conventional button electrode(Fig. 6), the gap is made larger (about 3mm) and N type connector is used. We tested it with e^- and e^+ beams and observed no abnormal signal.

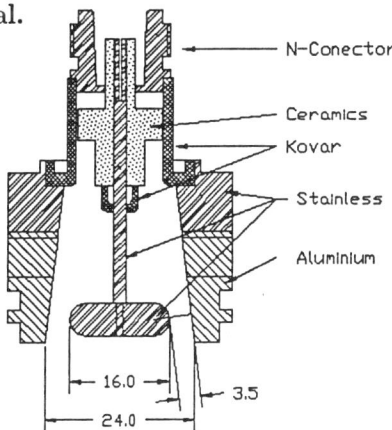

Fig. 5 New BPM electrode for test chamber.

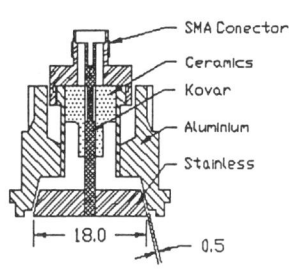

Fig. 6 Conventional BPM in TRISTAN MR.

3. We took out and examined four electrodes that had displayed a discharge in the RF section, as is shown Table 1. There was some discharge trace on the side wall of the electrode housing, as is shown in Fig. 7a,b.

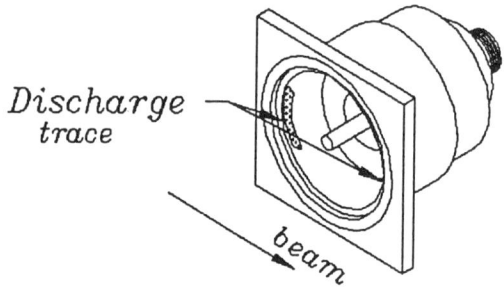

Fig. 7 Discharge trace on housing for button electrode.

4 Multipacting discharge

Judging from the above observation, we think that the origin of the abnormal signals is the multipacting discharge phenomenon between the button electrode and the chamber housing. In the TRISTAN ring when a short bunch beam passes through RF cavities or gate valves, wake fields are generated. The higher-order mode components can propagate in the beam pipe because the cutoff frequency of the TM01-cylindrical mode is 1.8 GHz. Then fields are induced on the button electrodes (standing wave TE11-coaxial mode), and when the frequency and voltage meet the condition of multipacting, the discharge occurs, as is shown in Fig. 8.

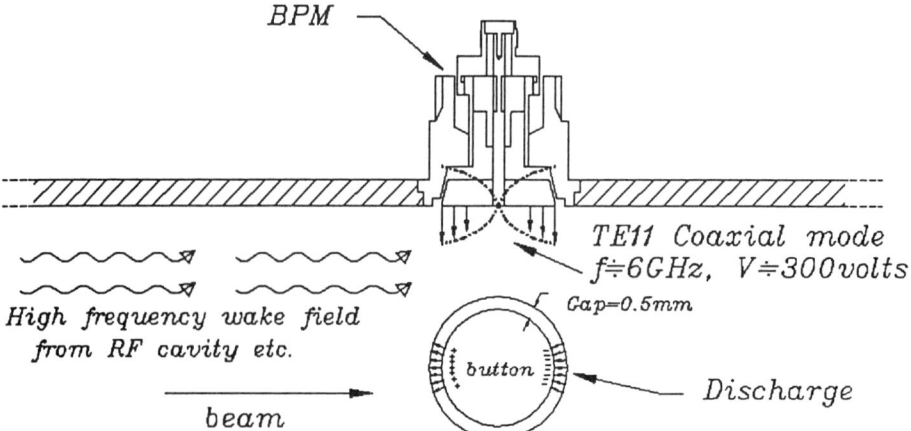

Fig. 8 Mechanism of multipacting discharge in the button electrode.

5 Development of new electrodes

Fig. 9 shows structure of our new electrodes designed to avoid discharge. They have connectors of two types, N-type or SMA. The ratio of the gap(b) and the diameter(d) of the button is in the range,

$$\frac{b}{d} < 6\pi \cdot 10^{-3} \quad or \quad \frac{b}{d} > 3\pi \cdot 10^{-2}$$

We are going to install a test chamber with SMA connectors during summer-autumn shutdown (1990), and we will test them with actual beams early next year(1991).

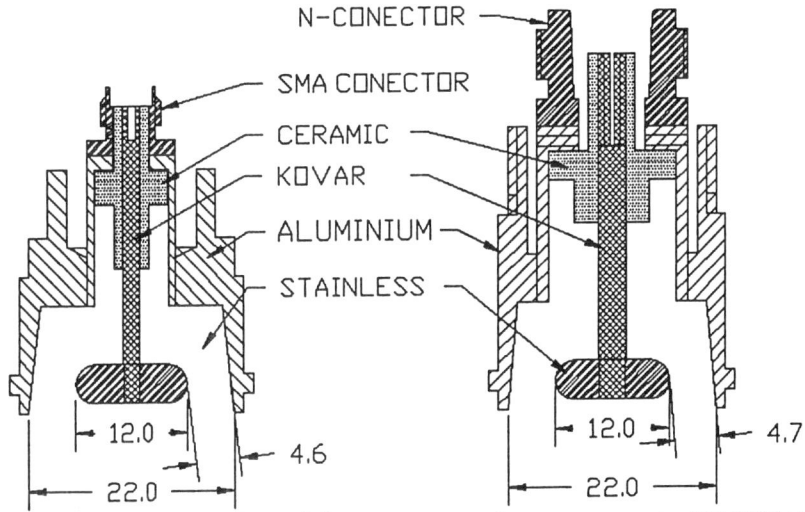

Fig. 9 New BPM electrodes with two types of connectors in TRISTAN.

6 Acknowledgment

We wish to thank the KEK Vacuum Group and K.Kamada, G.Mizuno, M.Yoshioka, K.Takata.

References

[1] M.Tejima, et.al., "Abnormal beam-induced signals in the button-electrodes of the beam position monitors of TRISTAN MR", presented at 7th Sympo. on Accelerator Science and Technology, Osaka, Japan,1989.

[2] A.J.Hatch, "Suppression of multipacting in particle accelerators" Nucl. Instrum. and Methods 41(1966) 261-271

THE BEAM LOSS MONITORS AT CEBAF*

John Perry
Continuous Electron Beam Accelerator Facility
12000 Jefferson Avenue, Newport News, VA 23606

REQUIREMENTS

The CEBAF beam will carry about 800 kW of power concentrated into a 200 μm diameter. Current estimates are that a stray beam will burn through the vacuum wall within 50–100 μsec. The CEBAF Fast Shutdown (FSD) system and the Beam Loss Monitor (BLM) system have been designed to prevent such an occurrence.

Initial estimates of the time constraints[1] dictated that FSD sensor functions would have about 10 μsec to detect and forward a fault to the FSD. The BLM is one of the sensor functions; therefore, we designed it to this specification. Reliability of the FSD was a critical concern that dictated a 5 MHz permission signal sent from the sensor function; permission exists when the signal is present, and is removed if the signal is not present for any reason, including system failure. A strong desire was expressed among the CEBAF design community for a short history of the radiation levels before a fault condition. All these considerations went into the design of the BLM system.

ARCHITECTURE

The BLM detects a beam loss fault by the radiation shower generated by the interactions between the beam electrons and surrounding matter. The critical time limit dictated by the vacuum wall burn through time placed strong constraints on the type of sensor used, on the signal conditioning method, and on the fault detect processing.

The BLM's fit into the FSD system as the leaves of the FSD tree (figure 1). This makes the fault trace fast and straightforward. Part of the FSD specification for permission inputs is that each input hold its fault state until queried by the FSD. The FSD only holds permission state information for the control system, and cannot distinguish among the various types of faults except as to their location on the tree.

Therefore, the BLM module must not only condition the sensor signal; it must also hold its own fault state, accept commands to set radiation trip levels, and provide this information to the control system upon command. We have successfully incorporated all these functions into a four channel CAMAC module which sends four independent FSD permission signals to the FSD interlock tree,

---

* This work was supported by the U.S. Department of Energy under contract DE-AC05-84ER40150.

accepts independent trip level commands for each channel, provides current radiation level for each channel, and keeps for access by the control system a −3 ms to +1 ms log of each channel's radiation level history around a fault.

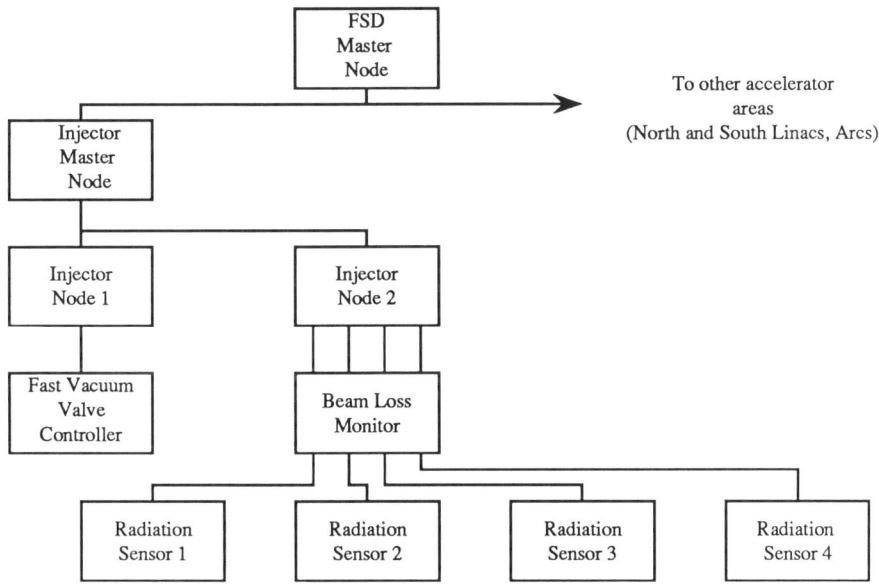

Figure 1. Beam Loss Monitors in Fast Shutdown system.

SENSORS

We investigated a number of types of radiation sensors for our use. Gas chambers were excluded quickly because they are mostly far too slow to satisfy the 10 μsec sense requirement. Some can react fast enough to remove permission within the limit, but they all require a high voltage power supply, and a great deal of manual intervention for construction and maintenance. We, therefore, investigated photoemissive and photoconductive devices for sensitivity and speed to see if their generally lower cost systems could satisfy our needs.

Preliminary research revealed[2,3] that photomultipliers would respond strongly to scintillation in their own envelopes, removing the need for scintillator material. Further, even the least expensive photomultipliers were sensitive enough to supply our needs. We looked at these as the first alternative to the very expensive Aluminum Cathode Electron Multipliers (ACEM's) used at CERN.[4] The question remaining was whether the envelope would darken during the life of the accelerator, making the photomultipliers blind to their own scintillation. The 10^5 rads of exposure required according to the literature to reduce their sensitivity to about half the initial value[5] will be sustained over a period of at least three years.[6] If the sensors respond well to visible wavelengths, that time will be stretched even further. The photomultipliers we are currently investigating are a factor of 10 less expensive than the ACEM's; hence, they are quite cost-effective. Furthermore, since the BLM was emphatically intended as an interlock system and not as a measurement system, we did not see gradual degradation in the response to fault level radiation as a handicap.

This still left us with the rather expensive high voltage power supply required for photomultipliers. We are presently investigating the use of vacuum photodiodes and silicon PIN diodes with scintillator material to eliminate the need for the high voltage. Preliminary results with vacuum photodiodes are very encouraging, and the lifetime considerations mentioned for photomultipliers apply equally to vacuum photodiodes.

NODE MODULE

The BLM module (figure 2) has been explicitly designed to permit a number of different signal conditioners to be used. This allowed us to design the fully specified control system interface immediately, leaving the less well defined analog system for further development. Since it was already established that the ACEM's and photomultipliers would fulfill our requirements at considerable cost, it was clear that the only questions lay in the kind of sensors we could use.

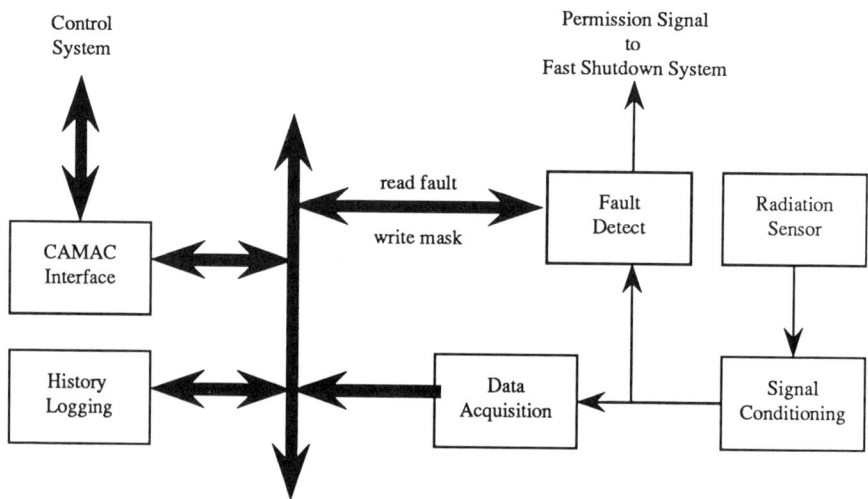

Figure 2. Beam Loss Monitor block diagram (one of four channels).

The module is therefore divided into a 3/4 CAMAC data acquisition board and a 1/4 replaceable analog signal conditioner. In principle, the BLM module could be used as a low resolution 8-bit data acquisition module with limited sequential storage.

The acquisition module interfaces the data acquisition blocks to the CAMAC bus, translating CAMAC function codes and addresses into command writes and data reads of the acquisition circuitry. The signal conditioner board allows us to continue to develop lower cost sensors, and fortuitously allows us to adapt a common interface to different sensors. This will permit us to back up whatever BLM sensors we choose with a slower gas-type detector. The present organization will allow us to use the same acquisition interface for both types of sensors, avoiding yet another software/hardware development cycle. The software interface provides the following functions:

Function F0 (CAMAC read)
A0–A3: Read current ADC data (8 bits plus fault bit)
A4–A7: Read current mask and fault status
Bit 0: Mask status
0 → channel may remove permission
1 → channel may not remove permission
Bit 1: Fault status
0 → channel has not tripped since last reset
1 → channel is in tripped state

Function F16 (CAMAC write)
A0–A3: Write current trip point (8 bit value for DAC)
A4–A7: Set mask and fault conditions
Bit 0: Mask command
0 → allow channel to remove permission
1 → prevent channel's removing permission
Bit 1: Fault command
0 → clear any existing fault
1 → generate a test fault

Setting both mask and fault in a command is not used (however, it results in safe operation if the module is tripped).

Function F2 (CAMAC destructive read)
A0–A3: Read channel history (Q-stop block protocol)

The data acquisition circuitry consists of an inexpensive 8-bit analog-digital converter with sequencing circuits that control both the ADC and the history block, which is a first-in, first-out storage chip. All the sequence control is done by two other chips: a dual monostable and a 20-pin programmable logic chip. This arrangement keeps the acquisition running constantly, updating the FIFO at each ADC cycle. When the channel detects a fault and the control system must read the history, it sends an F2 read command. The acquisition system does not stop for an F2 command; the channel must be faulted to get intelligible data. A test fault suffices.

Finally, the fault circuitry runs in parallel with the other functions. The fault circuitry monitors the conditioned analog voltage coming from the signal conditioning block without disturbing it in any way, until a fault is detected. Then the fault signal removes the 5 MHz permission from the output to the FSD, sets the fault bit in the CAMAC status register, and stops the acquisition cycle after approximately 1 millisecond.

EXPERIENCE

Prototypes of the BLM module in the FSD system have been running and successfully shutting down the accelerator during our recently concluded injector tests. Both photomultiplier tubes and vacuum photodiodes have been incorporated into the system, and have given full scale signal levels (5 volts out of the signal conditioners) for stray beam radiation levels. Placement of the sensors has not usually been particularly critical; however, we have as yet made no effort to characterize response vs. placement. Neither have we as yet made any attempt to measure radiation levels vs. signal strength in the accelerator. We have attempted to make such measurements at the cancer treatment center of a local hospital; however, in the limited time available we were not able to reduce the radiation level at the sensor enough to get linear readings suitable for obtaining numeric relations. When our beam, whose intensity can be controlled, turns on again, we expect to be able to make such measurements.

The CEBAF beam loss monitor design has shown itself to be a highly effective tool in protection for and analysis of the operation of the CEBAF injector. Further development of sensor technique is required to optimize cost and response.

REFERENCES

1. The CEBAF Fast Shutdown System, CEBAF PR 90-015, given at the 1990 LINAC Conference, Albuquerque, New Mexico, September 1990.

2. Young, Andrew T., "Cosmic Ray Induced Dark Current in Photomultipliers", *Review of Scientific Instruments,* Vol. 37 no. 11 (Nov. 1966), p. 1472.

3. Makeev, S. N., Petrov, V. A. Smelyanskii, I. M., Trubitsyn, N. M., "Variation of Photomultiplier Anode Current for Continuous and Pulsed Gamma Radiation", Trans. in *Instr. and Exp. Techn.* (USA), vol. 29 no. 1, pt. 2, (Jan–Feb 1986), pp. 182–185.

4. Agoritsas, V., Beck, F., Benincasa, G. P., Bovigny, J. P., "A Microprocessor-Based System for Continuous Monitoring of Radiation Levels Around the

CERN PS and PSB Accelerators", presented at the 2nd International Workshop on Accelerator Control System, Los Alamos, New Mexico, October 7–10, 1985.

5. Takasaki, F., "A Note on Radiation Damage to Photomultiplier Tubes", Appendix 15, *Radiation Effects at the SSC*, SSC-SR-1035, M. G. D. Gilchriese, Ed., Publ. by SSC Central Design Group, Lawrence Berkeley Laboratory, June 1988.

6. Stapleton, Geoffrey, *Summary of Radiation Damage Problems in Beam Enclosures*, CEBAF TN-0069, Dec. 1987.

A Longitudinal Emittance Measurement Program for the Fermilab Booster

V. Bharadwaj and M. Popovic
Fermi National Accelerator Laboratory
P.O.Box 500, Batavia, Il 60510

Abstract

We will describe a method of a longitudinal emittance measurement using TEK-DSA 602 scope and Accelerator Network (ACNET) console program. The scope is used to measure Booster bunch length and a total peak to peak RF voltage on the gaps of the cavities. The signal for bunch length comes from 2 GHz bandwidth Resistive Wall Current Monitor. An extension of this method to measure greater than 100 values of the longitudinal emittance in one accelerator cycle is proposed. In addition this device can act as digital Mountain Range.

1 Introduction

The variables we will use for description of longitudinal motion[1] are

$$q = \phi - \phi_s \quad \text{and} \quad y = \frac{E - E_s}{\omega_{rf}}, \tag{1}$$

where q is dimensionless and y has the units of eV-sec. The ϕ is the phase and E the energy of a particle that passes the accelerating gap. We use the convention that $\phi = 0$ when the RF voltage is zero and rising. The ϕ_s and E_s

are the phase and energy of the synchronous particle and ω_{rf} is the angular radio frequency.

In terms of these variables, the longitudinal motion of the particles in circular accelerators with a time varying electric field at the accelerating gap is described by the Hamiltonian,

$$H(q,t) \equiv -\frac{1}{2}Ay^2 + B(q\Gamma + \cos(q+\phi_s)) = C, \qquad (2)$$

where;

$$\Gamma = \sin(\phi_s) \quad , \quad \eta = \frac{1}{\gamma^2} - \frac{1}{\gamma_t^2}, \qquad (3)$$

$$A = (\frac{hc}{R})^2 \frac{\eta}{E_s} \quad \text{and} \quad B = \frac{eV}{2\pi h}, \qquad (4),$$

the radius of the Booster $R = 75.74$ meters, the harmonic number $h = 84$, V_0 is the peak RF voltage per turn, and the transition gamma $\gamma_t = 5.446$.

Each particle in the beam traces out a trajectory in phase space determined by the value of its Hamiltonian, C, and each particle has an emittance corresponding to the area enclosed by its trajectory. We will define the longitudinal beam emittance as the largest emittance of the particles in the beam.

The above discussion has been intended to clarify the terminology and introduce notation to be used here. In the next section we will present the formulae that are used to calculate beam parameters.

2 Calculational Method

The beam emittance ε_l as defined above is given by solving the Hamiltonian for y and integrating y as function of q

$$\varepsilon_l = \sqrt{\frac{2B}{A}} \int_{q_1}^{q_2} \sqrt{\cos(\phi_s + q) + \Gamma q + C} dy, \qquad (5)$$

where,

$$C = \cos(\phi_s + q_1) + \Gamma q_1 \qquad (6)$$

and $q_1 - q_2$ is the bunch length. Although it is possible to find q_1 and q_2 from of the bunch length and synchronous phase in the existing program we follow the way the longitudinal emittance is calculated traditionally at Fermilab[2]:

$$\varepsilon_l = (Moving\_Bucket\_Area) * \left(\frac{Bunch\_Length}{Bucket\_Length}\right)^2 * C_A. \qquad (7)$$

The C_A is the function displayed[3] in Figure 1.

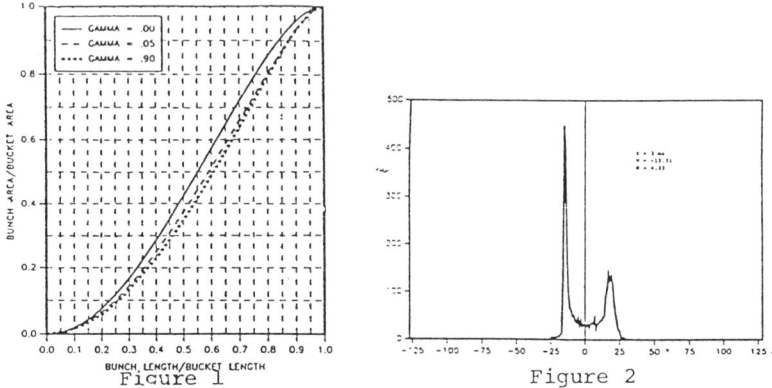

Figure 1   Figure 2

In the program the three curves are fitted to fifth order polynomials and for values of $\Gamma = \sin(\phi_s)$ in the indicated intervals linear interpolation is used[4]. The bucket length is defined as;

$$Bucket\_Length = \phi_L - \phi_R, \qquad (8)$$

where $\phi_R = \pi - \phi_s$ and ϕ_L is the second point where the separatrix crosses the $y = 0$ line and can be found as a solution of the following equation

$$\Gamma\phi_L + \cos(\phi_L) = \Gamma\phi_R + \cos(\phi_R). \qquad (9)$$

The ϕ_s can be calculated using the equation,

$$\phi_s = \frac{360}{2\pi} \arcsin(\frac{2\pi R}{V_0}\frac{dp}{dt}) \qquad (10)$$

and the fact that we know $p(t)$, the linear momentum of the Booster beam, as a function of time. The V_0 is the total peak RF voltage supplied to the beam by all cavities in one Booster turn and it is a measured quantity. The moving bucket area can be found as an integral of the equation for the separatrix;

$$Moving\_Bucket\_Area = \sqrt{\frac{2B}{A}} \int_{\phi_L}^{\phi_R} \sqrt{\cos(\phi) + \Gamma q + C_1} d\phi, \quad (11)$$

where,
$$C_1 = \cos(\phi_s + q_R) + \Gamma q_R \quad (12)$$

The $q_R = \pi - \phi_s$ and q_L is found from;

$$\cos(\phi_s + q_L) + \Gamma q_L = \cos(\phi_s + q_R) + \Gamma q_R \quad (13)$$

After integration

$$Moving\_Bucket\_Area = \alpha(\Gamma) \frac{16R}{h^2 c} \sqrt{\frac{heV_0 E_s}{2\pi\eta}}. \quad (14)$$

The program uses lookup table for $\alpha(\Gamma)$ which is indentical to the table in Bouvet at al[5].

3 Measurement of the bunch length and RF Voltage

As we have said above, the program uses the TEK DSA602 scope to measure the RF voltage, V_0, and the *Bunch_Length*. The RF voltage is the sum of the voltages coming from upstream gap monitors located in each RF cavity. These signals are properly phased by adjusting lengths of connecting cables and the summed signal is brought to the scope. The program uses a linear fit to compensate for the changes in the voltage as function of the frequency and at the same time to compansete for the loss due to the lossy cables and reduction factor of the pickup monitors. RF voltage and a bunch length are measured every 3 ms during the Booster cycle starting at 3 ms after injection. The signal for beam analyses comes from 2 GHz bandwidth Resistive Wall Current Monitor. For each point total of 2.56μs of the beam is analyzed. This portion of the beam is digitized in 5120 points and stored in the scope's memory. The stored waveform is then transferred trough ACNET and numerically analyzed. The bunch parameters are derived based on the sample

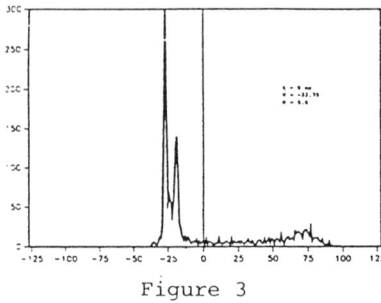
Figure 3

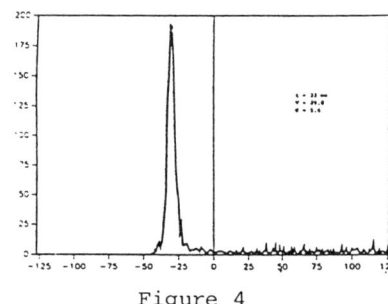
Figure 4

of 55 consecutive bunches. The baseline of the signal is defined as follows. First we find the voltage distribution, $N(V)$. The digitized values of the voltages are in an interval between -127 and +128 in some arbitrary units. Next we take cut at zero value and for purpose of defining baseline of the signal consider only region of negative voltage. We define the baseline of the signal, V_b as

$$V_b = \overline{V} + \sigma \qquad (15)$$

where

$$\overline{V} = \frac{\sum_{V=-127}^{0} V N(V)}{\sum_{V=-127}^{0} N(V)} \qquad (16)$$

and

$$\sigma^2 = \frac{\sum_{V=-127}^{0} (V - \overline{V})^2 N(V)}{\sum_{V=-127}^{0} N(V)} \qquad (17)$$

Figures 2, 3 and 4 are voltage distributions at 3, 9 and 33 ms.

4 Operational Experience and Future Development

A present the system described above will measure a value of the longitudinal emittance in approximately 2 minutes. An example of a set of measurements is given in Figures 5 and 6 where the longitudinal emittance is measured every three milliseconds in the acceleration cycle. We expect to improve the software such that we can take an arbitary number of emittance values, but

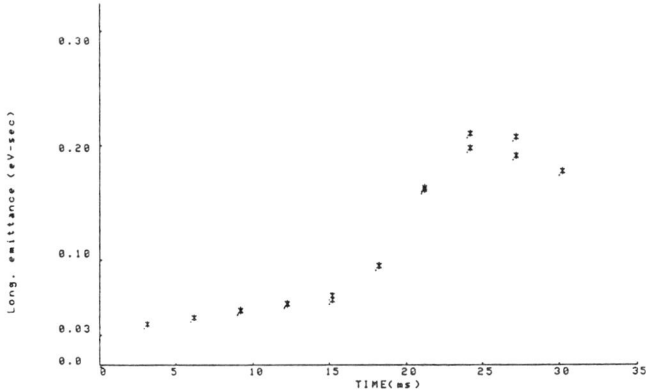

Figure 5, Longitudinal emittance during the Booster cycle.

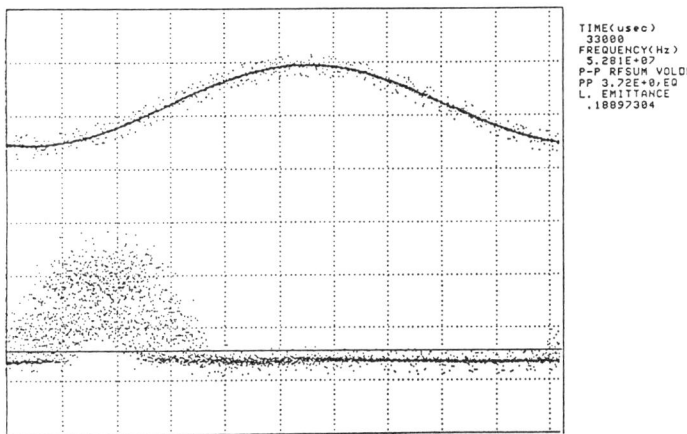

Figure 6, Typical display of 56 bunches overlapped; rf sum signal.

the limitations of our present system are obvious, ie. we cannot take multiple values of the emittance in one Booster beam cycle and the system cannot be made to readout any faster that the one point every 2 minutes. These limitations come from the nature of the TEK scope (only 32k samples of memory and 60 Hz maximum trigger rate) and the ACNET system that is used to readout and analyze the data. In a more powerful system we would propose to replace the hardware so that these limitations disappear. A possible system is shown in Figure 7. The heart of such a system would be a 2 Gs/s digitizer, with good analog bandwidth and dynamic range (eg. 1 Ghz and 8 bits) and a modest fast memory depth (8k samples). The digitizers should be able to operate on an external clock and we would like to be able to read out the fast memory into a large, regular, dual-ported memory in one millisecond. The system shown can then be operated in two distinct ways. If

Figure 7, A possible digitizing scheme for the stipline detectors.

the 4 digitizers are run in series then one can get 4 microseconds worth of data at 2 Gs/s into the regular memory every 1/4 millisecond. This information can then be analyzed to get a value of the longitudinal emittance and we will be able to get greater than 100 values on the longitudinal emittance in one Booster acceleration cycle. If the 4 digitizers are run in parallel then one can get an effective sampling rate of 8 Gs/s and by using the external clock feature of the digitizer one can subdivide the 8k samples of each digitizer arbitarily in the cycle. It is easy to see that one can use this method to generate a digital mountain range (8k samples = 128 * 64 samples, 64 samples @ 2 Gs/s = 32 ns). This mountain range scheme offers considerable advantages over the analog system now being used in that one has the option to analyze the longitudinal bunch data. In this system we are talking about a large amount of data generation and we propose using powerful microprocessors to do data reduction so that only a small amount of useful data is sent to the user consoles.

5 References

1. F. T. Cole, "Longitudinal Motion in Circular Accelerators", AIP Conf. Proc. No. 153 (US Particle School), 44-82.

2. S.Ohnuma, "The Beam Emittance", Fermilab, Exp-111, Nov. 1983.

3. P. S. Martin and S. Ohnuma, "Longitudinal Phase Space in Circular Accelerators", AIP Conf. Proc. No. 184 (1987-1988 U.S. School), 1941-1968.

4. J. Crisp Fermilab, unpublished.

5. C. Bouvet et.al., "A Selection of Formulae and Data Useful for the Design of A.G. Synchrotrons", CERN/MPS-SI/Int. DL/70/4, April 1970.

LOG-RATIO CIRCUIT FOR BEAM POSITION MONITORING*

F. D. Wells, R. E. Shafer, J. D. Gilpatrick and R. B. Shurter
MS: H808, Los Alamos National Laboratory
Los Alamos, NM 87545

Abstract

A synopsis is given of work in progress on a new signal processing technique for obtaining real-time normalized beam position information from sensing electrodes in accelerator beam pipes. The circuit employs wideband logarithmic amplifiers in a configuration that converts pickup electrode signals to position signals that are substantially independent of beam current. The circuit functions as a ratio detector that computes the logarithm of (A/B) as (Log A - Log B), and presents the result in a video (real-time analog) format representing beam position. It has potential benefits of greater dynamic range and better linearity than other techniques currently used and it may be able to operate at substantially higher frequencies.

The Log-Ratio Signal Processing Algorithm

Consider a pair of microstrip pickups in a circular beam pipe as shown in Fig. 1. When a short beam bunch, $I(r_o, \theta_o)$ travels past the electrodes in the z-direction, image currents are induced into the microstrips. The ratio of these currents can be expressed as[1],

$$\frac{I_A}{I_B} = \frac{1 + \frac{4}{\phi_o}\sum_{n=1}^{\infty}\frac{1}{n}(\frac{r_o^n}{R^n})\cos(n\theta_o)\sin(\frac{n\phi_o}{2})}{1 + \frac{4}{\phi_o}\sum_{n=1}^{\infty}\frac{1}{n}(\frac{r_o^n}{R^n})\cos(n\theta_o)\sin n(\pi + \frac{\phi_o}{2})} \quad (1)$$

where, $\emptyset_o$ is the angle subtended by the pickup elements,
R is the radius of the pickup aperture,
r_o and θ_o are the coordinates of the beam bunch,
$r_o \cos \theta_o = X$ is the beam displacement from the center.

Taking the first term of the series,

$$\frac{I_A}{I_B} \cong \frac{1 + \frac{4}{\phi_o}(\frac{X}{R})\sin(\frac{\phi_o}{2})}{1 - \frac{4}{\phi_o}(\frac{X}{R})\sin(\frac{\phi_o}{2})} = \frac{1+\varepsilon}{1-\varepsilon} \quad (2)$$

Converting this ratio to decibels gives,

$$20\log\frac{I_A}{I_B} \cong 20\log\frac{1+\varepsilon}{1-\varepsilon} = \frac{20}{\text{Ln }10}\text{Ln}\frac{1+\varepsilon}{1-\varepsilon} \quad (3)$$

$$\cong \frac{40}{\text{Ln }10}[\varepsilon + \frac{\varepsilon^3}{3} + \frac{\varepsilon^5}{5} + ...] \quad (4)$$

----

*Work supported by the US Department of Energy, Office of High Energy and Nuclear Physics.

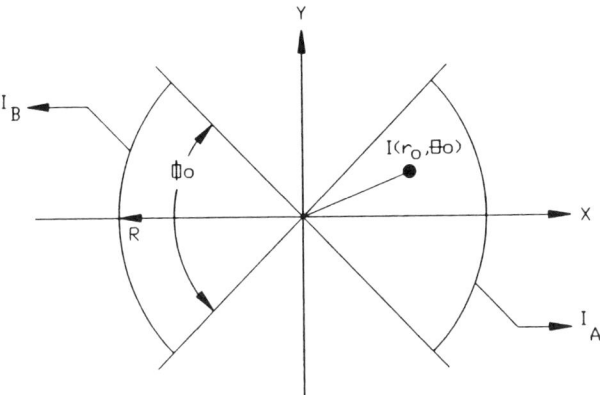

Fig. 1. Cross section of a pair of microstrip electrodes in a circular beam pipe.

Taking the first term of this expression gives,

$$20\log \frac{I_A}{I_B} \cong \frac{40}{\text{Ln }10}\left[\frac{4}{\phi_o}\left(\frac{X}{R}\right)\sin\frac{\phi_o}{2}\right] \quad (5)$$

Solving for X gives,

$$X \cong \frac{\text{Ln }10}{160}\left(\frac{R\phi_o}{\sin\frac{\phi_o}{2}}\right) 20\log \frac{I_A}{I_B} \quad (6)$$

Thus, the logarithmic ratio of the signal amplitudes provides a reasonably linear measurement proportional to beam position.

Fig. 2 compares the response of log-ratio processing to the two most commonly used processing techniques: difference-over-sum and amplitude-modulation to phase-modulation (AM/PM) conversion. Notice that log-ratio processing combined with the microstrip electrode beam position response theoretically provides the most linear response over the full aperture of the pickup[2].

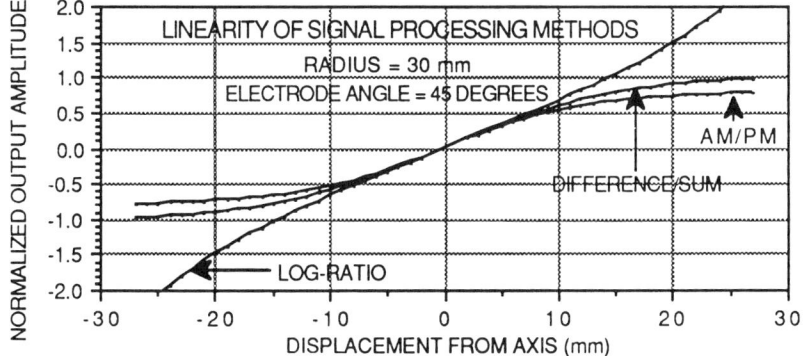

Fig. 2. Response curves of the log-ratio, difference-over-sum and AM/PM processing techniques.

310 Log-Ratio Circuit

Implementation

Detector-logarithmic video amplifiers are used extensively in radar and electronic warfare applications, and are available from a number of manufacturers. Input frequencies range from DC to tens of GHz and the input dynamic range can exceed 60 dB (typically -45 dBm to 15 dBm)[3].

In 1989 the model AD640 Logarithmic Amplifier became available from Analog Devices Corporation[4]. It is a monolithic logarithmic amplifier containing five cascaded dc-coupled amplifier/limiter stages, each having a small signal voltage gain of 10 dB and a -3 dB bandwidth of 350 MHz. Each stage has an associated fullwave detector that produces an output current that is proportional to the absolute value of the input voltage applied to the amplifier. The detector outputs are summed to produce a composite current that approximates a logarithmic transfer function. To complete the circuit, the current is converted to a voltage and filtered to provide a video envelope representing the input signal.

For the log-ratio application two AD640 chips have been cascaded as shown in Fig. 3. The current outputs are summed and converted to voltage by an amplifier and the video output is extracted from a 2 MHz low-pass filter.

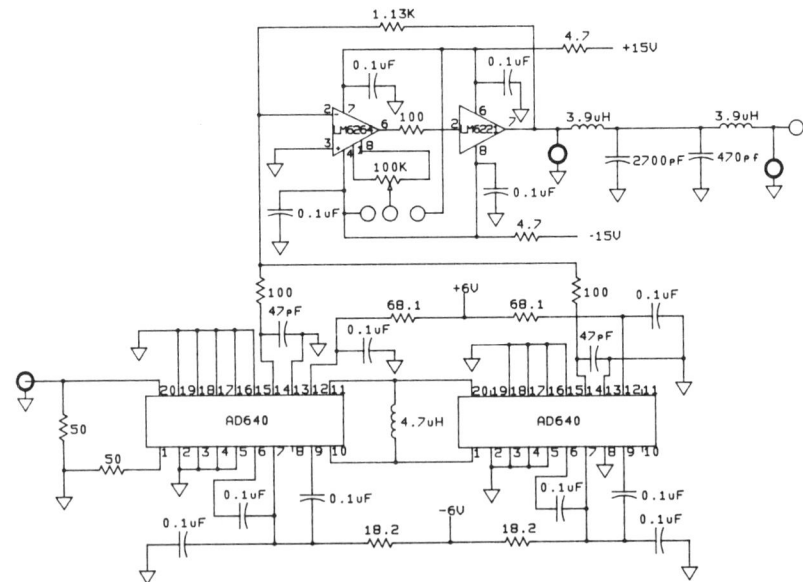

Fig. 3. Circuit diagram of a 70 dB dynamic range logarithmic amplifier for 50 MHz to 150 MHz operation.

Figs. 4 and 5 show the transfer and error curves for two of these amplifier circuits responding to 60 MHz RF input signals. Superimposed on the transfer curves are straight line fit curves. The slopes of the two amplifiers differ by less than 1% and the zero crossing points are nearly identical. Over the range from -10 dBm to -60 dBm the amplifier deviations are less than 1 dB from the straight line, as shown by the error curves.

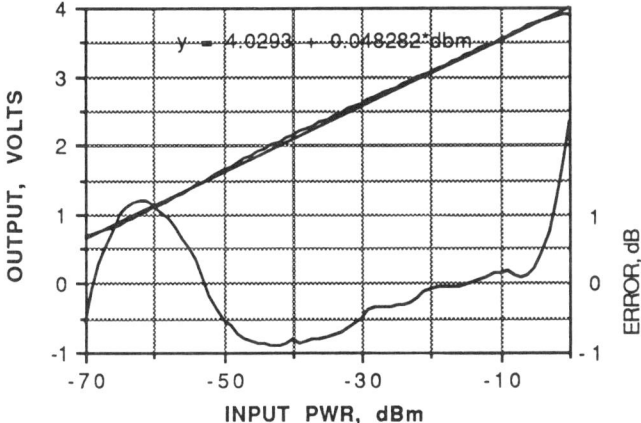

Fig. 4. Transfer curve, straight line fit and error curves for logarithmic amplifier number one.

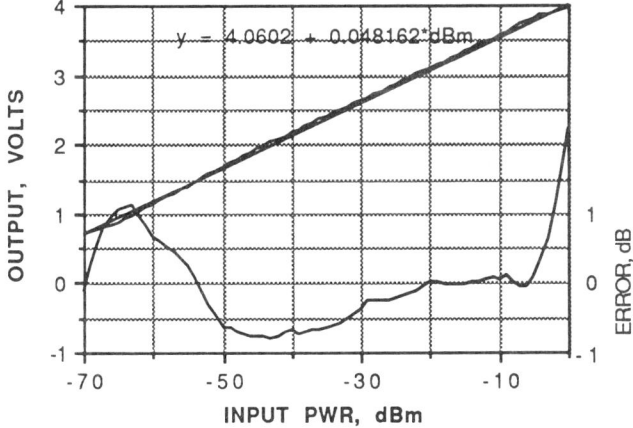

Fig. 5. Transfer curve, straight line fit and error curves for logarithmic amplifier number two.

Log-Ratio Circuit

The complete log-ratio circuit is shown in Fig. 6. Two logarithmic amplifiers are connected to a differencing amplifier to produce a video output proportional to Log (A/B).

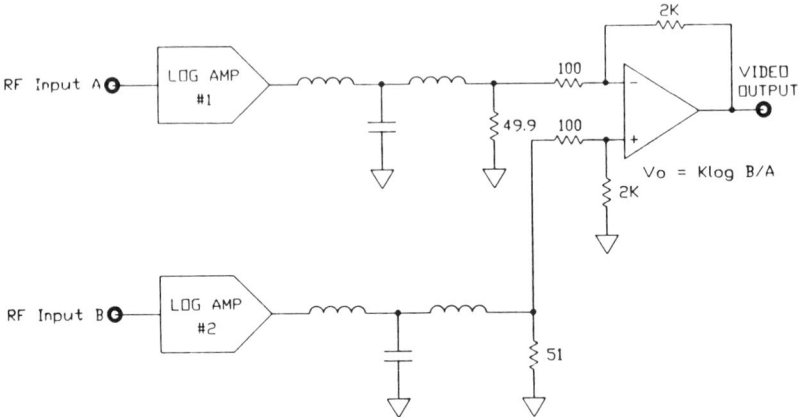

Fig. 6. The log-ratio circuit showing the two logarithmic amplifiers, the lowpass filters and the differencing amplifier.

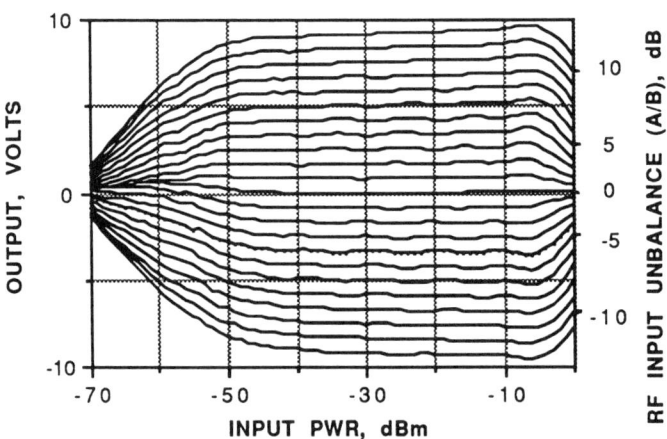

Fig. 7. Log-ratio circuit response curves to 60 MHz RF input signals.

The circuit response is shown by the curves of Fig. 7. On the horizontal axis the RF input power to the A and B inputs is plotted, ranging from 0 dBm to -70 dBm. This family of curves represents 23 position values corresponding to differential signal intensity changes of 1 dB/step to the two inputs. The center trace results when the two signals are equal (A = B). The upper traces correspond to A > B, while the lower traces result when A < B. Best operation occurs in the range of -10 dBm to -50 dBm. Ideally the traces should be straight horizontal lines. The variations from straightness are due to the deviation of the amplifier transfer curves from the straight line fit.

Fig. 8 is a plot of the output voltage versus the offset from center for an input power of -15 dBm. This illustrates the linearity characteristic of the log-ratio processing technique.

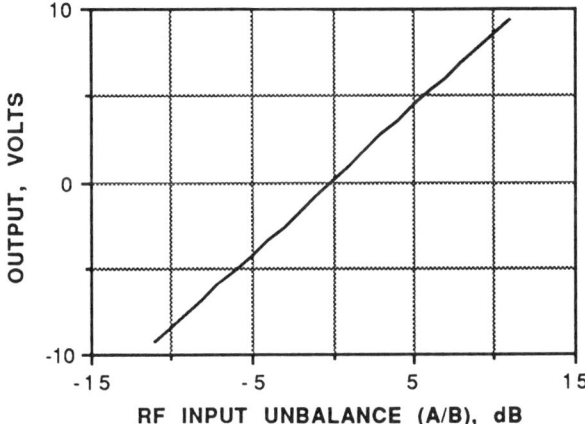

Fig. 8. Transfer curve of the log-ratio circuit at -15 dBm input power.

For the amplifier gains used in this investigation the average calibration factor over the input power range of -10 to -45 dBm is 0.853 volts/dB with individual factors differing by a maximum of ± 4.3%. If the circuit was connected to a microstrip pickup having a sensitivity of 6 mm/dB the overall calibration factor would be 7.03 ± 0.3 mm/volt.

Conclusions

1. The log-ratio circuit technique provides a more linear response characteristic than other types of position detection such as difference-over-sum and AM/PM processing.

2. The deviation of the logarithmic amplifier transfer curves from a straight line fit has, thus far, limited the accuracy and dynamic range.

3. The output is a real-time normalized position signal with good bandwidth.

4. A major advantage over AM/PM processing is that cables connecting the pickup electrodes to the processor do not need to be closely phase matched because the log-ratio circuit responds to amplitude differences and is not sensitive to phase differences.

5. The availability of commercially manufactured quasi-logarithmic response chips with wide dynamic range, high bandwidth and reasonable cost make log-ratio processing a technique that deserves continued investigation.

6. More work is planned to determine if the log-ratio process is a viable alternative to the difference-over-sum and AM/PM processing techniques. Improvements in the circuit response may be effected by close matching of the logarithmic amplifier chips. Other topics to pursue include noise considerations, drift sensitivity and cost effectiveness.

References

1. R. E. Shafer, "Characteristics of Directional Coupler Beam Position Monitors," IEEE Transactions on Nuclear Science, Vol. NS-32, No. 5, (Oct., 1985), pp. 1933-1937.

2. R. E. Shafer, "Beam Position Monitoring," AIP Conference Proceedings 212, Accelerator Instrumentation, (1989), p. 48.

3. R. S. Hughes, Logarithmic Amplification (Artech House, Inc., Dedham, MA 02026, 1986), p. 15.

4. "Monolithic DC-to-120-MHz Log-Amp is Stable and Accurate," Analog Dialogue, Vol. 23, No. 3, p. 3 (1989), (Analog Devices, Inc., Norwood, MA 02062-9106).

Imaging micron-sized beams with optical
transition radiation

D. W. Rule and R. B. Fiorito
White Oak Laboratory
Naval Surface Warfare Center
Silver Spring, Maryland   20903-5000

Optical transition radiation (OTR) can be used to obtain images of beam profiles and to measure beam divergence, and thus to determine beam emittance. We discuss the effects of diffraction on the limit of resolution of OTR images used as a diagnostic for micron-sized relativistic beams.

I. INTRODUCTION

In this paper, we will discuss the features of transition radiation production which have a bearing on its use as a beam diagnostic for relativistic beams with radii of the order of 1-100 μm. In particular, we will examine the limits of resolution of beam images made with optical transition radiation (OTR).

The early treatments of TR were based on the solution of the classical boundary value problem associated with a charge crossing the interface between media possessing different dielectric constants (see Ter-Mikaelian[1] for a review). The quantum mechanical treatment was carried out by Ritchie[2]. Pafomov[3] and more recently Wartski[4] have provided a treatment which utilizes the Fresnel transmission and reflection coefficients, replacing the Fourier transform of the particle's current by a line of Hertzian dipoles. This method thus uses the known results of reflection of plane waves by boundaries, rather than solving the boundary value problem, and, in essence, rederiving the Fresnel relations. All these methods are carried out in the frequency domain. In order to shed a different light on the TR production process and to emphasize the spacial and temporal properties of TR, we review the results of a simple treatment carried out in the time domain in Section II. In Section III we present the results of an analysis of the diffraction limit to the resolution of beam spot images. This analysis is based on a scalar diffraction treatment in which the pupil function has the angular dependence of transition radiation. We show that the limiting resolution is not significantly different from standard diffraction limits.

II. TRANSITION RADIATION FIELDS NEAR A BOUNDARY

This discussion is based on the tutorial treatment by Carron[5], whose results were previously derived by Maresca and Liboff[6] using a different approach. The radiation fields were derived in Ref. (5) for a particle crossing a perfectly conducting boundary plane. The method of image charges was used. The case of a particle emerging (forward TR) from the perfect conductor was examined. It was shown that the radiation component of the field propagated as a spherical wave on a hemispherical shell of radius R=ct as in Fig. 1. The radiation field components are[5]

$$E_r = 0$$
$$E_\theta = B_\phi = \frac{2\beta q \sin\theta}{R(1-\beta^2\cos^2\theta)} \begin{cases} \delta(ct-R) \text{ forward} \\ \delta(R-ct) \text{ backward,} \end{cases} \quad (1)$$

where $\beta=v/c$, velocity in units of c, the speed of light, q is the particle's charge. In Fig. 1, R is the magnitude of the vector from the point where the charge crosses the conducting boundary to the point where the TR is detected and θ is the angle measured from the direction parallel to the particle velocity (+z-direction) for forward TR and from the antiparallel direction (-z-direction) for backward TR, Fig. 2. While this wave is <u>spherical</u>, the field strength is a strong function of θ, which peaks at $\theta=\sin^{-1}(1/\beta\gamma)$, where γ is the usual Lorentz factor.

It has been suggested[7] that the strong variation in TR intensity with angle somehow relates to an uncertainty principle such that the peak angle ($\sim 1/\gamma$) times the uncertainty in the point of origin of the photon is of the order of the wavelength of the radiation. This would yield an uncertainty in position $\lambda\gamma$, which happens to be the characteristic radial scale length of the Lorentz transformed Coulomb field of a relativistic particle. Such an uncertainty effect would make it impossible to resolve micron sized beams, even at relatively modest energies.

However, according to Eq. (1), the origin of the TR is well defined: it is at R=ct=0, at the point where the particle crosses the boundary in Fig. 1. The angular dependence of the spherical wave's intensity, which peaks at $\theta\sim 1/\gamma$ does not correlate with an uncertainty in the point of origin of the TR photon. Such an uncertainty

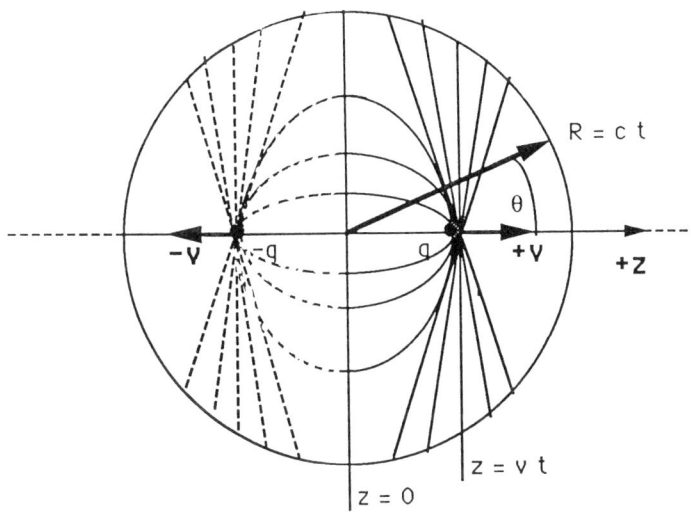

Fig. 1. Charge q emerging from a perfect conductor and its image, -q. Coulomb fields exist only inside radius R = ct and radiation fields exist on the shell R = ct.

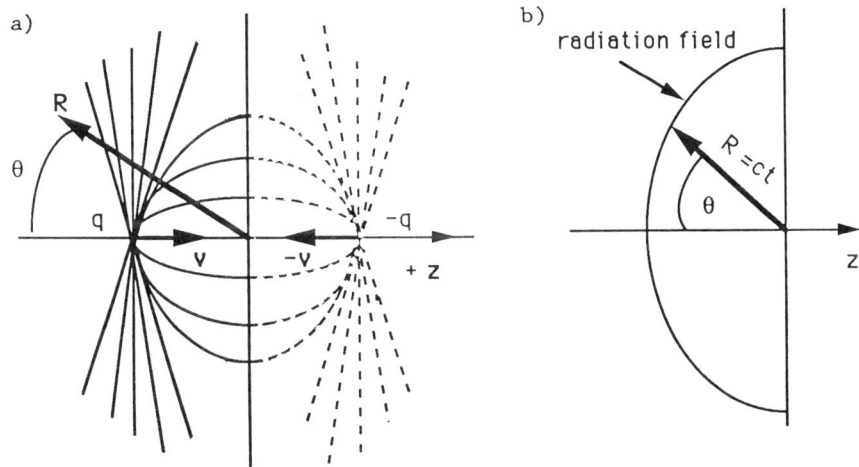

Fig. 2. (a). Particle entering a perfect conductor. At $t' = t - |\vec{R} - z\hat{z}|/c < 0$ only Coulomb fields exist. (b) When t' > 0, spherical radiation fields exist on shell R = ct.

effect conceptually might arise from a δ-function distribution in angle at θ=1/γ. In fact, considerable irradiance contributions come from angles larger than 1/γ.

Wartski[4] has made optical TR monitors for 1 GeV beams in the Orsay linear accelerator at the electron-positron conversion target, in order to optimize positron production. His observations of beam diameters of 1 mm were consistent with estimates based on the performance of the focusing triplet used and the beam emittance. For this measurement, $\gamma\lambda \approx 1$ mm, thus the effect of an uncertainty circle of diameter $2\gamma\lambda \approx 0.3$ mm should have been observed if this so-called uncertainty effect were valid.

Wartski observed, in fact, sharper beam images in TR, than from fluorescent zinc sulfide screens. More recently, our colleagues at the Boeing Physical Sciences FEL facility have found that only by replacing phosphor screens in the wiggler with TR foils, could they obtain sufficient beam spot resolution to characterize the beam transport in the wiggler and, in so doing, achieve a matched beam at the entrance to the wiggler.[8] Their beam diameter is ~1 mm and the beam energy is ~109MeV.

III. DIFFRACTION LIMIT FOR IMAGING TRANSITION RADIATION

It has been suggested by Jenkins[9] that the diffraction limit for TR imaging of ultrarelativistic beams is significantly larger than the usual limit from standard diffraction theory, for example the Rayleigh criterion. We have reexamined this question and come to a different conclusion: that the diffraction limit is not significantly larger than the usual situation.

Consider a spherical wavefront focused to a Gaussian image point on the optical axis of an optical system at a distance f from the aperture. This wavefront is limited by an aperture of radius a, as in Fig. 3. Following Born and Wolf's[10] treatment of the distribution of light near focus, we use scalar diffraction theory to determine the scalar function $U(\vec{r})$, at a distance $\vec{r}$ from the focal point in the image plane shown in Fig. 3. We take $\vec{r}$ along the x axis for simplicity and we assume that f >> a, a >> λ and f >> r.

The scalar diffraction can be expressed in terms of the Debye integral, which is an integration over plane waves differing in phase by $k\vec{q}\cdot\vec{r}$ where $\vec{q}$ is the unit vector shown in Fig. 3., which makes an angle α with respect to the optical axis and has an azimuthal angle φ in

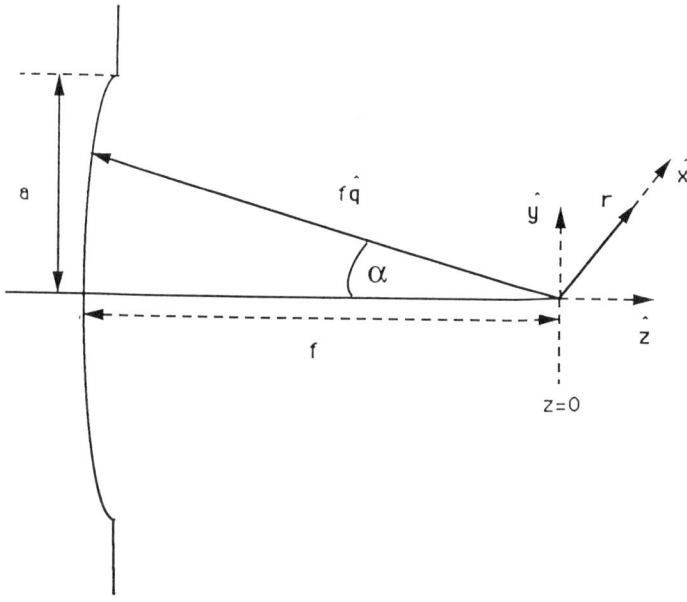

Fig. 3. Diffraction of converging spherical wave at a circular aperture.

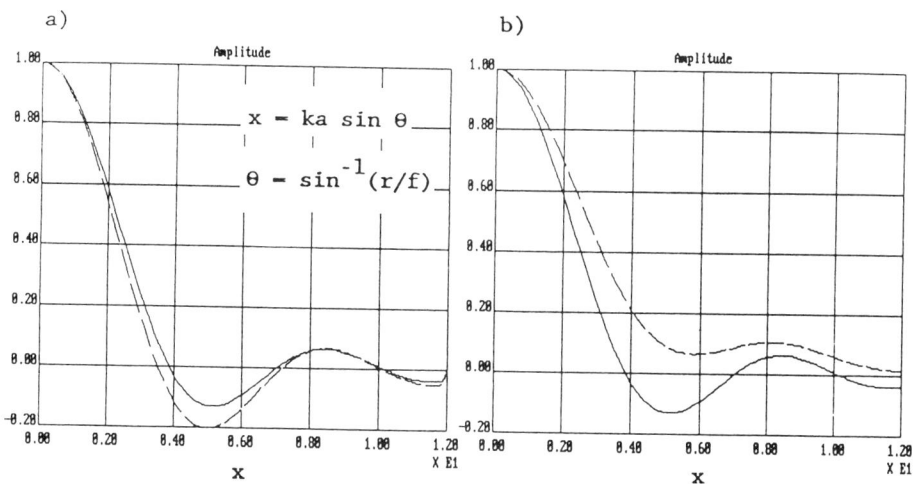

Fig. 4. Comparison of standard diffraction amplitudes (solid curve) with TR diffraction (dashed curve), calculated for: (a) an aperture angle $\alpha_m = 1/\gamma = 0.01$, and (b) an aperture angle $\alpha_m = 0.25$, i.e., $\alpha_m \gg 1/\gamma$ if $\gamma = 100$.

the plane perpendicular to the optical axis. The Debye integral for our case, in which the pupil function $A(\vec{q})$ is not constant, is

$$U(\vec{r}) = -\frac{i}{\lambda} \int_0^{2\pi} \int_0^{\alpha_m} A(\vec{q}) \, e^{-i\vec{kq} \cdot \vec{r}} \sin\alpha \, d\alpha \, d\phi. \quad (2)$$

The pupil function $A(\vec{q})$ is, in the case of focused TR,

$$A(\vec{q}) = A(\alpha) = A_0 \frac{\sin\alpha}{1 - \beta^2 \cos^2\alpha}, \quad (3)$$

where A_0 is a constant with appropriate units. The integral over ϕ gives the usual Bessel function: $2\pi J_0(kr \sin\alpha)$. The remaining integral can be performed numerically, where the limit is given by $\alpha_m = \sin^{-1}(a/f)$.

We considered two cases: $\alpha_m = 1/\gamma$, and $\alpha_m \gg 1/\gamma$. The results of numerically integrating (2) are given in Fig. 4. Fig. 4(a) shows the case where $\alpha_m = 1/\gamma$ (dashed curve). This result scales so that it is independent of the particular value of γ, if $\gamma \gg 1$. In Fig. 4(b) the diffraction amplitude for $a/f = 0.25 \gg 1/\gamma$ is shown for the case of $\gamma = 10^2$ (dashed curve), in comparison to standard diffraction theory (solid curve). The corresponding intensities are just the square of the amplitudes shown. We have also calculated the integral for $a/f = 0.25$ and $\gamma = 10^5$, or 51 GeV electrons. The result was only slightly different from Fig. 4(b).

The results of Fig. 4(a) for $\alpha_m = 1/\gamma$ show a slightly narrower diffraction pattern for TR than the standard pattern, but the second maximum is about twice the height of the usual value for the Airy disc. This case corresponds to the situation calculated by Jenkins[9]. Fig 4(b) shows that if $\alpha_m \gg 1/\gamma$, the diffraction pattern is slightly broader and resembles an apodized diffraction pattern; however, this effect is not large enough to seriously degrade imaging ultrarelativistic beams.

IV. DISCUSSION

Based on our understanding of the properties of optical transition radiation (OTR) produced by relativistc charged particles, we conclude that the concept of an uncertainty relation relating the angle of emission to an uncertainty in the position of the origin of the photon relative to the particle trajectory is not correct. We have also considered the diffraction of transition radiation by an optical system using scalar diffraction theory and have found that the limit of resolution for imaging beam spot profiles with OTR is not significantly different from the usual diffraction limits.

ACKNOWLEDGEMENT

This work was supported by the NAVSWC Independent Research Program.

1. M. L. Ter-Mikaelian, High Energy Processes in Condensed Media (Wiley-Interscience, 1972).

2. R. H. Ritchie and H. B. Eldridge, Phys. Rev., 126, 1935 (1962).

3. V. E. Pafomov, in Proc. P. N. Lebedev Phys. Inst., edited by D. V. Skobel'tsyn, 44, 25-127 [Consultants Bureau, New York, 1971].

4. L. Wartski, Thesis, Univ. de Paris-Sud, Centre d'Orsay (1976); J. Bosser et al., "Optical Transition Radiation Proton Beam Monitor," CERN/SPS/84-17.

5. N. J. Carron, Mission Research Corp., Report No. MRC-R-918 (1985).

6. Neal J. Maresca and Richard L. Liboff, J. Math. Phys., 16, 116 (1975).

7. K. T. McDonald and D. P. Russell, in Proc. of the Joint US-CERN School on Observation, Diagnosis, and Correction in Particle Beams, Capri (1988).

8. D. H. Dowell et al., Nuc. Instrum. & Methods in Phys. Res., A296, 351-357 (1990).

9. Edgar W. Jenkins, SLC, Single Pass Collider Memo CN-260 (1983).

10. Max Born and Emil Wolf, Principles of Optics, 2nd Ed. (Pergamon, 1964).

A TUNE MEASUREMENT SYSTEM FOR THE TEVATRON USING A PHASE-LOCKED LOOP

J. Fitzgerald, R. Gonzalez
Fermi National Accelerator Laboratory*
P.O. Box 500, Batavia, Il. 60510

ABSTRACT

A system for continuous real time measurement of betatron tune has been developed and installed for the Fermilab Tevatron and Accumulator ring. A phase-locked loop, using a considerable amount of input signal preconditioning, is used to lock to a betatron oscillation harmonic in the Schottky detector output signal. This system has demonstrated the capability of measuring the Tevatron fractional tune to an accuracy of $\pm.002$ at 100 Hz bandwidth. The desired tune spectrum line has a very low signal to noise ratio, undesired revolution frequency harmonics, frequency modulation with wide deviations, and amplitude modulation with levels to 100%. Shown are the techniques used to overcome some of these limitations, with examples of the systems accuracy and tracking performance.

INTRODUCTION

The tune number is described as the number of betatron oscillations around one revolution of the ring, which is normally in the range of 19.35 to 19.45 for the Tevatron. It is commonly referred to by the fractional part of the number, symbolized by ν. The signal originates from a Schottky detector as a 21 MHz signal, a harmonic of the Tevatron 47 KHz revolution frequency (F_{rev}). This signal is down-converted in two stages by mixing with multiples of F_{rev} leaving a difference frequency of $\nu(F_{rev})$ or nominally 19 KHz.[5] Measuring the frequency and scaling by a fixed value for F_{rev} produces the desired fractional tune value. This method allows the use of a low frequency spectrum analyzer to monitor the tune, however they are too slow for making fast time plots or for use in a possible feedback scheme. Another method is to use a phase-locked loop "tune tracker". Described in this report are the efforts to overcome some application problems and optimize the performance of a phase-locked loop (PLL) system.

DESIGN CONSIDERATIONS

The Tevatron Signal:

1. The tune range of .35 to .45 for the Tevatron gives a tune signal range of 16.7 to 21.5 KHz.
2. The output amplitude level is nominally -20dBm, but ranges from -50 to 0 dBm depending on acceleration mode.

---

*Operated by Universities Research Association Inc. under contract with the U.S. Department of Energy.

3. The signal is normally only 3 to 10 dB above the noise floor. The S/N ratio is calculated from the ratio of the power spectrums[2] and typically has a value less than one.
4. Revolution frequency and betatron oscillation harmonics and sidebands are present at every $N(F_{rev})$. The largest of these is the tune signal image frequency at $(1-\nu)F_{rev}$ (26-31 KHz), (Fig. 1b.) and is about equal to the desired signal.
5. The signal is frequency modulated (FM), with a gaussian distribution of frequency components of 0 to 1KHz.
6. The signal is also amplitude modulated (AM), (Fig. 2) with levels reaching 100%, and with frequency components to 1KHz BW.

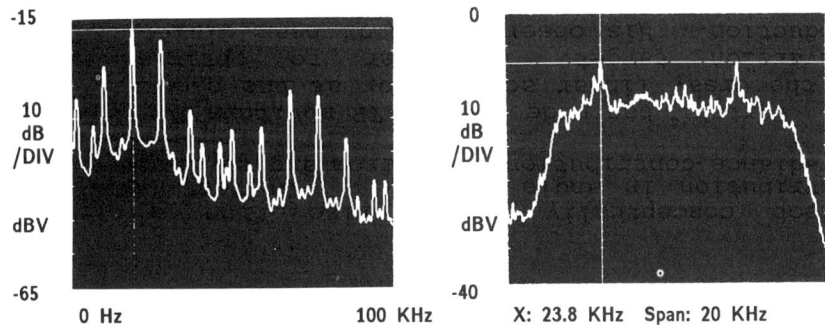

Fig.1. (a) Frequency spectrum of the tune signal, showing revolution frequency harmonics and sidebands present. (b) Spectrum showing the image frequency at $F_{rev}(1-\nu)$.

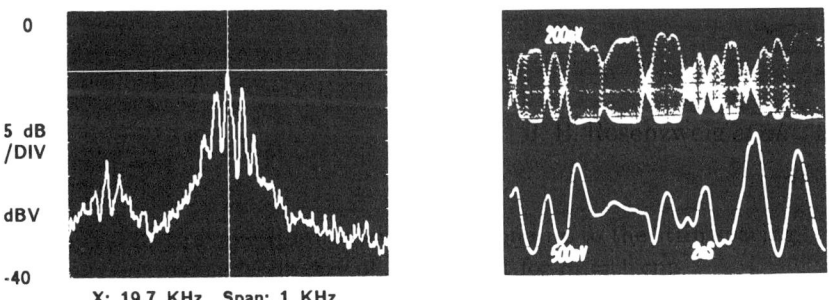

Fig. 2. (a) Frequency spectrum showing the vertical signal feedthrough on the horizontal signal. (b) Photo of a typical signal with 100% A.M. (top) and a PLL error signal showing loss of lock at the signal minimums (bottom).

For the Anti-Proton (AP) Accumulator ring a low frequency Schottky detector is used, with output at 245 KHz. A variation of the same PLL system is used, in this case without frequency conversion, but requiring a real time F_{rev} measurement. The AP signal is similar to the Tevatron signal except that it has a better signal to noise ratio and the undesirable harmonics are farther away.

A Tune Measurement System for the Tevatron

Performance Goals:

1. Use a PLL circuit to measure the fractional tune of the Tevatron in real time with resolution and accuracy to within .002 fractional tune.
2. The dynamic response characteristics of PLL, must enable tracking a tune signal containing FM of 500 Hz bandwidth (BW), at a 100 Hz Mod. rate, maintaining accuracy, and FM of 6 KHz BW maintaining lock.
3. The PLL must lock to and track the tune signal from an initial unlock condition, at turn on, or for any frequency step within the tune range.
4. Provide a noise free output to indicate a "locked" condition.

CIRCUIT DESIGN

A review of PLL design theory shows the relative merits of the various types of circuits. There are many types of phase-detectors, but all can be classified as either linear or digital. The PLL textbooks[1,2] strongly emphasize the use of a linear phase-detector for tracking filters, or for any application where the signal is buried in noise.

The digital detector, as its name implies, uses logic gates which are switched at threshold levels, zero crossings, or rising/falling edges. It is highly susceptible to any noise or jitter on the input signal at the threshold levels of the input comparator, while its internal logic functionally stores the resulting false triggers. The output uses logic levels with a varying duty cycle to indicate the phase error, again introducing uncorrelated digital noise. These detectors produce phase error noise which is totally unproportional to the original cause. In the worst case some detectors (the phase and frequency sensitive type) may loose the lead/lag relationship of the inputs, changing signs on the error signal, producing big transients and sometimes loss of lock.

The linear phase-detector is an analog device, usually a mixer or an integrated circuit multiplier, which actually multiplies the inputs together, giving an output product containing the phase error. Noise on the input signals has an effect on the output only in "linear" proportion to the relative noise power. Small glitches produce a corresponding small amount of phase noise, more readily reduced by the loop filter.

Several different types of phase-detectors were tried on the actual tune signals to check their performance. The above comments were verified, with the digital types able to maintain lock only about 75% of the time, compared to nearly 100% for the linear types. The linear detector however, is amplitude sensitive and since the input has a high level of AM and wide signal strength varation, additional conditioning circuitry is required. References[3,5] indicated that the best performance would be obtained by using a limiter section before the phase-detector. Comparisons to automatic gain circuits, and fast comparators showed that the limiter was indeed superior.

Shown in Fig. 4. is a block diagram of the Tevatron circuit. The overall circuit includes a narrow bandpass filter, a high gain limiter, and a linear phase-detector. The BPF, an 8th order Chebyshev, attenuates most undesirable harmonics by 40 to 70 dB. The problem of large amplitude variations and heavy AM was largely overcome by the amplifier-limiter stage. The limiter is a simple cascade of four clipping amps with an overall gain of 100 dB. The preamp, filter and the limiter combine to produce a dynamic range of about 80 dB, with the PLL section maintaining lock.

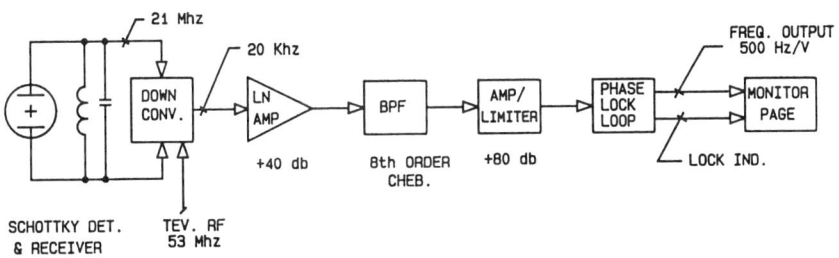

Fig. 4. Block diagram of the Tevatron tune measurement system.

The PLL circuit used in this system is a second-order type using an active loop filter. The circuit is constructed using separate IC's for the major components, to allow flexibility in the loop filter, loop gain and the output scaling. The loop filter natural frequency is set at 2 Khz which is as low as possible consistent with a damping factor of .7, and gain that allows sufficient tracking and lock in range. The PLL error voltage passes through final filters (post PLL BW limiting), and buffers to provide the outputs.

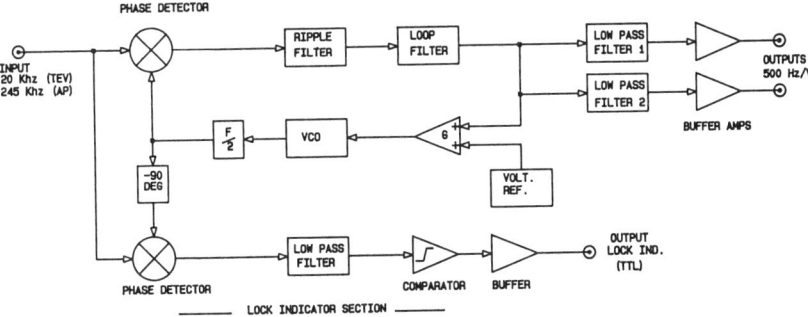

Fig. 5. Block diagram of the phase-locked loop circuit.

RESULTS

The performance was monitored during Collider operation and during a recent fixed target run where the "Tune" ranged from .45 at injection to .5 at extraction. Shown in Fig. 5 are fast time plots of the PLL output, for complete Tevatron cycles, showing a continuous tune measurement. Comparisons with data from simultaneous measurements with a spectrum analyzer indicate that the system is capable of accuracy to $\pm$ .002 (of fractional tune). The quality of the signals has a wide variation, depending on the mode of acceleration operation, and is not always sufficient to give this accuracy, or even a locked condition. In this application, vertical signal feedthrough on the horizontal signal (and vice-versa) is too close to be completely filtered out, and tends to pull the PLL off frequency, producing a measurement error. Further testing is indicated to better quantify and correlate the signal quality with the measurement accuracy.

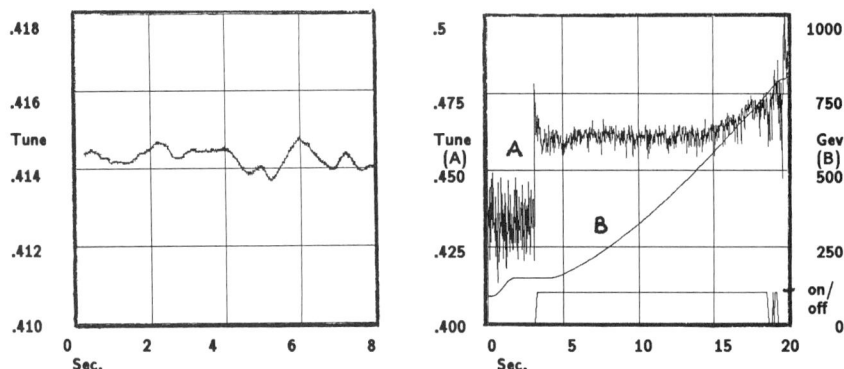

Fig. 6. Fast time plots of Tevatron tune measurement during collider operation, and (b) during fixed target run.

CONCLUSIONS

Because of the complex nature of the input signal, the high noise level, and the various harmonics and modulation present, a study of performance is highly statistical in nature. A phase-locked loop, connected as a tracking filter, will "lock" or center its VCO on the average of the frequency spectrum of the input signal, not necessarily the peak harmonic. The PLL will respond to all harmonics or FM noise on the input, in accordance with their relative power content, as modified by the roll-off characteristics of the loop filter.

The measurement accuracy is therefore somewhat dependent on the signal to noise ratio and the symmetry of the input spectrum. In general, the accuracy of $\pm$.002 tune requires a signal that is at least 5 dB above a relative flat noise floor. Although it is by no means perfect, and has not replaced the spectrum analyzer, this system has shown to be a useful tool for measuring the tune.

The system is under continuing development, in an effort to improve its performance when the signal quality is poor. Testing has shown that the input conditioning circuit is not yet optimized. Under consideration is a narrow-band tracking-filter (adaptive filter), that has shown considerable improvement in isolating the peak signal. Future developments will also include an expanded system, that will provide the capability of beam excitation[6] which greatly increases the detector signal strength when used.

REFERENCES

(1) Gardner, Floyd M., "Phaselock Techniques", 2nd ed., John Wiley and Sons, New York, 1978

(2) Best, Roland E., "Phase-locked Loops", McGraw-Hill, Inc., New York, 1978

(3) Jaffe, R. and Rechtin E., "Design and Performance of Phase Lock Circuits Capable of Near Optimum Performance Over a Wide Range of Input Signal and Noise Levels", IRE Trans. Information Theory, vol. IT-1, Mar., 1955
Reprinted by:
Lindsey, William C. and Simon, Marvin K., "Phase-Locked Loops & their Application", IEEE Press, pp. 20, New York, 1978

(4) Johnson, David E. and Hilburn, John L., "Rapid Practical Designs of Active Filters", John Wiley & Sons, pp. 145, New York, 1978

(5) Martin, D. et al., "A Schottky Receiver for Non-Perturbative Tune Monitoring in the Tevatron", Proceedings of the 1989 IEEE Particle Accelerator Conference, Chicago, IL

(6) Linnecar, T. et al., "Continuous Tune Measurements Using the Schottky Detector", IEEE Transactions on Nuclear Science, Vol. NS-30, No. 4, pp. 2185, 1983

Progress Towards a Turn-by-turn Beam Profile Monitor for the Fermilab Booster

J.B. Rosenzweig, V. Bharadwaj, J. Lackey and P. Zhou

Fermi National Accelerator Laboratory
P. O. Box 500, Batavia, Illinois 60510

Abstract

This paper describes the design and pre-installation testing of a turn-by-turn beam profile monitor for the Fermilab Booster. This non-intrusive monitor collects the ions created by beam particle collisions with the residual gas onto a microchannel plate (MCP) detector using a large (50-100 kV/m) clearing field, to obtain a projected image of the beam distribution in a given plane. The output of the MCP is an anode strip array which gives a 48 channel image with 1.5 mm resolution. The strip current signal is digitized on a turn-by-turn basis to potentially generate more than 16,000 beam profiles in one Booster cycle. The purpose of this device is to aid in understanding the effects which drive the transverse emittance growth in the Booster, which may have time scales as short as one turn.

Introduction

The idea of taking advantage of the residual gas ionization process to obtain images of the beam profile in a circular accelerator is not new. Profile monitors based on this concept and employing a microchannel plate (MCP) as the primary detector and amplifier have been implemented successfully at Fermilab[1] and KEK[2]. This type of monitor has been studied and is in the process of installation at the Fermilab Booster. Unique and challenging aspects of this application include: the largeness of the beam space-charge, which perturbs the ion drift motion and potentially causes loss of spatial resolution, and the desire to extract profiles on a turn-by-turn (1.6-2.7 μsec) basis. This time-resolution is necessary to observe the fast evolution of the space-charge driven transverse emittance growth at injection. Previous attempts to measure this with other techniques have set an upper limit on the time scale of this growth to < 100 μsec[3]. It should also be useful for diagnosing any fast transverse phenomena which may occur in the cycle, especially near transition where space-charge problems can reappear. The purpose of this paper is to describe the design considerations and progress towards implementation of the ionization-based microchannel plate profile monitor (MCP-PM) at the Booster.

Detecting Ions With the MCP

The beam profile monitor, with its MCP detector drift electrodes are shown schematically in Fig. 1. The gap between the anode and the collecting surface of the MCP (the negative drift electrode) is $g = 12$ cm, the active width of the MCP is 7.2 cm and its active length (in the direction of the beam axis) is $l = 10$ cm. The electrode hardware and MCP bias and protection electronics employed in the prototype described here is essentially the same as used by Krider in the BPM he developed for the Fermilab Debuncher. The current emanating from the back of the MCP is collected on 48 strips of width $w = 1.2$ mm separated by 0.3 mm gaps layed out on teflon circuit board material. The collected current signal from the strips is amplified and then digitized turn-by-turn to form desired images of the beam profile.

The reproduction of the beam transverse distribution in the detector image is dependent on drifting the ions in approximately straight lines normal to the MCP surface. This requires that the electric field be quite uniform inside the collecting region. Electric field nonuniformity arises from the effects of beam space charge and electrode geometry. Uniformity can be optimized by appropriate use of field shaping electrodes tied to a resistor chain. This has been done adequately for the prototype, but will be improved in the future. The effects of the space-charge derived field can be overcome by raising the drift field $E_d = V_d/g$ until it is much larger than the beam's transverse field. For a round bi-gaussian beam of N particles of radial size σ_b in a circular

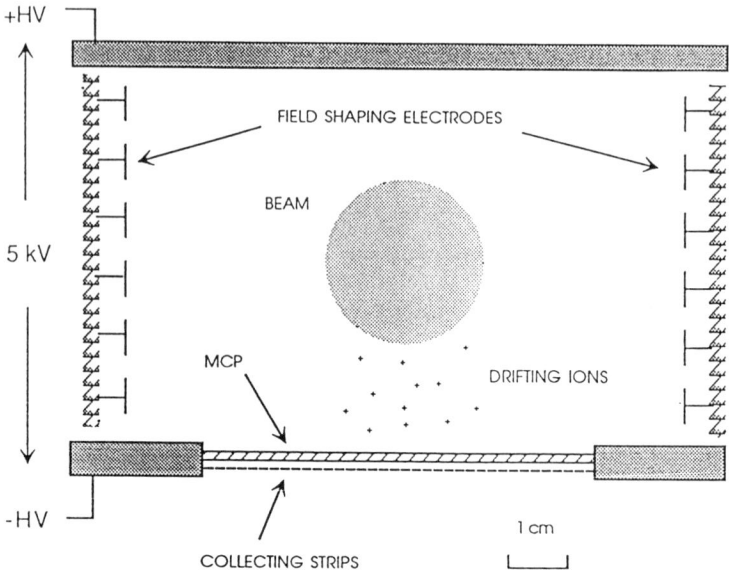

Figure 1: Schematic drawing of MCP-PM.

machine of radius R the maximum transverse electric force (the ions are slow $v_d \ll c$ and do not feel appreciable magnetic force) is approximately

$$eE_\perp \simeq \frac{0.9 N_b r_e m_e c^2}{\pi \sigma_b R}.$$

With a beam size of $\sigma_b = 1$ cm and $R = 75$ m and $N_b = 2.5 \times 10^{12}$, this gives a field of about 1.4 kV/m. The prototype allows a drift voltage of $V_d = 5$ kV, which gives a field of 42 kV/m. This is again adequate, but plans are to increase the voltage by a factor of two or three for eventual operation.

To simulate the operation of the prototype, a number of test particles equal to that expected in the detector per turn (see discussion below), set up initially in the same form as the beam transverse distribution, were tracked under the combined drift and beam fields derived from POISSON, with beam charge and radius as given above. The results allow us to calculate an inherent resolution $\Delta x = \sqrt{x_{rms,f}^2 - x_{rms,i}^2} = 3$ for this case, in which the space-charge effects are as severe as will ever be expected. The error induced in the measured rms beam size is about 5 percent, and therefore in the emittance is 10 percent. Improvements in the clearing voltage and field shaping should reduce Δx below the 1.5 mm level corresponding to the strip spacing, giving a resolution in the beam size of 1.1 percent for $\sigma_b = 1$ cm.

One other issue concerning the level of the drift voltage is the spread in the time-of-flight of the singly ionized particles of mass m_i created at the top and bottom of the beam distribution before collection

$$\Delta \tau_d \simeq \Delta y \frac{\partial \tau_d}{\partial y} \simeq \frac{2\sigma_b}{c}\sqrt{\frac{m_i c^2}{eV_d}} \simeq 65\sqrt{A_i(\text{amu})/V_d(\text{kV})}\,\text{nsec}.$$

Thus even for a heavy ion (e.g. CO+, $A_i = 28$) the difference in drift times is much smaller than a Booster turn.

The detection of sufficient numbers of ions to form an image in the few microseconds of a Booster turn requires that the residual gas density and the beam intensity be at a certain level. Since the expected vacuum pressure at the detector is about $p = 3 \times 10^{-7}$ Torr, the residual gas density is $n_g = p/kT \simeq 1.1 \times 10^{10} \text{cm}^{-3}$. The (species dependent) cross-section for ionization of the residual gas particles by beam protons is approximately $\sigma = 1.5 \times 10^{-22}$ m^2, and the mean distance travelled by the beam particles between ionizing events is $\lambda_i = (n_g \sigma_i)^{-1} \simeq 9 \times 10^8$ cm. The maximum ion current density on a plate parallel to the beam is (for a gaussian distribution)

$$J_m = \frac{I_b}{\sqrt{2\pi}\sigma_b \lambda_i} = \frac{\beta c e N_b n_g \sigma_i}{(2\pi)^{3/2} \sigma_b R} \simeq 0.94 \text{nA/cm}^2.$$

The maximum primary current per strip is approximately $I_m = 1.2\text{cm}^2 J_m = 1.12$ nA, and the maximum number of primary ions per strip per turn is $N_i = (2.7 \times 10^{-6}\text{sec}) I_m/e \simeq 1.95 \times 10^4$. The expected fractional statistical fluctuation in this signal is $N_i^{-1/2} = 0.7\%$. If better statistics are desired, or the beam current is lowered, the pressure at the MCP should be raised.

The microchannel plate inherited from the Debuncher used at present is a two-stage (chevron) capable of gains up to 10^8[4]. This is not necessary for the future, and a single stage with a gain of up to 10^4 will be employed. The maximum current density which can be supplied by the MCP before gain saturation onset is about 150 nA/cm^2[4]. Therefore the expected gain needed is only $G = 10^2$-10^3 and a single-stage MCP is perfectly adequate. Microchannel plates can only source a certain amount of total charge per unit area before the gain degrades due to accumulated damage to the microchannel surfaces[4]. For gain degradation less than ten percent $Q/A \leq 10^{-2}$ C/cm^2.

With the MCP run just below saturation the lifetime of the plate gain is about 66,000 seconds (18.5 hours) of beam-on time. The MCP bias voltage is only on during a measurement, and the bias circuit is interlocked with the output of the MCP[1], preventing damage and excessively fast aging of the plate. The MCP has been tested for uniformity of gain response using an ultraviolet source and is found to have 5.7 % rms variance in strip response, which will have negligible effect on our measurement. The unsaturated gain of the MCP was found to be exponential as a function of bias voltage over at least four orders of magnitude in voltage. The MCP detector hardware has recently been installed in the Booster.

Signal Amplification and Data Acquisition

Two stage amplifiers, with the initial stage configured as a transimpedance amplifier, capable of taking the (maximally) 150 nA current signals from the MCP output collector strips and producing a digitizable voltage output are currently under development. In order to resolve the tails of the profile distribution it is necessary that the signal-to-noise ratio be large enough to measure a few nA signal. Since turn-by-turn profiles are desired the bandwidth of the preamp has to be approximately DC to 1Mhz. To help alleviate noise problems the preamps are locate directly below the profile monitor current collection strips.

Since we would like to be able to read out the profile monitor turn-by-turn in the Booster the ADC system will need to digitize at a variable rate, *i.e.* 354 kHz at injection up to 628 kHz at Booster extraction. For the prototype test we will use a very simple system which uses LeCroy waveform digitizers (model 6810) to digitize the information coming out of the profile monitor amplifiers. These digitizers are capable of digitizing at the variable rate required and have a memory depth of 64k samples per channel which is more that adequate for our application. The digitizers will be read out through a LeCroy 6010 CAMAC crate controller, then acquired and analyzed by either the Accelerator Division Vaxes, or Sun workstations running UDAS (Unix Data Acquisition System) software, an option which has passed preliminary tests with our system.

Backgrounds

Primary beam particles impinging directly on the MCP or creating secondary particles which do so will form a background in this monitor. The shielding requirements needed to alleviate the background is being evaluated from measurement of the Booster beam loss. The beam particle loss rate at injection, which gives the number of primary particles leaving the machine, was measured to be $\dot{N}_b \simeq 2.7 \times 10^{13}$ Hz. The average flux through the torus defined by a radius $\rho = 5$ cm around the beam pipe is given by

$$\Psi_b \simeq \frac{\dot{N}_b}{4\pi^2 R\rho} = 1.87 \times 10^7 \text{cm}^{-2}\text{s}^{-1}.$$

Close to injection, therefore, the area of MCP above one strip would be expected to receive $(\Psi_b)(2.7 \times 10^{-6})(1.2\text{cm}^2) = 60$ primary particle hits per turn.

The beam loss monitor signal is the output of a log amp, which gives the integrated ion charge collected $Q = (10^{V/2.44} - 1.6)(4.88 \times 10^{-11})$ C where $V \simeq 5$ V and half of this charge is deposited in about 7 msec (as can also be seen in the decay of the circulating current signal). The calibration of the loss monitor is 7×10^8 C/rad[5]. Therefore the initial dose rate at injection is $\dot{D} = 0.2$ Gy/sec. The flux rate is given by

$$\Psi_i = \frac{6.24 \times 10^8 \dot{D}(\text{Gy/sec})}{dE/dx(\text{ MeV-m}^2/\text{kg})}$$

For the primary beam particles in argon $dE/dx \simeq 0.52\text{MeV} - \text{m}^2/\text{kg})$, about four times minimum ionizing. For slower primary or secondary particles dE/dx is even higher, and thus the value $\Psi_i = 2.37 \times 10^8 \text{cm}^{-2}\text{s}^{-1}$ should be taken as a lower limit. This rate corresponds to an average number of hits per turn per strip area of about 768, which is 12.6 times the rate of the primary flux. The signal-to-noise ratio in the strip with maximum signal is about 25:1 and is proportionally smaller in the tails. The background pedestal can be measured (with the drift field off, or from the strips away from the beam signal) and subtracted from the data. This background level, which we have probably overestimated due to uncertainty in dE/dx, is considered barely adequate, however, and lead shielding has been placed directly upstream of the MCP to improve the situation. The signal-to-noise ratio can also be helped by raising the local vacuum pressure.

Conclusions

The development of a residual gas ionization MCP-PM for turn-by-turn Booster measurements is well underway with first operational tests currently being performed this fall. The performance of the prototype will be evaluated and a final design, with better field shaping, higher drift voltage, a single stage MCP and console-based data acquisition system, will be implemented for use as permanent horizontal and vertical emittance monitors for the Booster. Turn-by-turn measurements should allow improved diagnosis of the causes of emittance growth in the Booster, whether it be space-charge tune shifting of particles down to the half-integer resonance, envelope instability[6] or some other fast mechanism. Corrective measures based on this information can then be taken. In addition, this type of instrumentation may be useful for other machines at Fermilab as well – at the end of the Linac, the Main Ring at injection, etc., wherever diagnosis of transverse emittance is needed. The performance of these devices is ultimately limited by the maximum space-charge field of the beam. For example, the Tevatron at 900 GeV gives ions a kick of $\Delta p \simeq 0.76$ MeV/c (300 eV of kinetic) energy for H+) in the 3 nsec time it takes the bunch to pass. A prohibitively large drift field would be required to suppress the ion cloud expansion caused by these space-charge kicks. One could immerse the profile monitor in a vertical magnetic field, which with $B \simeq 4$ T gives a Larmor radius, and thus a resolution, of 0.65 mm – approximately the beam size in the Tevatron. Use of a strong magentic field also complicates the operation of the MCP.[4]

References

[1] J. Krider, "Residual Gas Beam Profile Monitor", *Nucl. Instr. Meth. A* **278**, 660 (1989).

[2] T. Kawakubo, T. Adachi, E. Kadokura, H. Nakagawa, Y. Ajima, and T. Ishida, "Non-destructive Profile Monitor Using Microchannel Plate", Proceedings of the XIV International Conference on High Energy Accelerators, *Particle Accelerators* **30**,935 (1990).

[3] S. Machida, PhD Thesis, University of Tokyo, 1990.

[4] "Characteristics and Applications of Microchannel Plates" Hamamatsu Technical Manual RES-0795, also S. Matsura, S. Umebayashi, C. Okuyama, and K. Oba, "Characteristics of the Newly Developed MCP and its Assembly",*IEEE Trans. Nucl. Sci.* **NS-32**, 350 (1985).

[5] Robert E. Shafer, Rod E. Gerig, Alan E. Bambaugh and Carl R. Wegner, "The Tevatron Beam Position and Beam Loss Monitoring Systems", *Proceedings of the 12th International Conference on High Energy Physics Accelerators*, p. 609, Fermilab, 1983.

[6] P. Zhou and J.B. Rosenzweig, "Envelope Instability in Low Energy Proton Synchrotrons," to be published in *Nucl. Instr. Meth.*

Other Presentations

Each of the following topics was presented during the Workshop, but no contribution was made to these Proceedings.

Beam Profile Measurements Using Synchrotron Light
R. Nawrocky

Laser-Induced Neutralization Diagnostics Approach (Linda)
D. Sandoval

Recent Beam Instability and Impedance Measurements in the Fermilab Main Ring and Tevatron
G. Jackson

New Technologies of Beam Intensity Measurements Developed for the LEP Project.
K.B. Unser

Compact Cryogenic Toroid for Beam Current Measurements
J. F. Power

List Of Participants

Jeff Arthur
Fermilab, MS-308
PO Box 500
Batavia IL 60510
708-840-3162

Rosemary Baltrusaitis
EG & G
130 Robin Hill RD
Goleta CA 93117
805-681-2380

Alan Band
Los Alamos National Laboratory
H838, MP-5
PO Box 1663
Los Alamos NM 87545
505-665-1444

Walter Barry
CEBAF
12000 Jefferson Ave
Newport News VA 23606
804-249-7236

Ed Barsotti
Fermilab, MS-308
P.O. Box 500
Batavia IL 60510
708-840-2127

Edward Beadle
Brookhaven National Laboratory
Bldg 911B
Upton, NY 11973
516-282-7649

Gerald Bennett
Brookhaven National Laboratory AGS-911B
Upton, NY 11973
516-282-7590

Julien Bergoz
Bergoz, Inc
01170
Crozet France
33-50-41-00-89

Vinod Bharadwaj
Fermilab, MS-341
PO Box 500
Batavia IL 60510
708-840-4728

C.M. Bhat
Fermilab, MS-341
PO Box 500
Batavia, IL 60510
708-840-4821

Richard Biwer
Argonne National Laboratory
Bldg 362, Rm 157
9700 Cass Ave
Argonne IL 60439
708-972-4134

S. Alex Bogacz
Fermilab, MS-345
P.O. Box 500
Batavia IL 60510
708-840-3873

Eric Bong
SLAC, BIN 12
2575 Sand Hill Road
Menlo Park, CA 94025
415-926-3457

Jean Borer
CERN, CH-1211
Division SL-A1
Geneve 23 Switzerland
41-22-767-61 11

Bud Botkin
Advent Assoc. Ltd
5808 N. St. Louis Ave.
Chicago IL 60659-4408
312-583-7979

David Brown
Los Alamos National Laboratory, MS-H838
Los Alamos NM 87545
505-667-3277

List of Participants

Ernest Buchanan
Fermilab, MS-341
P.O. Box 500
Batavia IL 60510
708-840-4842

John Byrd
Wilson Synchrotron Laboratory, Cornell
Dryden Road
Ithaca NY 14853

Stephen Caldwell
Los Alamos National Laboratory, MS-D-410
P.O. Box 1663
Los Alamos NM 87545

Oscar Calvo
Bates Linear Accelerator Center
P.O. Box 846
Middletown MA 01949
617-245-6600

Youngjoo Chung
Argonne National Laboratory
C121, Bldg 362
9700 S. Cass Ave
Argonne IL 60439
708-972-4601

Dominic Ciardullo
Brookhaven National Laboratory
Bldg 911 B
Upton, NY 11973
516-282-5264

Rick Coleman
Fermilab, MS-340
PO Box 500
Batavia IL 60510
708-840-3030

Roger Connolly
Los Alamos National Laboratory
MS H808
Los Alamos NM 87545
505-665-2740

Le Croy Corporation
#415
4811 South 76th Street
Greenfield WI 53220
414-281-7300

Jim Crisp
Fermilab, MS-306
PO Box 500
Batavia IL 60510
708-840-4460

Sam Crivello
SSCL, MS-1046
2550 Beckleymeade
Dallas TX 75237
214-708-3129

Vernon Cupps
Fermilab, MS-119
PO Box 500
Batavia IL 60510
708-840-3582

Bill Curry
Argonne National Laboratory, EP/207
9700 S Cass Ave
Argonne IL 60439
708-972-4007

Roy Cutler
SSCL
2550 Beckleymeade Ave.
Dallas TX 75237
214-708-3070

Glenn Decker
Argonne National Laboratory
Bldg 362 APS
9700 S. Cass Ave
Argonne IL 60439
708-972-6635

Bob Demaat
Fermilab, MS-220
PO Box 500
Batavia IL 60510
708-840-2188

Terry Dillahunty
Kaman Instr Corp
P.O. Box 7463
1500 Garden Of The Gods Rd
Colorado Springs, CO 80933-7463
719-599-1500

Gerry Dugan
Fermilab, MS-306
PO Box 500
Batavia, IL 60510
708-840-4668

Curt Dunnam
Wilson Lab, Cornell University
Ithaca NY 14853
607-255-5749

Electronic Design INC
1776 E Washington St
Urbana, IL 61801

Brian Fellenz
Fermilab, MS-308
PO Box 500
Batavia IL 60510
708-840-2512

Kenneth Fertner
Advanced Tech Labs
1111 Street Road
Southampton PA 18966
215-355-8111

Charles Fink
Argonne National Laboratory
Bldg 207
9700 S. Cass Ave
Argonne IL 60439
708-972-6611

Ralph Fiorito
Naval Surf Warfare Center, MS-R42
10901 New Hampshire Ave
Silver Springs MD 20903
202-394-1908

James Fitzgerald
Fermilab, MS-308
PO Box 500
Batavia IL 60510
708-840-4978

Jacob Flanz
Bates Linear Accelerator Center
PO Box 846
Middleton MA 01946
617-245-6600

Ray Fuja
Argonne National Laboratory
Bldg 371-T APS DIV
9700 S. Cass Ave
Argonne IL 60439
708-972-6442

George Gabor
Lawrence Berkeley Laboratory
Bldg 46-149B
1 Cyclotron RD
Berkeley CA 94720
415-486-6711

Hank Gerwers
Scientific International
141 Snowden Lane, PO Box 143
Princeton NJ 08540
609-924-3011

Ed Gill
Brookhaven National Laboratory
Bldg 911A
Upton NY 11973

J. Gilpatrick
Los Alamos National Laboratory, H808
PO Box 1663
Los Alamos NM 87545
505-473-1373

David Goldberg
Lawrence Berkeley Laboratory MS-47/112
1 Cyclotron RD
Berkeley CA 94720
415-451-7222

Elizabeth Hansen
Hewlett-Packard
1200 E Diehl Road
Naperville IL 60566
708-505-8800 X2649

Fady Harfoush
Fermilab, MS-345
PO Box 500
Batavia IL 60510
708-840-2411

Michael Harms
SLAC, PO Box 4349
Stanford CA 94309
415-926-2915

Kim Hartman
Tektronix, Inc
5350 Keystone CT.
Rolling Meadow IL 60008
708-259-7580

Jay Heefner
CEBAF
12000 Jefferson Ave
Newport NEWS VA 23606
804-249-7271

Greg Heinen
USA SDC, C SSD-DE-C
P.O. Box 1500
Huntsville AL 35807-3801
205-895-3941

James Holt
KEK
Oho 1-1 Tsukuba-Shi
Ibaraki Japan 305
298-64-1171, X4126

Gary Horner
Cyclotron Lab, Michigan State University
S. Shaw Lane
E. Lansing MI 48824-1321
517-355-8532

Takao Ieiri
Accelerator Division, KEK
Oho1-1 Tsukuba-Shi
Ibaraki-Ken, Japan 305
(81)298-64-1171 X6403

Stephen Jachim
Los Alamos National Laboratory
MS-H827
PO Box 1663
Los Alamos NM 87545
505-667-9692

Gerald Jackson
Fermilab, MS-308
PO Box 500
Batavia IL 60510
708-840-2317

Georges Jamieson
SSCL, MS-1046
2550 Beckleymeade Ave
Dallas TX 75237
214-708-3128

Keith Jobe
SLAC, P.O. Box 4349
Stanford CA 94309
415-926-2084

Alan Jones
SSCL
2550 Beckleymeade Ave
Dallas TX 75237
214-708-3107

Kevin Jones
Los Alamos National Laboratory
MS-H812
PO Box 1663
Los Alamos NM 87545
505-667-4974

Tomotaro Katsura
KEK
1-1 Oho, Tsukuba
Ibaraki 305 Japan
0298-64-1171

List of Participants   343

Kinetic Systems
Kinetic Systems Corp
11 Maryknoll Drive
Lockport IL 60441
815-838-0005

Eugene Klein
Advanced Tech Labs
1111 Street Road
Southampton PA 18966
215-355-8111

Kevin Kleman
Synchotron Radiation Center
3731 Schneider Dr.
Stoughton WI 53589
608-873-6651

Stephen Kramer
Brookhaven National Laboratory
Bldg. 725C
Upton NY 11973
516-282-4925

John Krider
Fermilab, MS-220
P.O. Box 500
Batavia IL 60510
708-840-3647

T. Kuzay
Argonne National Laboratory, APS 360
Argonne IL 60439
708-972-3084

James Lackey
Fermilab, MS-341
PO Box 500
Batavia IL 60510
708-840-4403

Sharon Lackey
Fermilab, MS-340
P.O. Box 500
Batavia IL 60510
708-840-4453

Ray Larsen
Analytek Ltd
365 San Aleso Ave
Sunnyvale CA 94086
408-745-1114

Gordon Leifeste
SSCL
2550 Beckleymeade Ave
Dallas TX 75237
214-708-3126

Frank Lenkszus
9700 S Cass Ave
Argonne IL 60439
708-972-6972

Andy Ligeti
Insulator Seal
23874 B Cabot Blvd
Hayward CA 94545
415-887-8664

James Logan
Pearson Electronics
1860 Embarcadero Rd
Palo Alto CA 94303
415-494-6444

Alex Lumpkin
Los Alamos National Laboratory
Group P-15, MS-D406
Los Alamos NM 87545
505-667-7726

George Mackenzie
TRIUMF
4004 Wesbrook Mall
Vancouver Canada V6T 2A3
604-222-1047 X 248

Eugene Marsh
Charles Evans & Associates
301 Chesapeake Drive
Redwood City CA 94063
415-369-4567

William Marsh
Fermilab, MS-347
Po Box 500
Batavia IL 60510
708-840-4018

Felix Marti
Cyclotron Lab, Michigan State University
S Shaw Lane
East Lansing MI 48824
517-353-8726

Donald Martin
SSCL, MS-1046
2550 Beckleymeade Ave.
Dallas TX 75237
214-708-3066

Manuel Martin
Fermilab, MS-352
P.O. Box 500
Batavia IL 60510
708-840-3859

Philip Martin
Fermilab, MS-306
Po Box 500
Batavia IL 60510
708-840-4547

Randy McConeghy
SSCL
2550 Beckleymeade Ave
Dallas TX 75237
214-708-3061

Elliott McCrory
Fermilab, MS-307
PO Box 500
Batavia, IL 60510
708-840-4808

David McGinnis
Fermilab, MS-341
P.O. Box 500
Batavia IL 60510
708-840-2789

Robert Melen
Analytek, Ltd
365 San Aleso AVE
Sunnyvale CA 94086
408-745-1114

L. Mestha
SSCL, MS-1046
2550 Beckleymeade Ave
Dallas TX 75237
214-708-3048

Bradley Micklich
Argonne National Laboratory, EPB-207
9700 S. Cass Ave
Argonne IL 60439
708-972-4849

Scott Miller
SSCL, MS-1046
2550 Beckleymeade Ave.
Dallas TX 75237
214-708-3054

Ron Nawrocky
Brookhaven National Laboratory
MS-725B
Upton NY 11973
516-282-4449

King Yuen Ng
Fermilab, MS-345
P.O. Box 500
Batavia IL 60510
708-840-4597

Mark Nyman
Lawrence Barkeley Laboratory
1 Cyclotron Road
Berkeley CA 94720
415-486-7512

Marvin Olson
Fermilab, MS-308
Po Box 500
Batavia IL 60510
708-840-4445

Russell Parker
Loma Linda University Medical Center
SAIC
1123 Via Barcelona
Redlands CA 92374
714-824-4135 X2016

Ralph Pasquinelli
Fermilab, MS-341
Po Box 500
Batavia IL 60510
708-840-4724

Jerry Paul
Los Alamos National Laboratory
MS-H838
Los Alamos NM 87545

Steve Peggs
Fermilab, MS-345
P.O. Box 500
Batavia IL 60510
708-840-4946

John Perry
CEBAF
12000 Jefferson Ave.
Newport News VA 23606
804-249-7249

David Peterson
Fermilab, MS-341
P.O. Box 500
Batavia IL 60510
708-840-3073

Michael Plum
Los Alamos National Laboratory, MS-H838
MP-5
Po Box 1663
Los Alamos NM 87545
505-667-7547

Milorad Popovic
Fermilab, MS-341
Po Box 500
Batavia IL 60510
708-840-4478

John Power
Los Alamos National Laboratory
MS-H808 AT-3
Po Box 1663
Los Alamos NM 87545
505-667-7045

William Rawnsley
TRIUMF
4004 Wesbrook Mall
Vancouver BC, Canada V6T2A3
604-222-1047 X 285

R. Kenneth Reece
Brookhaven National Laboratory
Bldg 911A
Upton NY 11973
516-282-4767

Luigi Rezzonico
Ch-5232 Villigen PSI
Switzerland
011-4156993377

Joseph Rogers
Brookhaven National Laboratory
Bldg 725C
Upton NY 11973
516-282-3887

James Rosenzweig
Fermilab, MS-341
PO Box 500
Batavia IL 60510
708-840-2981

Marc Ross
SLAC, Bin 66
P.O. Box 4349
Stanford CA 94309
415-926-3526

Don Rule
Naval Surface Warfare Center, MS-R42
10901 New Hampshire Ave
Silver Springs MD 20903-5000
301-394-2260

Darryl Sandoval
Los Alamos National Laboratory
MS H808
Los Alamos NM 87545
505-665-1832

Todd Satogata
Fermilab, MS-345
P.O. Box 500
Batavia IL 60510
708-840-2766

Grant Schrag
Insulator Seal
23874 B Cabot Blvd
Hayward CA 94545
415-887-8664

Robert Shafer
Los Alamos National Laboratory, MS-H808
Po Box 1663
Los Alamos NM 87545
505-667-5877

Tom Shea
Brookhaven National Laboratory
Bldg 1005
Upton NY 11973
516-282-2435

R. Brad Shurter
Los Alamos National Laboratory, MS-H808
Po Box 1663
Los Alamos NM 87545
505-665-1122

W. Sims
Brookhaven National Laboratory
Bldg 911-B
Upton NY 11973
516-282-3271

Gary Smith
Brookhaven National Laboratory
Bldg 911C
Upton NY 11973
516-282-3473

Vernon Smith
SLAC, Bin 50
Po Box 4349
Stanford CA 94309
415-926-3519

Lief Solensten
Grumman Corp, MS A01-26
Bethpage NY 11714
516-346-8977

Dave Spence
Argonne National Laboratory
9700 S. Cass Ave
Argonne IL 60439
708-972-6611

Arnold Stillman
Brookhaven National Laboratory
Bldg 911-B
Upton NY 11973
516-282-4944

Greg Stover
Lawrence Berkeley Laboratory
MS 64-121
1 Cyclotron Rd
Berkeley CA 94720
415-486-7706

Gianni Tassotto
Fermilab, MS-222
P.O. Box 500
Batavia IL 60510

Masaki Tejima
KEK
1-1 Oho, Tsukuba-Shi
Ibaraki-Ken Japan 305
0298-64-1171

Robert Traller
SLAC
PO Box 4329
Palo Alto, CA 94309
415-926-4063

List of Participants 347

Klaus Unser
CERN SL-Division
Ch-1211
Geneva Switzerland
41-22767-3784

Olin Van Dyck
Los Alamos National Laboratory, MS-H844
PO Box 1663
Los Alamos NM 87545
505-667-7323

VG Instruments Corp
32 Commerce Center
Cherry Hill DR.
Danvers, MA 01923

Gregory Vogel
Fermilab, MS-308
Po Box 500
Batavia IL 60510
708-840-4942

Robert Webber
SSCL, MS-1046
2550 Beckleymeade Ave.
Dallas TX 75237
214-708-0002

Frank Wells
Los Alamos National Laboratory
MS H808 AT-3
PO Box 1663
Los Alamos NM 87545
505-665-0956

W. Weng
Brookhaven National Laboratory
Bldg 911A
Upton NY 11973
516-282-2135

Steve Williams
SLAC, Bin 60
PO Box 4349
Stanford CA 94309
415-926-2276

Richard Witkover
Brookhaven National Laboratory
Bldg. 911B
Upton, NY 11973
516-282-4607

Guan-Hong Wu
Fermilab, MS-341
PO Box 500
Batavia, IL 60510

Chong-Guo Yao
CEBAF
12000 Jefferson Ave
Newport News VA 23606
804-249-7592

Yan Yin
TRIUMF
4004 Wesbrook Mall
Vancouver CANADA V6T 2A3
604-222-1047

David Zack
Andrew Corporation
10500 W 153rd ST
Orland Park IL 60462
708-349-5440

Jim Zagel
Fermilab, MS-308
Po Box 500
Batavia IL 60510
708-840-4076

Christopher Ziomek
Los Alamos National Laboratory, MS-H827
Po Box 1663
Los Alamos NM 87545
505-667-0688

AIP Conference Proceedings

		L.C. Number	ISBN
No. 210	Production and Neutralization of Negative Ions and Beams (Brookhaven, NY, 1990)	90-55316	0-88318-786-8
No. 211	High-Energy Astrophysics in the 21st Century (Taos, NM, 1989)	90-55644	0-88318-803-1
No. 212	Accelerator Instrumentation (Brookhaven, NY, 1989)	90-55838	0-88318-645-4
No. 213	Frontiers in Condensed Matter Theory (New York, NY, 1989)	90-6421	0-88318-771-X 0-88318-772-8 (pbk.)
No. 214	Beam Dynamics Issues of High-Luminosity Asymmetric Collider Rings (Berkeley, CA, 1990)	90-55857	0-88318-767-1
No. 215	X-Ray and Inner-Shell Processes (Knoxville, TN, 1990)	90-84700	0-88318-790-6
No. 216	Spectral Line Shapes, Vol. 6 (Austin, TX, 1990)	90-06278	0-88318-791-4
No. 217	Space Nuclear Power Systems (Albuquerque, NM, 1991)	90-56220	0-88318-838-4
No. 218	Positron Beams for Solids and Surfaces (London, Canada, 1990)	90-56407	0-88318-842-2
No. 219	Superconductivity and Its Applications (Buffalo, NY, 1990)	91-55020	0-88318-835-X
No. 220	High Energy Gamma-Ray Astronomy (Ann Arbor, MI, 1990)	91-70876	0-88318-812-0
No. 221	Particle Production Near Threshold (Nashville, IN, 1990)	91-55134	0-88318-829-5
No. 222	After the First Three Minutes (College Park, MD, 1990)	91-55214	0-88318-828-7
No. 223	Polarized Collider Workshop (University Park, PA, 1990)	91-71303	0-88318-826-0
No. 224	LAMPF Workshop on (π, K) Physics (Los Alamos, NM, 1990)	91-71304	0-88318-825-2
No. 225	Half Collision Resonance Phenomena in Molecules (Caracus, Venezuela, 1990)	91-55210	0-88318-840-6
No. 226	The Living Cell in Four Dimensions (Gif sur Yvette, France, 1990)	91-55209	0-88318-794-9
No. 227	Advanced Processing and Characterization Technologies (Clearwater, FL, 1991)	91-55194	0-88318-910-0
No. 228	Anomalous Nuclear Effects in Deuterium/Solid Systems (Provo, UT, 1990)	91-55245	0-88318-833-3
No. 229	Accelerator Instrumentation (Batavia, IL 1990)	91-55347	0-88318-832-1